Occupational Safety and Health Law Handbook

authors

Lesa L. Byrum
Peter L. de la Cruz
Michael C. Formica
Lawrence P. Halprin
Michael T. Heenan
Scott E. Kauff
Chris S. Leason
Margaret S. Lopez
Martha E. Marrapese

Jennifer E. McCadney
Kathryn McMahon-Lohrer
Marshall L. Miller
Frank F. Murtha
John B. O'Loughlin, Jr.
Manesh K. Rath
William J. Rodgers
Arthur G. Sapper
Stanley M. Spracker

Government Institutes
An imprint of The Scarecrow Press, Inc.
Lanham, Maryland • Toronto • Oxford

Government Institutes

Published in the United States of America
by Government Institutes, an imprint of The Scarecrow Press, Inc.
A wholly owned subsidiary of
The Rowman & Littlefield Publishing Group, Inc.
4501 Forbes Boulevard, Suite 200
Lanham, Maryland 20706
www.govinstpress.com

Estover Road, Plymouth PL6 7PY, United Kingdom

Copyright © 2001 by Government Institutes

All rights reserved. No part of this publication may be reproduced, stored in a retrieval system, or transmitted in any form or by any means, electronic, mechanical, photocopying, recording, or otherwise, without the prior permission of the publisher.

The reader should not rely on this publication to address specific questions that apply to a particular set of facts. The author and the publisher make no representation or warranty, express or implied, as to the completeness, correctness, or utility of the information in this publication. In addition, the author and the publisher assume no liability of any kind whatsoever resulting from the use of or reliance upon the contents of this book.

British Library Cataloguing in Publication Information Available

Library of Congress Cataloging-in-Publication Data

OSHA law handbook / authors, Marshall Lee Miller...[et al.]. p. cm.
Includes bibliographical references and index.
ISBN 10: 0-86587-640-1 (cloth : alk. paper)
ISBN 13: 0-86587-640-8 (cloth : alk. paper)
    1. Industrial safety—Law and legislation—United States. 2. Industrial hygiene—
       Law and legislation—United States. I. Miller, Marshall Lee. II. United States.
       Occupational Safety and Health Administration.

KF3570.O844 2001
344.73'0465—dc21

∞™ The paper used in this publication meets the minimum requirements of American National Standard for Information Sciences—Permanence of Paper for Printed Library Materials, ANSI/NISO Z39.48-1992.
Manufactured in the United States of America.

Occupational Safety and Health Law Handbook

Summary Contents

Chapter 1   **Occupational Safety and Health Act**
Marshall Lee Miller 1

Chapter 2   **The Rulemaking Process**
Lawrence P. Halprin 53

Chapter 3   **The Duty to Comply**
Arthur G. Sapper 77

Chapter 4   **The General Duty Clause**
Peter L. de la Cruz 93

Chapter 5   **Recordkeeping**
Martha E. Marrapese 111

Chapter 6   **Employers' and Employees' Rights**
Stanley M. Spracker and John B. O'Loughlin, Jr. 141

Chapter 7   **Refusal to Work and Whistleblower Protection**
Frank F. Murtha and William J. Rodgers 151

Chapter 8   **Hazard Communication and Employee Right-to-Know**
Chris S. Leason and Scott E. Kauff 165

Chapter 9   **Self-Audits**
Michael T. Heenan and Margaret S. Lopez 197

Chapter 10   **Inspections**
Lawrence P. Halprin 215

Chapter 11   **Contesting Citations and Penalties**
Manesh K. Rath 235

Chapter 12   **Criminal Enforcement of Violations**
Michael C. Formica and Marshall Lee Miller 259

Chapter 13 Judicial Review of Enforcement Actions
   Stanley M. Spracker and John B. O'Loughlin, Jr. 267

Chapter 14 Imminent Dangers Inspections
   Kathryn McMahon-Lohrer and Jennifer E. McCadney 279

Chapter 15 State Plans
   Lesa L. Byrum 299

Contents

Preface .... *xvii*

About the Authors .... *xix*

Chapter 1   Occupational Safety and Health Act

Overview 1
    Comparison of OSHA and EPA 2
    OSHA, the Organization 2

Legislative Framework 3
    Purpose of the Act 4
    Coverage of the Act 5
    Exemptions from the Act 5
    Telecommuting and Home Workplaces 6

Scope of OSHA Standards 7
    Areas Covered by the OSHA Standards 7
    Overview of Standards 8
    Overview of Health Standards 9
    Overview of Safety Standards 10

Standard Setting 11
    Consensus Standards: Section 6(a) 11
    Standards Completion and Deletion Processes 12
    Permanent Standards: Section 6(b) 13
    Emergency Temporary Standards 15
    General Duty Clause, 5(a)(1) 15
    Feasibility and the Balancing Debate 16

Variances 19
    Temporary Variances 20
    Permanent Variances 20

Compliance and Inspections 20
    Field Structure 21
    Role of Inspections 21
    Training and Competence of Inspectors 22
    Citations, Fines, and Penalties 23
    OSHA Citation and Penalty Patterns 24
    Communicating and Enforcing Company Rules 26
    Warrantless Inspections: The *Barlow* Case 27

Recordkeeping 28
    Accident Reports 28
    Monitoring and Medical Records 29
    Hazard Communication 30
    Access to Records 30

Programmatic Standards 31
Refusal to Work and Whistleblowing 31
    Refusal to Work 31
    Protection of Whistleblowing 32

Federal and State Employees 34
    Federal Agencies 34
    State Employees 34

State OSHA Programs 34
    Concept 35
    Critiques 35

Consultation 36

Overlapping Jurisdiction 37

Occupational Safety and Health Review Commission 39
    OSHRC Appeal Process 39
    Limitations of the Commission 39

National Institute of Occupational Safety and Health 40

Hazard Communication Regulations 41
    Reason for the Regulation 41
    Scope and Components 42
    Hazard Evaluation 43
    Trade Secrets 44
    Federal Preemption Controversy 45

Ergonomics Issues 47
    Overview 47
    Scope of the Standard 48
    Initial Actions and Action Triggers 48
    The Required Ergonomics Program 49

Legislation 49

Chapter 2   The Rulemaking Process

Overview 53
    The Decision to Initiate a Rulemaking 54
    The Traditional Rulemaking Process 54
    Negotiated Rulemaking 59

Safety and Health Standards 59
    Basic Framework 59
    OSHA's Statutory Authority to Issue Standards 60
    Other Controlling Statutes and Executive Orders 61
    Characterization of the Safety and Health Standards 66
    Emergency Temporary Standards 70

State Jurisdiction and Plans 71

Challenging OSHA Standards 73

Variances 73
    Temporary Variances 74
    Permanent Variances 74
    Other Variances 74
    Interim Order 75

Chapter 3   The Duty to Comply

Overview 77

Applicability of OSHA Standards 77
    The General Principle of Preemption 77
    Special Applicability Problems 78

General Principles of the Duty to Comply 79
    The Exposure Rule 79
    To Whose Employee Does the Duty Run? 81

Actual or Constructive Knowledge 84

Additional Elements That OSHA Must Sometimes Prove 85

The Employer's Substantive Affirmative Defenses 86
    Infeasibility 86
    The Greater Hazard Defense 89
    Unpreventable Employee Misconduct 90
    Invalidity of the Standard 91
    De Minimus 92

Chapter 4   The General Duty Clause

Overview 93

Guidance Provided by Congress 94

Basic Framework 96

Recognized Hazards 98
    Employer Recognition 99
    Industry Recognition 100
    "Common Sense" Recognition 102

Unforeseeable Hazards 102

Employee Misconduct and the General Duty Clause 103

Demonstration of Serious Harm 104

Feasible Hazard Abatement 105

Replacement of the General Duty Clause with OSHA Standards 106
Conclusion.. 108

Chapter 5 Recordkeeping

Overview.. 111
Statutory Basis for OSHA Records 111
Reasons for Requiring Records.. 112
Types of Records.. 114
Recordkeeping Requirements for Chemical-Specific Standards........ 114
Recordkeeping Requirements for Safety Standards 118
Illness and Injury Recordkeeping Requirements 120
    Those Subject to the Rule 121
    Exemptions... 122
    Recordkeeping Forms 123
    Posting.. 124
    Maintaining and Retaining These Records............................. 124
    Access 125
    Information to Be Recorded... 126
    Musculoskeletal Disorders ("MSDs") 129
Recordkeeping Requirements for the Hazard Communication
    Standard 132
Enforcement of OSHA Recordkeeping Requirements 135
Access to Records 136
Conclusion... 139

Chapter 6 Employers' and Employees' Rights

Overview.. 141
Employers' Rights 142
    Inspections and Warrants... 142
    Challenging Citations and Civil Penalties.............................. 143
    Judicial Review 144
    Participation in Rulemakings 144
    Protection of Trade Secrets 145
Employees' Rights 146
    Complaints... 146
    Refusal to Work... 147
    Protection from Discrimination 147
    Participation in Inspections and Enforcement 148

    Access to Information 149

    Conclusion 150

Chapter 7   **Refusal to Work and Whistleblower Protection**

    Overview 151

    Refusal to Work 152
        Federal Statutes 152
        State Statutes 160
        Common Law 160

    Whistleblowing 161
        Federal Statutes 161
        State Statutes 162
        Common Law 163

    Conclusion 164

Chapter 8   **Hazard Communication and Employee Right-to-Know**

    Overview 165

    Assessing the Hazard 168
        The Hazard Determination 168
        Employee Exposure 171
        Mixtures 171

    Written Hazard Communication Program 173

    Labeling and Other Forms of Warning 174

    MSDS 176

    Employee Information and Training 181

    Exemptions 184
        Partially Exempted Work Places/Conditions 184
        Partially Exempted Materials (Chemicals Subject to the Labeling
            Requirements of Other Statutes or Regulations 186
        Fully Exempted Materials 187

    Trade Secrets 189

    Multiple Employer Worksites 190

    Preemption and Approval of State Plans 191

    The Future of HAZCOM-The Globally Harmonized
        System ("GHS") 193

Chapter 9 Self-Audits

Overview 197

The Significance of Voluntary Safety and Health Auditing 198
    Overview of Audits 199
    Auditing Tips 201

OSHA's Voluntary Self-Audit Policy 203
    Purpose 203
    Scope 203
    Provisions 204
    Limitations 205
    Critique 206

Privileges and Protections from Disclosure of Audit Information 207
    Introduction 207
    The Self-Audit Privilege 207
    The Attorney/Client Privilege 211
    Attorney Work Product Doctrine 213

Conclusion 213

Chapter 10 Inspections

Overview 215

OSHA's Authority to Inspect Places of Employment 216

The 4th Amendment's Prohibition against Unreasonable Search and Seizure, Warrants, and Probable Cause 216

Consensual Searches v. Warrant Requirement 219
    Background 219
    Challenging a Warrant 221
    Warrant Requirement Exceptions 222

Types of Inspections 223

Inspection Procedures 223
    General Requirements 223
    Opening Conference 224
    Records Inspection/Subpoenas 227
    The Walkaround 229
    Pictures, Documents and Other Evidence 230
    Employee Interviews 231
    Closing Conference 232

General Rules on Managing OSHA Inspections 233

Special Consideration for Construction Sites and Other Multi-Party Worksites 233

Chapter 11 Contesting Citations and Penalties

Overview 235

The OSHA citation 236

Classification of OSHA Violations 236
    De minimus Violations 237
    Other-than-Serious Violations 237
    Serious Violations 237
    Willful Violations 237
    Repeat Violations 238
    The Egregious Penalty Policy 238
    Increasing Size of Penalties 239
    OSHA Penalties in Light of Its Budget 239

Challenging a Citation 240
    Settlement Agreement with Area Director 240
    The Notice of Contest 241
    Extending the Abatement Period 242
    Employee Participation 242
    Opportunity for Settlement after Filing the Notice of Contest .. 243

Proceedings before the Administrative Law Judge 244
    The Complaint 244
    The Answer 244
    Defenses 245
    Discovery 249
    Hearing 254
    Simplified Proceedings 255

Review by the Commissioners 255
    Interlocutory Review 255
    Final Decisions and Reports of Judges 256
    Discretionary Review 256
    Commission Review 256
    Stay of Commission Order 257

Chapter 12 Criminal Enforcement of Violations

Overview 259

Federal Prosecution 260
    Definition of "Employee" 260
    Willful Violations Causing Death to Employee 261
    False Statements and Advance Notice 262

State Enforcement 263

Prosecution under Environmental Statutes 264

Recent Legislation 265

Chapter 13 Judicial Review of Enforcement Actions

Overview.. 267

Jurisdiction 268
    Parties That Have Standing to Bring an Appeal....................... 268
    Courts That Have Jurisdiction over Appeals.............................. 270

Timing... 270
    Final Commission Orders 270
    Exhaustion of Administrative Remedies..................................... 271
    Constitutional Challenges 273

Scope of Judicial Review.. 274
    Procedural Matters.. 274
    Standard of Review for Conclusions of Law 275
    Standard of Review for Findings of Fact................................... 276
    Precedential Effect of Judicial Decisions 278

Conclusion.. 278

Chapter 14 Imminent Danger Inspections

Overview.. 279
    Background on Imminent Danger Inspections 280

Statutory Framework and Regulatory Guidance 282
    Imminent Dangers Defined 282

Inspection Procedure... 283
    OSHA Investigation... 283
    Secretary of Labor's Response.. 284
    Employer Response 286

Remedying Imminent Dangers-Injunction Proceedings................... 286

Employee Rights 288
    Right to Inform and Assist With Inspection 288
    Requesting a Writ of Mandamus 289
    Refusing Dangerous Work 290

Penalties... 291

Imminent Danger Inspections in Practice.. 292

Comparison of Federal Standards with MSHA and State OSHA
    Programs... 294

Legislation... 296

Chapter 15 State Plans

    Overview 298

    Approval of State Plans 298
        Introduction 298
        Program Content 300
        Steps for Approval 301

    Complaints against State Plans 302

    Approved State Plans 303

    State Standards 304

    Preemption 305

    Appendix: State Agency Contact Information 307

Appendix A Occupational Safety and Health Act *311*

Index *351*

Preface

We are happy to present this inaugural edition of the *OSHA Law Handbook*. Occupational safety and health is a fundamental concern of all employers, and is among the most pervasive of regulatory obligations. This book serves as a comprehensive reference for reliable and practical information concerning occupational safety and health compliance issues.

In creating the *OSHA Law Handbook*, we drew on the goals and framework that have made its sister publication, the *Environmental Law Handbook*, so highly respected. We have endeavored to provide users with accurate compliance information presented by the most respected experts in the field. We have also tried to present this information in a clear, concise manner with a minimum of legal jargon, but with enough detail and references that readers can readily immerse themselves more deeply into any subject if the situation warrants.

We begin with an overview of the Occupational Safety and Health Act (OSH Act), which creates the framework for this regulatory landscape. This is complemented with a discussion of rulemaking practices that incorporates both general administrative law concepts as well as those features unique to the Occupational Safety and Health Administration (OSHA). We then address universal compliance issues involving OSHA standards, the general duty clause, and OSHA's seemingly ubiquitous recordkeeping and reporting requirements.

To be well understood, the OSH Act needs to be viewed within the broader context of employment law. This framework is aptly provided by chapters on employers' and employees' rights, including thorough discussions of issues relating to refusal to work and whistle blowing.

Because hazard communication and employee right-to-know issues are both broadly applicable and the source of many citations, we have dedicated a separate chapter to this standard. From there, we transition from compliance to enforcement by reviewing OSHA's audit policy and OSHA inspections. We then progress through various enforcement scenarios with chapters on contesting OSHA violations and penalties, criminal prosecutions, and civil litigation.

To complete the picture, we discuss imminent dangers and the special criteria governing these situations. Finally, we conclude by recognizing the important place that states play in this regulatory setting of shared jurisdiction with a discussion of state plans and authority.

We invite readers who want additional occupational safety and health information to consult Government Institutes' website at www.govinst.com. Because the field is ever-changing, we will revise this book periodically to provide the most accurate and relevant information possible. Particularly because this is a first edition, we welcome comments from readers on how this publication might be improved to help all employers comply with both the spirit and the letter of an increasingly complex regulatory system.

Peter de la Cruz, Keller and Heckman
Contributing Author

Charlene Ikonomou, Government Institutes
Editor

About the Authors

Lesa L. Byrum

Ms. Byrum is an attorney in the Washington, D.C. law firm of Keller and Heckman LLP, where she practices exclusively in the areas of occupational safety and health law and employment law. Ms. Byrum develops safety and health programs, as well as negotiates OSHA citations and litigates the citations should pre-litigation negotiations fail. She also advises clients on a day-to-day basis with regard to various employment issues, including terminations, disciplinary procedures and wage and hour issues. Ms. Byrum also has significant litigation experience in defending employers against a variety of employment discrimination claims. She lectures and writes on a wide range of safety and employment issues. Ms. Byrum holds a B.S. in Political Science from James Madison University and a J.D. from George Mason University.

Email
byrum@khlaw.com

Website
www.khlaw.com

Peter de la Cruz

Peter de la Cruz is a partner in the Washington, D.C. law firm of Keller and Heckman LLP. From 1976 to 1980, he served as a trial attorney with the Antitrust Division of the United States Department of Justice and concentrated on the briefing and oral argument of appellate cases before the United States Courts of Appeal. Based on his work for the National Commission for the Review of Antitrust Laws and Procedures in 1978 and 1979, Peter received the Justice Department's Special Achievement Award for sustained superior performance. Peter received an A.B. from Cornell University in 1969, and a J.D. from the University of Toledo College of Law in 1976, where he was Associate Editor of the law review. Since joining Keller and Heckman in 1980, his principal areas of practice have included occupational safety, environmental, food and drug, antitrust and trade association matters.

Email
delacruz@khlaw.com

Website
www.khlaw.com

Michael C. Formica

Email
mformica@dslaw.net

Website
www.dclaw.net

Michael C. Formica is an attorney with the Washington, D.C. law firm of Baise & Miller, P.C., where he practices in the areas of environmental, health, and safety law. Mr. Formica has experience with all the major environmental statutes, and additional experience in matters concerning international trade, foreign policy, intellectual property, and criminal law. Prior to joining Baise & Miller, P.C., Mr. Formica clerked in both the Environment and Natural Resources Division of the U.S. Dept. of Justice, and in the General Counsel's Office of Blue Cross / Blue Shield, Vermont. Mr. Formica was educated at the University of Tennessee, University of Rhode Island, and Vermont Law School.

Lawrence P. Halprin

Email
halprin@khlaw.com

Website
www.khlaw.com

Lawrence P. Halprin is a partner in the Washington, D.C. law firm of Keller and Heckman LLP. Mr. Halprin draws on his technical and business background to counsel individual companies and trade associations in a broad range of workplace health and safety, environmental, and product safety and quality issues. He has a hands-on familiarity with the manufacturing environment and has represented clients in the chemical, construction, electronics, food, paper, pharmaceuticals, plastics, steel, telecommunications and transportation industries. He attended the University of Pennsylvania (B.S.Ch.E., *with honors,* 1974); Duquesne University (J.D., 1977, Law Review); George Washington University (M.B.A. in Finance and Investments, 1984, Beta Gamma Sigma).

Michael T. Heenan

Michael T. Heenan is a founding partner of the law firm of Heenan, Althen & Roles, LLP in Washington, D.C. Mr. Heenan has focused on workplace safety and health law for thirty years. He is the author of two books on mine safety and has written various chapters for books and treatises on OSHA and MSHA. As legal editor for *Pit & Quarry* magazine he writes a monthly column on occupational safety issues and related topics and he also speaks regularly on these matters. He serves on the Board of Trustees of the Energy and Mineral Law Foundation and is a member of the Council of Counsel of the National Stone, Sand and Gravel Association.

Email
mheenan@harlaw.com

Website
www.harlaw.com

Scott E. Kauff

Scott Kauff is an attorney in the Washington, D.C. office of McKenna & Cuneo, L.L.P, where he specializes in environmental, health and safety law. He has extensive experience counseling clients regarding health and safety compliance, with a concentration in chemical hazard compliance. Mr. Kauff also conducts audits for environmental, health and safety compliance and aids clients in self-reporting violations to federal agencies. Previously, Mr. Kauff was a Water Resources Associate with the New York State Water Resources Institute and clerked with the U.S. Environmental Protection Agency. Mr. Kauff received his J.D. from the University of Pennsylvania Law School and his B.A., cum laude, from Cornell University.

Email
scott_kauff@
mckennacuneo.com

Website
www.mckennacuneo.com

Chris S. Leason

Email
chris_leason@mckennacuneo.com

Website
www.mckennacuneo.com

Chris Leason is a partner in McKenna & Cuneo's Washington, D.C. office where he practices environmental and occupational safety and health law. He has counseled clients on compliance with numerous OSHA standards, such as the HAZCOM Standard, Process Safety Management Standard, and Air Contaminants Standard, and has represented companies in OSHA enforcement actions for alleged violations of the Occupational Safety and Health Act and OSHA's implementing regulations. Mr. Leason has also represented clients in legal challenges to OSHA final rules, including, most recently, OSHA's controversial Ergonomics Standard. Mr. Leason received his J.D. degree, with honors, from the George Washington University National Law Center and his B.S. in chemical engineering from the Pennsylvania State University.

Margaret S. Lopez

Email
mlopez@harlaw.com

Website
www.harlaw.com

Margaret S. Lopez is an attorney in the Washington, D.C. office of Heenan, Althen & Roles, LLP. Ms. Lopez regularly advises companies in heavy industries on matters related to occupational safety and health. Ms. Lopez is the author of a law review article titled *Application of the Audit Privilege to Occupational Safety and Health Audits: Lessons Learned from Environmental Audits* for the University of Kentucky College of Law's *Journal of Natural Resources & Environmental Law*, and she regularly speaks on topics related to workplace safety and health and employment law. Ms. Lopez received her law degree with Honors from The George Washington University Law School where she was an editor of *The George Washington Journal of International Law and Economics*.

Martha E. Marrapese

Martha Marrapese is employed with the law firm of Keller and Heckman in Washington, DC. Ms. Marrapese's principal areas of practice are environmental and occupational safety and health law and regulatory reform. She has co-authored numerous publications relating to environmental, health, and safety matters. Ms. Marrapese is the current editor of Keller and Heckman's Environmental Regulatory Advisory. Prior to joining Keller and Heckman, Ms. Marrapese served as a legislative assistant to former Congressman (now U.S. Senator) Robert C. Smith (R-NH) and subsequently as law clerk to the Chief Judicial Officer of the U.S. Environmental Protection Agency.

Email
marrapese@khlaw.com

Website
www.khlaw.com

Jennifer E. McCadney

Jennifer E. McCadney is an Environmental Associate with Collier Shannon Scott, PLLC. Ms. McCadney was awarded her J.D. degree with Honors from The George Washington University Law School. Prior to joining Collier Shannon, she served as a part-time law clerk at Conservation International. From 1995 to 1997, she tracked U.S.-Japan trade and economic issues as a research associate at the Japan Information Access Project. She also served as an intern for the U.S. Department of Commerce, Japanese Official Development Assistance Group. Ms. McCadney has authored articles for *Business American* and *Public Contrac Journal*, including an article titled, "The Green Society? Leveraging the Government's Buying Power to Create Markets for Recycled Products."

Email
JMccadney@
colliershannon.com

Website
colliershannon.com

Kathryn McMahon-Lohrer

Kathryn McMahon-Lohrer is employed with the Washington, D.C. law firm of Collier Shannon Scott, PLLC. Ms. McMahon has practiced in the areas of environmental and occupational safety and health law for nearly ten years. In the environmental area, she represents large and mid-size manufacturing concerns and trade associations in a broad range of regulatory compliance matters, ranging from RCRA permitting to Clean Water Act compliance. She also handles judicial and administrative environmental litigation, including CERCLA contribution actions, Department of Justice enforcement matters, and federal and state administrative actions against her clients. Ms. McMahon received her B.A. from Syracuse University in 1988 and her J.D. from Cornell Law School in 1992.

Email
kmcmahon@
colliershannon.com

Website
colliershannon.com

Marshall Lee Miller

Marshall Miller is a partner in the Washington, D.C. office of the law firm of Baise & Miller, where he specializes in the areas of environmental law, occupational health and safety, and international transactions. Mr. Miller was previously special assistant to the first administrator of the U.S. Environmental Protection Agency, chief EPA judicial officer, associate deputy attorney general in the U.S. Department of Justice, and deputy administrator and acting head of the Occupational Safety and Health Administration. He was educated at Harvard, Oxford, Heidelberg, and Yale.

Email
envirlaw@aol.com

Website
www.dclaw.net

Frank Murtha

Frank Murtha is Of Counsel to the law firm of Collier Shannon Scott, PLLC in Washington, DC, where he specializes in Employment Law. Prior to joining Collier Shannon, Mr. Murtha was Labor Counsel for American Can Company. For many years he was also an Adjunct Law Professor at Pace University School of Law, where he taught Advanced Labor Law and Occupational Safety and Health Law. He is former president of the National Foundation for Occupational and Environmental Health Research. A graduate of Columbia Law School, he is a member of the New York State Bar and is admitted to practice before various federal courts.

Email
fmurtha@colliershannon.com

Website
www.colliershannon.com

John B. O'Loughlin, Jr.

John B. O'Loughlin Jr. is a senior associate in the Washington, DC, office of Weil, Gotshal & Manges, LLP. Mr. O'Loughlin's practice is devoted to environmental, safety and health counseling, with a particular emphasis on product safety and workplace health and safety issues. Mr. O'Loughlin focuses on matters under the jurisdiction of EPA, OSHA, FDA, CPSC, and other regulatory agencies. Mr. O'Loughlin was educated at Towson University, Richmond College in London, Boston University, and the University of Maryland School of Law.

Email
john.o'loughlin@weil.com

Website
www.weil.com

Manesh K. Rath

Email
rath@khlaw.com

Website
www.khlaw.com

Manesh K. Rath is an attorney at Keller and Heckman LLP's Washington, D.C. office. Mr. Rath engages in labor law, employment law and OSHA law, representing management. He has authored several articles on OSHA and employment law issues, including articles in *The Labor Lawyer*, *South Texas Law Review* and *Compliance Magazine*. Mr. Rath has served as chair of the Virginia Bar Association Young Lawyers Division Mentor Program and has been active in several other committees on behalf of the Virginia Bar Association. He is also a member of the Society of Human Resources Managers, where he serves on the Northern Virginia Chapter Education Committee. Mr. Rath earned his juris doctor at The College of William and Mary School of Law and his bachelor of sciences degree in business finance at Virginia Polytechnic Institute and State University (Virginia Tech).

William J. Rodgers

Email
wrodgers@
colliershannon.com

Website
www.colliershannon.com

William J. Rodgers is employed with the Washington, D.C. law firm of Collier Shannon Scott, PLLC. Mr. Rodgers has over 30 years' experience in the field of labor relations law and litigation. His practice includes ERISA and employment law litigation; contract negotiations, grievance handling and arbitration; and administrative litigation and representation before all national and local agencies. Mr. Rodgers assists clients regarding OSHA compliance and provides counsel and advice to corporate human resource departments regarding labor law compliance and corporate personnel policies and procedures. Mr. Rodgers received his B.S. from Cornell University in 1965 and his J.D. from Fordham University in 1968.

Arthur G. Sapper

Arthur G. Sapper is a partner in the OSHA Practice Group of McDermott, Will & Emery in Washington, D.C. Mr. Sapper regularly represents employers and major trade associations in OSHA cases before the federal appellate courts and the Occupational Safety and Health Review Commission and has participated in ground-breaking cases in the field. Mr. Sapper was an adjunct professor of OSHA law at Georgetown University Law Center for nine years. Previously, he was the Deputy General Counsel of the Occupational Safety and Health Review Commission and the Special Counsel to the Federal Mine Safety and Health Review Commission. Mr. Sapper has testified before Congress on occupational safety and health issues, published numerous articles in the field of OSHA law, and is a contributing editor of *Occupational Hazards* magazine. Mr. Sapper was awarded a J.D. degree from Georgetown University Law Center.

Email
asapper@mwe.com

Website
www.mwe.com

Stanley M. Spracker

Stanley M. Spracker is a partner in the Washington, DC, office of Weil, Gotshal & Manges, LLP. Mr. Spracker's principal areas of practice are environmental and workplace health and safety law, focusing on EPA and OSHA. In that regard, Mr. Spracker has extensive experience in litigation under all federal environmental statutes including Superfund litigation, district court and administrative proceedings under the Resource Conservation and Recovery Act, and insurance litigation involving claims for coverage for costs of hazardous waste cleanups. He is a graduate of the University of Wisconsin and received his J.D. degree from the University of Chicago Law School.

Email
stanley.spracker@weil.com

Website
www.weil.com

Chapter 1

Occupational Safety and Health Act

Marshall Lee Miller, Esq.
Baise & Miller, LLP
Washington, D.C.

1.0 Overview

The U.S. Occupational Safety and Health Administration (OSHA) was once called the most unpopular agency in the federal government. It was criticized for its confusing regulations, chronic mismanagement, and picayune enforcement. With somewhat less accuracy, business groups likened it to an American gestapo, while labor unions denounced it as ineffective, unresponsive, and bureaucratic.

Most damning of all, OSHA was often simply ignored. It no longer is. Although OSHA still has its weaknesses and many of its standards are outmoded, its penalties have sharply increased in severity. This has caught the attention of labor and management alike. Moreover, the agency has gradually improved its general reputation. The prestigious Maxwell School of Government at Syracuse University has graded a number of federal agencies and has given OSHA a B-minus, the same grade as EPA. A decade or two ago, the grade would likely have been D-plus or C-minus, so this is a step up.[1]

It is not often recognized, however, that OSHA is also perhaps the most important environmental health agency in the government. Even the Environmental Protection Agency (EPA), with far greater resources and public attention, deals with

---

[1] "Report Card In on Government Agencies," AP, 2 February 1999. Nevertheless, in another study, OSHA tied with the Internal Revenue Service for the lowest ranking among federal agencies in terms of "customer approval." University of Michigan Business School, "American Customer Satisfaction Report," 15 December 1999. Not everything has improved. A detailed critique of OSHA prepared by an outgoing senior official a quarter-century ago could regrettably be re-issued today with relatively few changes. See "Report on OSHA: Regulatory and Administrative Efforts to Protect Industrial Health," Jan. 1977, 108 pp., by the author of this chapter.

a smaller range of much less hazardous exposures than does OSHA. After all, individuals are more likely to be exposed to high concentrations of dangerous chemicals in their workplaces than in their backyards.

1.1   Comparison of OSHA and EPA

There are several distinct differences between OSHA and EPA, besides the obvious occupational jurisdiction.

First, OSHA has major responsibility over safety in the workplace as well as health. Second, OSHA is essentially an enforcement organization, with a majority of its employees as inspectors, performing tens of thousands of inspections a year. This "highway patrol" function, inspecting and penalizing thousands of businesses large and small, has been the major reason for OSHA's traditional unpopularity. At EPA, on the other hand, inspections and enforcement are a relatively smaller part of the operation.

Third, whereas EPA is an independent regulatory agency, albeit headed by presidential appointees, OSHA is a division of the Department of Labor. This organizational arrangement provides not only less prestige and less independence for OSHA, but also poses an internal conflict of whether OSHA should be primarily a health or a labor-oriented agency. Nevertheless, OSHA and EPA regulate different aspects of so many health issues—asbestos, vinyl chloride, carcinogens, hazard labeling, and others—that it is reasonable to regard them both as overlapping environmental organizations.[2]

1.2   OSHA, the Organization

OSHA has a staff of 2,200 throughout the country in ten regional offices and scores of area offices. Almost exactly half of the personnel are safety and health inspectors. Around 600 workers are located at OSHA headquarters in Washington, D.C., near Capitol Hill. The budget is $336 million.

The organization is administered by the Assistant Secretary of Labor for Occupational Safety and Health, currently R. Davis Layne (with John L. Henshaw nominated as press time). Before joining OSHA in late 1997, he was an official in the North Carolina Department of Labor and the department's chief lobbyist on occupational safety and health matters with the state legislature.

---

[2]   To prevent this overlap from causing jurisdictional confusion, the two agencies developed a Memorandum of Understanding (MOU) in 1990 to delineate and coordinate their respective activities. OSHA-EPA MOU, 23 November 1990.

The head of OSHA is aided by one to three Deputy Assistant Secretaries, as well as by a number of other senior personnel who head offices such as health standards, safety standards, enforcement, policy planning, and federal programs.

This chapter emphasizes the health aspects of OSHA, because most press attention and the agency's own public emphasis since the mid-1970s has been on toxic hazards. Nevertheless, OSHA is predominantly an occupational safety organization. The two parts of the organization are quite distinct: there are separate inspectors and standards offices for each, and the two groups are different in terms of background, education, and age. There are also far more safety inspectors than health inspectors.

In the most recent fiscal year, the agency conducted 34,000 inspections and proposed penalties of around $90 million. Over half of the inspections were in the construction area, and a quarter were in manufacturing. The number of inspections is only about half of what it was in some earlier years, but this fact alone is not a particularly reliable indicator of agency effectiveness.

State OSHA inspections average a little less than 60,000 a year, but with only $50 million in proposed penalties.

2.0 Legislative Framework

OSHA was created in December 1970—the same month as EPA—with the enactment of the Occupational Safety and Health Act (OSH Act),[3] and officially began operation in April 1971. When compared with other environmental acts, the OSH Act is very simple and well drafted. This does not mean that one necessarily agrees with the provisions of every section, but it is clearly and concisely written so that details can be worked out in implementing regulations. And unlike the other environmental laws which have been amended several times, becoming more tangled each time, the OSH Act has scarcely been amended or modified since its original passage.[4]

---

[3] Occupational Safety and Health Act of 1970, PL 91-596, 84 Stat. 1590.

[4] This lack of change could obviously also be considered a negative factor, but a comparison with some of EPA's ponderously detailed legislation shows the benefits of keeping the basic statute simple. OSHA annual appropriations legislation, however, has been modified several times to restrict OSHA authority over small businesses, farming, hunting, and other subjects.

2.1 Purpose of the Act

The act sets an admirable but impossible goal: to assure that "*no* employee will suffer material impairment of health or functional capacity" from a lifetime of occupational exposure.[5] It does not require—or even seem to allow—a balancing test or a risk-benefit determination.[6] The supplementary phrase in the OSH Act, "to the extent feasible," was not meant to alter this. This absolutist position, comparable only to one provision in the Clean Air Act,[7] reflects Congress' displeasure at previous overly permissive state standards, which traditionally seemed always to be resolved against workers' health. In fact, the concession to *feasibility* was added almost as an afterthought.

Business groups did obtain two provisions in the law as their price for support. First, industry insisted that states should be encouraged to assume primary responsibility for implementation, in order to minimize the role of the federal OSHA. Second, because of their distrust for the allegedly pro-union bias of the Department of Labor, responsibility for first-level adjudication of violations would be vested in an independent Occupational Health and Safety Review Commission (OHSRC) with a three-member panel of judges named by the President and approved by the Senate.

Congress did reject an industry effort to separate the standard-setting authority from the enforcement powers of the new organization, but it gave a special role to the National Institute for Occupational Safety and Health (NIOSH), located in another government department, the Department of Health and Human Services, in the standard-setting process.

Thus, the three main roles of OSHA are these:
1. setting of safety and health standards,
2. their enforcement through federal and state inspectors, and
3. employee and employee education and consultation.

---

[5] OSH Act § 6(b)(5), emphasis added.
[6] This issue will be discussed in detail in Section 4.6 of this chapter.
[7] Clean Air Act § 112, 42 U.S.C. § 1857, the National Emission Standards for Hazardous Air Pollutants (NESHAP).

2.2 Coverage of the Act

In general, coverage of the act extends to all employers and their employees in the fifty states and all territories under federal government jurisdiction.[8] An *employer* is defined as any "person engaged in a business affecting commerce who has employees but does not include the United States or any State or political subdivision of a State."[9] Coverage of the act was clarified by regulations published in the Federal Register in January 1972.[10] These regulations interpret coverage as follows:

1. The term *employer* excludes the United States and states and political subdivisions.
2. Any employer employing one or more employees is under its jurisdiction, including professionals, such as physicians and lawyers; agricultural employers; and nonprofit and charitable organizations.
3. Self-employed persons are not covered.
4. Family members operating a farm are not regarded as employees.
5. To the extent that religious groups employ workers for secular purposes, they are included in the coverage.
6. Domestic household employment activities for private residences are not subject to the requirements of the Act.
7. Workplaces already protected by other federal agencies under other federal statutes (discussed later) are also excluded.

In total, OSHA directly or indirectly covers approximately 100 million workers in 6 million workplaces.

2.3 Exemptions from the Act

The OSH Act and regulations exempt a number of different categories of employees. The most important exemption is for workplaces employing 10 or fewer workers. What often is not recognized is that this exemption is only partial; these smaller establishments are still subject to accident and worker complaint investigations and the hazard communication requirements (discussed below).

---

[8] OSH Act § 4(a)–4(b)(2).

[9] OSH Act § 3(5). Congress' annual appropriations language has excluded several "peripheral" categories of employers in the past few years.

[10] 37 FR 929, 21 January 1972, codified at 29 C.F.R. § 1975.

Federal and state employees are also exempted from direct coverage by OSHA. As discussed below, however, the former are subject to OSHA rules under OSH Act Section 19 and several presidential Executive Orders, and most states having their own state OSHA plans also cover their state and local government workers.

Workers are also exempted if they are covered under other federal agencies, such as railroad workers under the Federal Railroad Administration or maritime workers subject to Coast Guard regulations. This exemption has sometimes generated intergovernmental friction where the other agency has general safety and health regulations but not the full coverage of OSHA regulations. In other words, is the exemption absolute or only proportional?

Under OSH Act Section 9, OSHA is supposed to defer to the other agency if it can better protect the workers and, similarly, the other agency is expected to recede when the situation is reversed. Of course, considerations of turf and politics are often paramount.[11]

2.4 Telecommuting and Home Workplaces

Workplaces are workplaces, even if they are in a private home. That was at least the principle OSHA relied on in 1999 to attempt to exert its authority over the growing number of white collar workers who use their modems rather than their motor cars to commute to work. Of course, OSHA had always claimed (if rarely exercised) jurisdiction over "sweat shops" and other industries even if operated from someone's home. Therefore, when OSHA was asked for a simple interpretation about its coverage of home office workers, it applied the same logic. In an interpretative ruling from the Office of Compliance Programs in November 1999, the Agency stated that OSHA would hold employers responsible for injuries to employees at home.[12] This triggered a political explosion.

---

[11] EPA learned this lesson back in 1984 when Deputy Administrator James Barnes quite properly deferred to OSHA on certain asbestos workplace matters. Congressional critics, who believed OSHA would not treat the matter seriously or competently, raised such furor that EPA retained jurisdiction. Even earlier, in 1973, OSHA and EPA had an acrimonious dispute over which agency should have primary jurisdiction over protecting farm workers from pesticides. EPA won.

[12] Richard Fairfax, Director of the Office of Compliance Programs, Opinion letter to CSC Credit Services of Houston, Texas, November 15, 1999. Lest one think this was merely a hasty OSHA response, note that the company's request for an opinion was submitted in August 1997, 27 months before.

The National Association of Manufacturers declared, "We see this as the long arm of OSHA coming into people's homes."[13] The chairman of a powerful congressional committee warned that the policy would put "home workers in the position of having to comply with thousands of pages of OSHA regulations."[14]

What OSHA had failed to realize was these new workers were not someone's employees needing protection from exploitive bosses. They were their own bosses, or they certainly saw themselves as such. And they saw OSHA intervention not as protective but intrusive.

On January 5, 2000, the Secretary of Labor, Alexis Herman, announced the cancellation of the short-lived OSHA policy.

3.0 Scope of OSHA Standards

To give the reader an idea of the areas covered by the standards, the following is a subpart listing from the Code of Federal Regulations, Part 1910, Occupational Safety and Health Standards. They are mostly safety standards. The health standards are all contained in Subpart Z, except for Subparts A, C, G, K and R, which cover both categories.

3.1 Areas Covered by the OSHA Standards

- **Subpart A**: General (purpose and scope, definitions, applicability of standards, etc.)
- **Subpart B**: Adoption and Extension of Established Federal Standards (construction work, ship repairing, long shoring, etc.)
- **Subpart C**: General Safety and Health Provisions (preservation of records)
- **Subpart D**: Walking-Working Surfaces (guarding floor and wall openings, portable ladders, requirements for scaffolding, etc.)
- **Subpart E**: Means of Egress (definitions, specific means by occupancy, sources of standards, etc.)
- **Subpart F**: Powered Platforms, Manlifts, and Vehicle-Mounted Work Platforms (elevating and rotating work platforms, standards, organizations, etc.)

---

[13] Jenny Krese, Director of NAM's employment policy, BNA, *OSHA Reporter*, January 6, 2000, p. 5.
[14] Rep. Pete Hoekstra (R-Mich.), chairman of the House Oversight and Investigations Subcommittee of the House Education and Workforce Committee, *id.*, January 13, 2000, p. 22.

- **Subpart G:** Occupational Health and Environmental Control (ventilation, noise exposure, radiation, etc.)
- **Subpart H:** Hazardous Materials (compressed gases, flammables, storage of petroleum gases, effective dates, etc.)
- **Subpart I:** Personal Protective Equipment (eye / face, respiratory, electrical devices, etc.)
- **Subpart J:** General Environmental Controls (sanitation, labor camps, safety color code for hazards, etc.)
- **Subpart K:** Medical and First Aid (medical services, sources of standards)
- **Subpart L:** Fire Protection (fire suppression equipment, hose and sprinkler systems, fire brigades, etc.)
- **Subpart M:** Compressed Gas and Compressed Air Equipment (inspection of gas cylinders, safety relief devices, etc.)
- **Subpart N:** Materials Handling / Storage (powered industrial trucks, cranes, helicopters, etc.)
- **Subpart O:** Machinery and Machine Guarding (requirements for all machines, woodworking machinery, wheels, mills, etc.)
- **Subpart P:** Hand and Portable Powered Tools and Other Hand-Held Equipment (guarding of portable power tools)
- **Subpart Q:** Welding, Cutting and Brazing (definitions, sources of standards, etc.)
- **Subpart R:** Special Industries (pulp, paper and paperboard mills, textiles, laundry machinery, telecommunications, etc.)
- **Subpart S:** Electrical (application, National Electrical Code)
- **Subpart T:** Commercial Diving Operations (qualification of team, pre- and post-dive procedures, equipment, etc.)
- **Subpart U-Y:** [Reserved]
- **Subpart Z:** Toxic and Hazardous Substances (air contaminants, asbestos, vinyl chloride, lead, benzene, etc.)

3.2 Overview of Standards

When OSHA was created, Congress realized that the new agency would require years to promulgate a comprehensive corps of health and safety standards. The

OSH Act therefore provided that for a two year period ending in April 1972 the Agency could adopt as its own the standards of respected professional and trade groups. These are the *consensus standards* issued under Section 6(a) of the statute.[15] Nobody could have imagined that three decades later these imperfect and outdated standards would still form the overwhelming majority of OSHA regulations.

3.3   Overview of Health Standards

Health issues, notably environmental contaminants in the workplace, have increasingly become OSHA's concern over the past few years. Health hazards are much more complex, more difficult to define, and because of the delay in detection, perhaps more dangerous to a larger number of employees. Unlike safety hazards, the effects of health hazards may be slow, cumulative, irreversible, and complicated by nonoccupational factors.

If a machine is unequipped with safety devices and maims a worker, the danger is clearly and easily identified and the solution usually obvious. However, if workers are exposed for several years to a chemical that is later found to be carcinogenic, there may be little help for those exposed.

In the nation's workplaces there are tens of thousands of toxic chemicals, many of which are significant enough to warrant regulation. Yet OSHA only has a list of fewer than 500 substances, and these are mostly simple threshold limits adopted under Section 6(a) from the recommended lists of private industrial hygiene organizations back in the 1960s and early 1970s. This list is being updated now but with glacial slowness.

The promulgation of health standards involves many complex concepts. To be complete, each standard needs medical surveillance requirements, recordkeeping, monitoring, and multiple physical reviews, just to mention a few. At the present rate, promulgation of standards on every existing toxic substance could take centuries.

Ironically, an attempt to update the health standards for hundreds of substances in one regulatory action by borrowing newer figures from respected health professional organizations was opposed by the labor unions (and industry) and struck down by an appellate court in 1992.[16]

---

[15]   These consensus standards are discussed below in Sections 3.4 and 4.1 of this chapter.

[16]   *AFL-CIO v. OSHA*, 965 F.2d 962 (11th Cir. 1992).

3.4 Overview of Safety Standards

Safety hazards are those aspects of the work environment that, in general, cause harm of an immediate and sometimes violent nature, such as burns, electrical shock, cuts, broken bones, loss of limbs or eyesight, and even death. The distinction from health hazards is usually obvious; mechanical and electrical are considered safety problems, while chemicals are considered health problems. Noise is difficult to categorize; it is classified as a health problem.

The Section 6(a) adoption of national consensus and other federal agency standards created chaos in the safety area. It was one thing for companies to follow industry or association guidelines that, in many cases, had not been modified in years; it was another thing for those guidelines actually to be codified and enforced as law. In the two years the act provided for OSHA to produce standards derived from these existing rules, the agency should have examined these closely, simplified them, deleted the ridiculous and unnecessary ones, and promulgated final regulations that actually identified and eliminated hazards to workers. But in the commotion of organizing an agency from scratch, it did not happen that way.

Nor did affected industry groups register their objections until later. During the entire two year comment period, not a single company or association filed an objection with OSHA.

Almost all of the so-called "Mickey Mouse" standards were safety regulations, such as the requirements that fire extinguishers be attached to the wall exactly so many inches above the floor. Undertrained OSHA inspectors often failed to recognize major hazards while citing industries for minor violations "which were highly visible, but not necessarily related to serious hazards to workers' safety and health."[17]

Section 6(g) of the OSH Act directs OSHA to establish priorities based on the needs of specific "industries, trades, crafts, occupations, businesses, workplaces, or work environments." The Senate report accompanying the OSH Act stated that the agency's emphasis initially should be put on industries where the need was determined to be most compelling.[18] OSHA's early attempts to target inspections, however, were sporadic and, for the most part, unsuccessful. The situation has improved somewhat in recent years, for both health and safety, in part because of

---

[17] Statement of Basil Whiting, Deputy Assistant Secretary of Labor for OSHA, before the Committee on Labor and Human Resources, U.S. Senate, 21 March 1980, pp. 5–6.

[18] For the legislative history of the act, *see especially* the Conference Report 91-1765 of 16 December 1970, as well as H.R. 91-1291 and S.R. 91-1282.

the recent requirement that some priority scheme be used that could justify search warrants.[19] But, as we shall see, that has brought its own problems.

4.0 Standard Setting

Setting standards can be a complex and protracted process. There are thousands of chemical substances, electrical problems, fire hazards, and many other dangerous situations prevalent in the workplace for which standards needed to be developed.

To meet the objectives defined in the act, three different standard-setting procedures were established:

1. Consensus Standards, under Section 6(a).
2. Permanent Standards, under Section 6(b).
3. Emergency Temporary Standards, under Section 6(c).

4.1 Consensus Standards: Section 6(a)

Congress realized that OSHA would need standards to enforce while it was developing its own. Section 6(a) allowed the agency, for a two-year period that ended on 25 April 1973, to adopt standards developed by other federal agencies or to adopt consensus standards of various industry or private associations.[20] This resulted in a list of around 420 common toxic chemicals with maximum permitted air concentrations specified in parts per million (ppm) or in milligrams per cubic meters (mg / M^3).

There are several problems inherent in these standards. First, these threshold values are the only elements to the standard. There are no required warning labels, monitoring, or medical recordkeeping, nor do they generally distinguish between the quite different health effects in 8-hour, 15-minute, peak, annual average, and other periods of exposure. Second, being thresholds, they are based on the implicit assumption that there are universal *no-effect levels*, below which a worker is safe. For carcinogens, this assumption is quite controversial.

Third, most of the standards were originally established not on the basis of firm scientific evidence but, as the name implies, from existing guidelines and limits of various industry, association, and governmental groups. Before OSHA's creation,

---

[19] *See Marshall v. Barlow's Inc.*, 436 U.S. 307 (1978), which will be discussed later in this chapter.
[20] 39 FR 23502, 27 June 1974.

they were intended to be general, nonbinding guidelines, and had been in circulation for a number of years with no urgency to keep them current. Consequently, neither industry nor labor bothered to comment when OSHA first proposed the consensus standards. Many of these "interim" standards were out of date by the time they were adopted by OSHA, and they are now frozen in time until OSHA goes through the full Section 6(b) administrative rulemaking process.

Fourth, OSHA consensus standards often involve "incorporation by reference," especially in the safety area. In some cases, these pre-1972 publications were not standards or even formal association guidelines but mere private association pamphlets that are no longer in print and not easily obtainable. For example, the general regulation on compressed gases merely states that the cylinders should be in safe condition and maintained "in accordance with Compressed Gas Association pamphlet P-1-1965" and several similar documents.[21]

Fifth, not all of these "toxics" are on the list because they really pose a health hazard. Although that has been the unquestioned assumption of certain later rulemakings, such as the requirement for Material Safety Data Sheets, some chemicals such as carbon black were listed because of "good housekeeping" practices—a facility with even a small amount of this intrusive black substance will look filthy—and not because it was hazardous at the levels set.

Nevertheless, Congress was undoubtedly correct in requiring the compilation of such a list. Otherwise, there would have been no OSHA health standards at the beginning. There are virtually no others even now.

4.2 Standards Completion and Deletion Processes

The agency has attempted to deal with one of the shortcomings of the consensus standards by what is called the Standards Completion Process. Over a number of years, OSHA has taken some threshold standards and added various medical, monitoring, and other requirements.[22] At least a broader range of protection is offered to exposed workers.

The agency has also sought to reduce the number of safety standards. This is done by eliminating the so-called "Mickey Mouse" standards that accomplish little, but impose voluminous requirements. More often, the simplification has come by

---

[21] 29 C.F.R. § 1910.101.

[22] Since the 6(a) process ended in April 1972, the standards promulgated thereunder cannot be modified or revised without going through the notice and comment administrative procedures under Section 6(b).

removing redundant sections and cross references. This eliminates pages, but not a lot more.

Nevertheless, OSHA is proud of its compliance with the presidential directive that federal agencies review and remove duplicative or repetitive regulations.[23]

4.3 Permanent Standards: Section 6(b)

Permanent standards must now be developed pursuant to Section 6(b). This is the familiar standard-setting and rule-making process followed by most other federal agencies under the Administrative Procedure Act.[24]

Permanent standards may be initiated by a well-publicized tragedy, court action, new scientific studies, or the receipt of a criteria document from the National Institute of Occupation Safety and Health (NIOSH), an organization described later in this chapter. The criteria document is a compilation of all the scientific reports on a particular chemical, including epidemiological and animal studies, along with a recommendation to OSHA for a standard. The recommendation, based supposedly only on scientific health considerations, includes suggested exposure limits (8-hour average, peaks, etc.) and appropriate medical monitoring, labeling, and other proscriptions.

Congress assumed that NIOSH would be the primary standard-setting arm of OSHA, although the two are in different government departments—Health and Human Services (HHS) and Labor, respectively. According to this model, OSHA would presumably take the scientific recommendations from NIOSH, factor in engineering and technical feasibility, and then promulgate as similar a standard as possible. However, the system has never worked this way. Instead, OSHA's own standards office has generally regarded NIOSH's contribution as just one step in the process—and not one entitled to a great deal of deference.[25]

Following receipt of the criteria document, or some other initiating action, OSHA will study the evidence and then possibly publish a proposed standard. Most candidate standards never get this far: the hundreds of NIOSH documents, labor

---

[23] *See*, for example, the OSHA press release of 19 June 1998: "OSHA Eliminates Over 1,000 Pages of Regulations to Save Employers Money, Reduce Paperwork, and Maintain Protection."

[24] 5 U.S.C. § 553 *et seq.*

[25] NIOSH criteria documents vary considerably in quality, depending in part on whom they were subcontracted to, but another problem is that too often they are insufficiently discriminating in evaluating questionable studies. That is, one scientific study is regarded as good as any other study, without regard to the quality of the data or the validity of the protocols. Of course, another factor in OSHA's attitude just might be the "not invented here" syndrome.

union petitions, and other serious recommendations have resulted in only a few new health standards since 1970.[26]

The proposed standard is then subjected to public comment for (typically) a 90-day period, often extended, after which the reactions are analyzed and informal public hearings are scheduled. In a few controversial instances, there may be more than one series of hearings and comments. Then come the post hearing comments, which are perhaps the most important presentations by the parties. After considerable further study, a final standard is eventually promulgated. The entire process might theoretically be accomplished in under a year, but in practice it takes a minimum of several years or, as with asbestos, even decades.

The following is a list of some of the final health standards which OSHA has promulgated to date:

1. Asbestos
2. Fourteen carcinogens

   - 4-Nitrobiphenyl
   - alpha-nephthylamine
   - methyl chloromethyl ether
   - 3,3'-dichlorolenzidine
   - bis-chloromethyl ether
   - beta-naphthylamine
   - 4-aminodiphenyl

   - benzidine
   - ethyleneimine
   - beta-propiolactone
   - 2-acetylaminofluorene
   - 4-dimethylaminozaobenzene
   - N-nitrosodimethylamine
   - (MOCA stayed by court action)

3. Vinyl chloride
4. Inorganic arsenic
5. Lead
6. Coke oven emissions
7. Cotton dust
8. 1,2-dibromo-3-chloropropane (DBCP)
9. Acrylonitrile

---

[26] This meager number of chemicals does *not* reflect OSHA's scientific judgment that the other candidates are unworthy or that the agency has sharply different priorities, although these may be partial factors. More important reasons are poor leadership, technical inexperience, and a bit of politics.

10. Ethylene oxide (EtO)
11. Benzene
12. Field Sanitation

The list is obviously incredibly short for three decades of OSHA standard setting.

4.4 Emergency Temporary Standards

The statute also provides for a third standard-setting approach, specified for emergency circumstances where the normal, ponderous rulemaking procedure would be too slow. Section 6(c) gives the agency authority to issue an *emergency temporary standard* (ETS) if necessary to protect workers from exposure to *grave danger* posed by substances "determined to be toxic or physically harmful or from new hazards."[27]

Such standards are effective immediately upon publication in the *Federal Register*. An ETS is only valid, however, for six months. OSHA is thus under considerable pressure to conduct an expedited rulemaking for a permanent standard before the ETS lapses. For this reason, a quest for an emergency standard has been the preferred route for labor unions or other groups seeking a new OSHA standard. These ETSs have not fared well, however, when challenged in the courts; virtually all have been struck down as insufficiently justified.

4.5 General Duty Clause, 5(a)(1)

There is actually a fourth type of enforceable standard, one that covers situations for which no standards currently exist.

Since OSHA has standards for only a few hundred of the thousands of potentially dangerous chemicals and workplace safety hazards, there are far more situations than the rules cover. Therefore, inspectors have authority under the *General Duty Clause* to cite violations for unsafe conditions even where specific standards do not exist.[28] Agency policy has shifted back and forth between encouraging the use of "Section 5(a)(1)," as the clause is often termed, since this ensures that unsafe conditions will be addressed, and discouraging its use on the theory that employers should be liable only for compliance with specific standards of which they are given knowledge.

---

[27] OSH Act § 6(c)(1).
[28] OSH Act § 5(a)(1); 29 U.S.C. § 654(a)(1).

However, the agency has acknowledged that many of the standards which do exist are woefully out of date and thus cannot be relied upon for adequate protection of worker safety and health. The traditional notion was that compliance with an existing specific standard—even if demonstrably unsafe—precluded an OSHA citation.[29] This has been called into question by the courts. In April 1988, a federal appellate court allowed OSHA to cite for violations of the General Duty Clause even where a company was in full compliance with a specific numerical standard on the precise point in question.[30] Bare compliance with the standards on the books, therefore, might not be responsible management.

4.6 Feasibility and the Balancing Debate

There has been a continuing debate over feasibility and balancing in OSHA enforcement. The important issues include the following:

- Can OSHA legally consider economic factors in setting health or safety standards levels?

- If so, is this consideration limited only to extreme circumstances?

- Does the Occupational Safety and Health Act provide for a balancing of costs and benefits in setting standards?

- Can OSHA mandate engineering controls although they alone would still not attain the standard?

- And, can OSHA require engineering controls even if personal protective equipment (such as ear plugs) could effectively, if often only theoretically, reduce hazards to a safe level and at a much lower cost?

---

[29] This is exemplified by *Phelps Dodge Corporation* (OSHRC Final Order, 1980), 9 OSHC 1222, which found no violation of the act to expose workers to "massive amounts of sulphur dioxide for short periods of time" since there was no maximum ceiling value in the standard and the employer was complying with the eight-hour average value required in the specific sulphur dioxide regulation. The citation for violation of § 5(a)(1) was therefore vacated.

[30] *International Union, UAW v. General Dynamics Land System Division*, 815 F.2d 1570, 13 OSHC 1201 (CADC 1988). The Court held that employer's knowledge was the crucial element; if he knew that the OSHA standard was not adequate to protect workers from a hazard, he could not claim he was maintaining a safe workplace within the meaning of § 5(a)(1), even if he were adhering to a standard he knew was outmoded. The Court thereby dismissed the argument that the employer would not know what is legally expected of him; he was expected to maintain a safe workplace, specific regulations notwithstanding. There was no specific provision in the statute which prevented a general duty citation when a specific standard existed. Note, however, that no other court has since used this rationale.

These questions have been extensively litigated before the Occupational Safety and Health Review Commission (OSHRC) and the courts. Most of the debate has been over the interpretation of *feasibility* in Section 6(b)(5) of the act.

One must remember that OSHA legislation was originally seen by Congress in rather absolutist terms: any standard promulgated should be one "which most adequately assumes...that no employee will suffer material impairment of health." Only late in the congressional debate was the Department of Labor able to insert the phrase "to the extent feasible" into the text. This was intended to prevent companies having to close because unattainable standards were imposed on them, but it was not spelled out to what extent economic as well as technical feasibility was included.[31]

Since the term *feasibility* was not clearly defined, there has been much confusion over how to interpret what Congress intended, as the earlier cases show. In *Industrial Union Department, AFL v. Hodgson*, the D.C. Circuit accepted that economic realities affected the meaning of *feasible*, but only to the extent that "a standard that is prohibitively expensive is not 'feasible.'"[32] It was Congress' intent, the court added, that this term would prevent a standard unreasonably "requiring protective devices unavailable under existing technology or by making financial viability generally impossible." The court warned, however, that this doctrine should not be used by companies to avoid needed improvements in their workplaces:

> Standards may be economically feasible even though, from the standpoint of employers, they are financially burdensome and affect profit margins adversely. Nor does the concept of economic feasibility necessarily guarantee the continued existence of individual employers.[33]

A similar view was adopted in 1975 by the Second Circuit in *The Society of the Plastics Industry v. OSHA*, written by Justice Clark, who cited approvingly the case above.[34] He held that *feasible* meant not only that which is attainable technologically and economically now, but also that which might reasonably be achievable in the future. In this case, which concerned strict emissions controls on vinyl chloride, he

---

[31] This account of the behind-the-scenes machinations is based on this author's personal discussions with the late Congressman William Steiger (R-Wisc.), a principal author of the act, and Judge Lawrence Silberman, now of the Court of Appeals for the District of Columbia Circuit who was then Solicitor of Labor. The legislative history is relatively unhelpful on this subject. *See*, for example, hearings before the Select Subcommittee on Labor, Committee on Education and Labor, "Occupational Safety and Health Act of 1969," two vols., 1969.

[32] 499 F.2d 467, 1 OSHC 1631 (D.C. Cir., 1974).

[33] 1 OSHC 1631 at 1639.

[34] 509 F.2d 1301, 2 OSHC 1496 (2nd Cir., 1975), *cert. den.* 421 U.S. 922.

declared that OSHA may impose "standards which require improvements in existing technologies or which require the development of new technology, and...is not limited to issuing standards based solely on devices already fully developed."[35]

Neither court undertook any risk-benefit analysis, such as attempting to compare the hundreds of millions of dollars needed to control vinyl chloride with the lives lost to angiosarcoma of the liver. Those who have attempted to develop such equations have generally concluded the task is undoable, at least for most such chronic health effects.[36]

A third federal appeals court, however, took a strongly contrary position in a case involving noise. In *Turner Co. v. Secretary of Labor*, the Seventh Circuit Court of Appeals decided that the $30,000 cost of abating a noise hazard should be weighed against the health damage to the workers, taking into consideration the availability of personal protective equipment to mitigate the risk.[37]

This holding is not unreasonable, but is based on a highly tenuous interpretation of the law. The court, without providing any clear rationale for its view, held that "the word 'feasible' as contained in 29 CFR § 1910.95(6)(1) must be given its ordinary and common sense meaning of 'practicable.'" (This may be so, but is of no analytical value.) From this the court concluded:

> Accordingly, the Commission erred when it failed to consider the relative cost of implementing engineering controls...versus the effectiveness of an existing personal protective equipment program utilizing fitted earplugs.[38]

This interpretation does not follow from the analysis. In fact, since the Turner Company had both the financial resources and the technical capability to abate the noise problem, compliance with the regulation would appear to be "practicable." The court, however, considered this term to mean that a cost-benefit computation should be made.

In 1982, the U.S. Court of Appeals for the Ninth Circuit, in the case of *Donovan v. Castle & Cooke Foods and OSHRC*,[39] also held that the Noise Act and the

---

[35] 509 F.2d at 1309, 2 OSHC at 1502 (2nd Cir., 1975).

[36] *See*, for example, the conclusions of the National Academy of Sciences report, "Government Regulation of Chemicals in the Environment," 1975.

[37] 561 F.2d 82, 5 OSHC 1970 (7th Cir., 1977). The Occupational Safety and Health Review Commission (OSHRC) decisions on *Turner* and the related *Continental Can* case can be found at 4 OSHC 1554 (1976) and 4 OSHC 1541 (1976), respectively.

[38] 5 OSHC 1790 at 1791.

[39] 692 F.2d 641, 10 OSHC 2169 (1982).

regulations permit consideration of relative costs and benefits to determine what noise controls are feasible.

OSHA gave the plant a citation on the grounds that, although Castle & Cooke required its employees to wear personal protective equipment, its failure to install technologically feasible engineering and administrative controls[40] constituted a violation of the noise standard, and that the violation could only be abated by the implementation of such controls. OSHA argued that engineering and administrative controls should be considered economically infeasible only if their implementation would so seriously jeopardize the employer's economic condition as to threaten continued operation.

On appeal, OSHA argued that neither the OSHRC nor the courts are free to interpret *economic feasibility*, because its definition is controlled by the Supreme Court's decision in *American Textile Manufacturers Institute, Inc. v. Donovan*.[41] The appeals court, however, decided that the Supreme Court's interpretation of the term *feasible* made in American Textile was not deemed controlling for the noise standards. It also affirmed that economic *feasibility* should be determined through a cost-benefit analysis, and that in the case of *Castle & Cooke* the costs of economic controls did not justify the benefit that would accrue to employees. Thus, the decision to vacate the citation was upheld.

5.0 Variances

Companies that complain that OSHA standards are unrealistic are often not aware that they might be able to create their own version of the standards. The alternative proposed has to be at least as effective as the regular standard, but it can be different.

---

[40] Engineering controls are those that reduce the sound intensity at the source of that noise. This is achieved by insulation of the machine, by substituting quieter machines and processes, or by isolating the machine or its operator. Administrative controls attempt to reduce workers' exposure to excess noise through use of variable work schedules, variable assignments, or limiting machine use. Personal protective equipment includes such devices as ear plugs and ear muffs provided by the employer and fitted to individual workers.

[41] 101 S. Ct. 2478, 9 OSHC 1913 (17 June 1981). In this case, representatives of the cotton dust industry challenged proposed regulations limiting permissible exposure levels to cotton dust. Section 6(b)(5) of the act requires OSHA to "set the standard which most adequately assures, to the extent feasible...that no employee will suffer material impairment of health...." The industry contended that OSHA had not shown that the proposed standards were economically feasible. However, the Supreme Court upheld the cotton dust regulations, holding that the "plain meaning of the word 'feasible' is capable of being done, executed, or effected," and that a cost-benefit analysis by OSHA is not required.

5.1 Temporary Variances

Section 6(b)(6)(A) of the OSH Act establishes a procedure by which any employer may apply for a "temporary order granting a variance from a standard or any provision thereof." According to the act, the variance will be approved when OSHA determines that the requirements have been met and establishes:

- that the employer is unable to meet the standard "because of unavailability of professional or technical personnel or of materials and equipment," or because alterations of facilities cannot be completed in time;
- that he is "taking all available steps to safeguard" his workers against the hazard covered by the standard for which he is applying for a variance; and
- that he has an "effective program for coming into compliance with the standard as quickly as practicable."[42]

This temporary order may be granted only after employees have been notified and, if requested, there has been sufficient opportunity for a hearing. The variance may not remain in effect for more than one year, with the possibility of only two six-month renewals.[43] The overriding factor an employer must demonstrate for a temporary variance is good faith.[44]

5.2 Permanent Variances

Permanent variances can be issued under Section 6(d) of the OSH Act. A permanent variance may be granted to an employer who has demonstrated "by a preponderance" of evidence that the "conditions, practices, means, methods, operations or processes used or proposed to be used" will provide a safe and healthful workplace as effectively as would compliance with the standard.

6.0 Compliance and Inspections

OSHA is primarily an enforcement organization. In its early years both the competence of its inspections and the size of the assessed fines were pitifully inadequate; they were the primary reason OSHA was not taken seriously by either labor unions or the business community. That picture has now changed significantly.

---

[42] OSH Act § 7(b)(6)(A).
[43] Id.
[44] E. Klein, Variances, in *Proceedings of the Occupational Health and Safety Regulations Seminar* (Washington, D.C.: Government Institutes, 1978), p. 74.

6.1 Field Structure

The Department of Labor (DOL) has divided the territory subject to the OSH Act into ten federal regions, the same boundaries that EPA also uses. Each region contains from four to nine area offices. When an area office is not considered necessary because of a lack of industrial activity, a district office or field station may be established. Each region is headed by a regional administrator, each area by an area director. In the field, compliance officers represent area offices and inspect industrial sites in their vicinity, except in situations where a specialist or team might be required.

6.2 Role of Inspections

The only way to determine compliance by employers is inspections, but inspecting all the workplaces covered by the OSH Act would require decades. Each year there are tens of thousands of federal inspections, and as many or more state inspections, but there are several million workplaces. Obviously, a priority system for high-hazard occupations is necessary, along with random inspections just to keep everyone "on his toes."

Inspections are supposed to be surprises; there are penalties for anyone alerting the sites beforehand. The inspections may occur in several ways: they may be targeted at random, triggered by worker complaints, set by a priority system based on hazardous probabilities, or by events such as a fatality or explosion. Inspectors expect admittance without search warrants, but a company has the Constitutional right to refuse admittance until OSHA obtains a search warrant from a federal district court.[45] Such refusal is frankly not a good idea except in very special circumstances, such as when the additional delay would allow a quick cleanup of the workplace to bring it into compliance.

---

[45] *See* a later section in this chapter on the Supreme Court's *Barlow's* decision interpreting the Fourth Amendment to the U.S. Constitution.

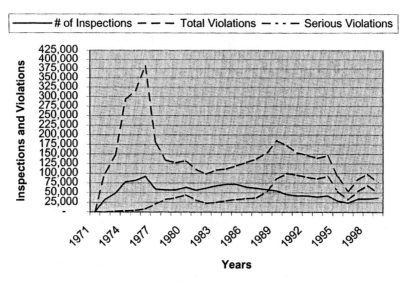

Figure 1.1 Inspections and Violations

6.3 Training and Competence of Inspectors

There has been a major problem with OSHA inspectors in the past—the training program did not adequately prepare them for their tasks, and the quality of the hiring was uneven. In the early days there was tremendous pressure from the unions to get an inspection force on the job as soon as possible, so recruitment was often hurried and training was minimal. Inspectors would walk into a plant where, for example, pesticide dust was so thick workers could not see across the room, yet, because there was no standard as such, the inspectors would not think there was a problem.[46] Yet, had there been a fire extinguisher in the wrong place, and had the inspector been able to see it through the haze, he would have cited the plant for a safety violation.

Competence among staff has markedly improved since the early days of the program. Both in-house training efforts by OSHA and increased numbers of professional training programs conducted by colleges and universities have contrib-

---

[46] This happened with Kepone in the notorious Hopewell, Virginia, incident in 1975, and with asbestos for years at a plant in Tyler, Texas.

uted to these improvements. There is also a greater sensitivity towards workers and their representatives.[47]

6.4 Citations, Fines and Penalties

If the inspector discovers a hazard in the workplace, a citation and a proposed fine may be issued. Citations can be serious, nonserious, willful, or repeated. By one count, there are at least nine types of penalty findings under the OSH Act. They are as follows:

- **De minimis**—A technical violation, but one posing insignificant risk and for which no monetary penalty is warranted.
- **Non-Serious**—The basic type of penalty. No risk of death or serious injury is posed, but the violation might still cause some harm.
- **Serious**—The hazard could lead to death or serious injury.
- **Failure to correct**—Violations when found must be remediated within a certain period of time. If a subsequent reinspection finds this has not been done, or has been allowed to recur, this fairly serious citation is in order.
- **Repeated**—Continuous violations, discussed below.
- **Willful**—Intentional violations, discussed below.
- **Criminal**—Applicable under the OSH Act only for cases involving death.
- **Egregious**—Supposedly heinous situations, discussed below.
- **Section 11(c)**—Penalties for company retaliation against complainers and whistleblowers, discussed at length below.

---

[47] Statement of Lane Kirkland, President, AFL-CIO, before the Senate Committee on Labor and Human Resources on Oversight of the Occupational Safety and Health Act, 1 April 1980.

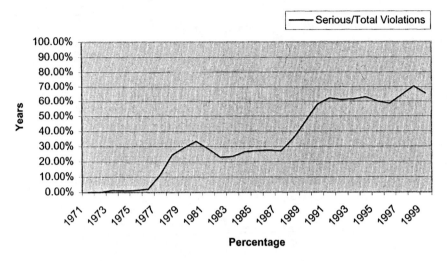

Figure 1.2 Serious Violations as a Proportion of Total Violations

6.5 OSHA Citation and Penalty Patterns

OSHA now averages about 35,000 inspections a year. These are focused on the industries and sectors where statistics indicate greater potential hazards. Contrary to the common assumption that most inspections are in manufacturing, in fact that sector accounts for only about one-fourth of the inspections. Over half are in the construction industry, with another quarter distributed over all other types of workplaces.

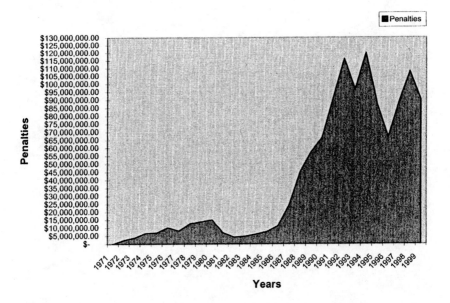

Figure 1.3 Penalties

Specially targeted sectors in the manufacturing area, with four-digit SIC codes, have most recently been designated:

- plastic products (3089)
- sheet metalwork (3444)
- fabricated structural metal (3441)
- metal stampings (3469)
- fabricated metal products (3499)
- motor vehicle parts (3714)
- construction machinery (3431)
- shipbuilding and repair (3731)[48]

---

[48] "Top Ten Federal OSHA Targeted SIC Codes," *Manufacturing Sector*, 4th Quarter 1998, OSHA.

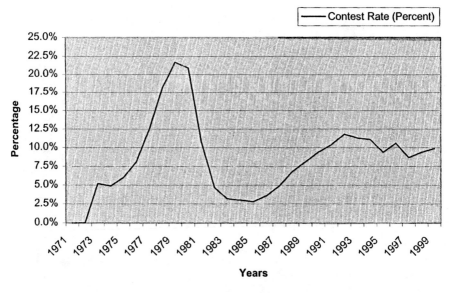

Figure 1.4 Contest Rate

6.6 Communicating and Enforcing Company Rules

Many, arguably, even most, accidents are due to human negligence, often involving an act which is contrary to company policy. Merely claiming a company policy, however, is not enough, for OSHA does not look very favorably upon this defense. For employers to plead employee misconduct as a defense to an OSHA citation, the company must first demonstrate three things:

- First, of course, is to prove the existence of such rules.[49]

- Second, an employer must prove that these rules were effectively communicated to the employees. Proof can include written instructions, evidence of required attendance at education sessions, the curriculum of training programs, and other forms that should be documented.[50]

---

[49] *The Carborundum Company* (OSHRC Judge, 1982), 10 OSHC 1979.
[50] *Schnabel Associates, Inc.* (OSHRC Judge, 1982), 10 OSHC 2109. Moreover, employers should retain copies of training curriculum, tests, and other evidence of the educational program, recommends Susan M. Olander, counsel for the Federated Rural Electric Insurance Exchange. BNA, *OSHA Reporter*, 19 October 200, p. 933-34.

- Third, many companies that can demonstrate the above two principles fall short on the third, namely that there should be evidence the policies are effectively enforced.[51] For this, evidence of disciplinary action taken against infractions of the rules, though not necessarily the precise rule that would have prevented the accident under investigation, is necessary. The closer to the actual circumstance, of course, the more that proof of active company enforcement is dispositive.[52]

If the above three principles can all be demonstrated, they constitute a reasonable defense to charges of violating the regulations, even in cases of death or serious injury.

Note that this defense is not limited to the misconduct of a low-ranking employee. Misconduct of a supervisor, although it may suggest inadequate company policy and direction, can also be shown as an isolated and personal failing. According to an appellate court, the proper focus of a court is on the effectiveness of the employer's implementation of his safety program, and not on whether the unforeseeable conduct was by an employee or by supervisory personnel.[53]

6.7 Warrantless Inspections: The *Barlow* Case

Litigants have challenged OSHA's constitutionality on virtually every conceivable grounds from the First Amendment to the Fourteenth.[54] The one case that has succeeded has led to the requirement of a search warrant, if demanded, for OSHA inspectors.

The Supreme Court in *Marshall v. Barlow's Inc.*[55] decided that the Fourth Amendment to the Constitution, providing for search warrants, was applicable to OSHA, thereby declaring unconstitutional Section 8(a) of the Act, in which Congress had authorized warrantless searches.[56]

---

[51] *Galloway Enterprises, Inc.* (OSAHRC Judge, 1984), 11 OSHC 2071.
[52] *Bethlehem Steel Corporation, Inc.* (OSAHRC Judge, 1985), 12 OSHC 1606. *Dover Electric Company, Inc.* (OSAHRC Judge, 1984), 11 OSHC 2175.
[53] *Brock v. L. E. Myers Company*, 818 F.2d 1270, 13 OSHC 1289 (6th Cir., 1987).
[54] A good, if dated, summary of these challenges is found in Volume I of Walter B. Connolly and David R. Cromwell, II, *A Practical Guide to the Occupational Safety and Health Act* (New York: New York Law Journal Press, 1977).
[55] 436 U.S. 307 (1978).
[56] There are circumstances in which warrants are not required, such as federal inspection of liquor dealers, gun dealers, automobiles near international borders, and in other matters with a long history of federal involvement.

While the court held that OSHA inspectors are required to obtain search warrants if denied entry to inspect, it added that OSHA must meet only a very minimal *probable cause* requirement under the Fourth Amendment in order to obtain them. As Justice White explained:

> Probable cause in the criminal sense is not required. For purposes of an administrative search such as this, probable cause justifying the issuance of a warrant may be based not only on specific evidence of an existing violation but also on a showing that "reasonable legislative or administrative standards for conducting an...inspection are satisfied with respect to a particular [establishment]."[57]

Moreover, if too many companies demanded warrants so that the inspection program was seriously impaired, the Court indicated it might reconsider its ruling. This ironically would make enjoyment of a Constitutional right partly contingent on few attempting to exercise it. It is therefore not surprising that commentators, both liberals and conservatives, were critical of the decision. Conservative columnist James J. Kilpatrick declared flatly:

> If the Supreme Court's decision in the *Barlow* case was a "great victory," as Congressman George Hansen proclaims it, let us ask heaven to protect us from another such victory anytime soon.[58]

7.0 Recordkeeping

For an agency that seems grounded in practical workplace realities, OSHA's regulations increasingly emphasize recordkeeping and paperwork requirements. Moreover, recent OSHA enforcement efforts have been directed heavily toward paperwork violations.

7.1 Accident Reports

Any workplace accident requiring treatment or resulting in lost work time must be recorded within six working days on an *OSHA Form 200*. This is officially entitled the Log and Summary of Recordable Occupational Injuries and Illnesses, although no one uses that longer term. This document is supposed to provide insight into accident types and causes for both the company and OSHA inspectors. It must be

---

[57] *Marshall v. Barlow's Inc., supra*, quoting *Camara v. Municipal Court*, 387 U.S. 523 at 538 (1967).
[58] *Washington Star*, 2 June 1978.

retained for five years. Criminal penalties apply to any "knowing false representation" on these and other required records.[59]

A second document is the *OSHA Form 101*, which describes in detail the nature of each of the recorded accidents. All the supporting information does not have to be on this one form, provided that the material is available in the file. This form is officially called the Supplementary Record of Occupational Injuries and Illnesses.

A third required document is the *Annual Summary* of accidents and illnesses, statistics based on the Form 200 data. This summary must be signed by a responsible corporate official and posted in some conspicuous place by the following 1st of February each year.[60]

7.2 Monitoring and Medical Records

OSHA's health standards increasingly contain provisions calling for medical records, monitoring of pollution, and other information. Safety, as well as health standards, may also require periodic inspections of workplaces or equipment, and these inspections must be recorded. These medical and exposure records must be retained for a staggering *30 years*. A company going out of business or liquidating must transfer these records to NIOSH.[61]

For example, the OSHA noise standards mandate baseline and periodic hearing tests,[62] the lead standard requires measuring of blood-lead levels and other data which can be the basis for removal from the workplace until the levels go down, and the ionizing radiation regulation requires careful recording of exposure and absorption information.

A host of safety (and some health) regulations requires (1) written safety programs, (2) specified training, (3) documented routine inspections, or combinations of all three.

There is no clear pattern to these requirements; they must be checked separately for each regulation. For example, the safety standard on derricks does not require the first but does require the second and third, while cranes require only the third.[63]

---

[59] OSH Act § 17(g).
[60] Recordkeeping requirements are set forth generally in 29 CFR 1904.
[61] 29 C.F.R. § 1910.20.
[62] OSHA's noise monitoring and recordkeeping requirements for hearing loss and *standard threshold shift* (STS, previously "significant threshold shift") are particularly complex and have been subject to considerable litigation.
[63] 29 C.F.R. § 1910.181, and § 1910.178–179.

Some safety standards, such as fire protection, lockout / tagout, process safety management, and employee alarms, require all three.[64] The health standards tend to require all three as well, including those for bloodborne pathogens and for hazard communications.[65]

OSHA has increasingly levied substantial fines for failure to comply with these recordkeeping regulations. For example, in October 2000 a Texas steelmaker was fined $1.7 million, much of it for "purposefully" not recording workplace injuries and illnesses.[66]

7.3 Hazard Communication

The OSHA hazard communication (hazcom) program, which is described more fully later in this chapter, requires companies making or using hazardous chemicals to provide information to their workers on possible exposure risks. The program provides for these measures:

1. labeling of toxic chemicals,
2. warning signs and posters,
3. material safety data sheets (MSDS) on hazardous chemicals,
4. a written policy setting forth the company's handling of issues under the hazcom program; and
5. a list of hazardous chemicals on premises.[67]

7.4 Access to Records

Employees and their designated legal or union representatives have the right to obtain access to their records within 15 working days. They may not be charged for duplication or other costs. Former employees are also given this access.

There are certain limited exceptions to disclosure dealing with psychiatric evaluation, terminal illness, and confidential informants. Otherwise the view is that even the most secret chemical formulas and business information must be revealed to the employees or former employees if they are relevant to exposure and toxicity.

---

[64] 29 C.F.R. §§ 1910.156 *et seq.*, 1910.147, 1910.119, and 1910.165.
[65] 29 C.F.R. § 1910.1030, and 29 C.F.R. § 1200.
[66] BNA, *OSHA Reporter*, 26 October 2000, p. 951.
[67] 29 C.F.R. 1910.1200.

This could be a godsend for industrial espionage, but so far there have been few claims that this is a practical problem.

OSHA inspectors also have access to these records. From time to time some company challenges this access as a violation of the Fourth Amendment, but an inspector has little difficulty in obtaining a search warrant.

7.5 Programmatic Standards

OSHA is giving more attention to programmatic standards. The controversial ergonomics standard, for example, is based on companies providing evidence that they have set up a specific program rather than having OSHA dictate what the detailed content of that program should be. Although these are perhaps not recordkeeping in the absolute sense, their reliance on paperwork and documentation merits their mention in this recordkeeping section.

8.0 Refusal to Work and Whistleblowing

Employees have a right to refuse to work when they believe conditions are unsafe. OSHA rules protect them from discrimination based on this refusal. And if employees see unsafe or unhealthy workplace conditions, they have a right to report them to OSHA without fear of reprisals or discrimination.

8.1 Refusal to Work

OSHA has ruled, and the Supreme Court has unanimously upheld, the OSHA principle that workers have the right to refuse to work in the face of serious injury or death.[68] The leading case was a simple one. Two workers refused to go on some wire mesh screens through which several workers had fallen part way and, two weeks before, another worker had fallen to his death. When reprimanded, the workers complained to OSHA. The Supreme Court had no difficulty in finding that the workers had been improperly discriminated against by their employer in this case.

How a court would rule in less glaring circumstances is harder to predict. Interestingly, there has not been a swarm of such cases in the two decades since this decision, despite dire predictions of wholesale refusal and consequent litigation.

---

[68] 29 C.F.R. § 1977.12 (1979); *Whirlpool Corp. v. Marshall*, 445 U.S. 1, 8 OSHC 1001 (1980). This case also stands as a textbook example of when not to appeal a lower court's ruling.

8.2 Protection of Whistleblowing

If a worker is fired or disciplined for complaining to governmental officials about unsafe work conditions, he has a legal remedy under the OSH Act for restoration of his job or loss of pay.[69] Similar provisions, administered also by OSHA's "11(c)" staff, have been inserted into other federal statutes, including EPA's Emergency Planning and Community Right-to-Know Act (EPCRA) in the Superfund legislation.[70]

Congress assumed that the employees in a given workplace would be best acquainted with the hazards there. It therefore statutorily encouraged prompt OSHA response to worker complaints of violations.[71] Since this system could be undermined if employers penalized complaining employees, the act in Section 11(c) provides sanctions against such retaliation or discrimination:

> No person may discharge or in any manner discriminate against any employee because such employee has filed any complaint or instituted or caused to be instituted any proceeding under or related to this Act or has testified or is about to testify in any such proceeding or because of the exercise by such employee on behalf of himself or others of any right afforded by this Act.[72]

If discrimination occurs, particularly if an employee is fired, a special OSHA team intervenes to obtain reinstatement, back wages, or—if return to the company is undesirable—a cash settlement for the worker. If agreement cannot be reached, the agency resorts to litigation.

This entire system has not worked as expected. First, worker complaints have surprisingly not been a very fruitful source of health and safety information. Far too many of the complaints came in bunches, coinciding with labor disputes in a

---

[69] OSH Act § 11(c), 29 U.S.C. § 660; 29 C.F.R. § 1977.

[70] Title III of the Superfund Amendment and Reauthorization Act of 1986 (SARA). These amendments are designed to "prevent future Bhopals" (the chemical disaster that killed thousands of residents of the city of Bhopal, India) by informing community fire and emergency centers what chemicals a company has on-site.

[71] OSH Act § 8(f)(1).

[72] OSH Act § 11(c)(1).

particular plant.[73] OSHA therefore finally had to abandon its policy of trying to investigate every single complaint. [74]

Second, the Title 11(c) process has worked slowly and uncertainly, so even though an employee may receive vindication, the months (or even years) of delay and anguish are a strong disincentive for workers to report hazards.

Third, it is often difficult to determine whether a malcontented worker was fired for informing OSHA or for a number of other issues which might cloud the employer-employee relationship. Does the complaint have to be the sole cause of dismissal or discrimination, or can some (fairly arbitrary) allocation be made?

Fourth, there is continuing controversy over whether 11(c) should protect workers complaining of hazards to those other than OSHA, even if the direct or indirect result is an OSHA inspection.

In the famous Kepone tragedy of 1975, an employee complained of hazardous chemicals to his supervisor, was fired, and only then went to OSHA. Not only was he declared unprotected by the act, but his complaint, no longer a worker complaint, was not even investigated at the time. Although agency officials have sworn not to repeat that mistake, the issue of what triggers 11(c) protection, (a) a complaint of unsafe workplace conditions, or (b) reporting that matter to OSHA, is a continuing one.

A related current issue is whether an employee who reports a hazard to the press, whose ensuing publicity triggers an OSHA investigation, is protected by 11(c). In one notable instance, OSHA regional officials decided in favor of the worker and won the subsequent litigation in federal district court. The Solicitor of Labor, however, disagreed and attempted to withdraw the agency from a winning position.[75]

For all these reasons, therefore, a worker can never really predict how he might fare if he does complain.

---

[73] A contrary view by Peg Seminario, AFL-CIO's director of Health and Safety, is that "Historically OSHA inspections conducted as a result of a complaint produce just as significant results in identifying serious violations and uncovering hazards as the general scheduled inspections." Quoted in BNA, *OSHA Reporter*, 16 March 2000, p. 202.

[74] OSHA has nevertheless strengthened the workers' role in the on-site consultation process. 29 C.F.R. § 1908 (December 2000).

[75] *Washington Post*, "About Face Considered in OSHA Suit," 20 October 1982.

9.0 Federal and State Employees

The exclusion of federal and state employees has been the topic of much discussion and debate.

9.1 Federal Agencies

Federal employees are not covered directly by OSHA, at least not to the extent that federal agencies are subject to fines and other penalties. However, the presumption was that the agencies would follow OSHA regulations in implementing their own programs. Section 19 of the OSH Act designates the responsibility for providing safe and healthful working conditions to the head of each agency. A series of Presidential Executive Orders has emphasized that this role should be taken seriously. Nevertheless, many commentators feel the individual agencies' programs are inadequate and inconsistent.

In 1980 the leading Presidential Executive Order[76] was issued, which broadened the responsibility of federal agencies for protecting their workers, expanded employee participation in health and safety programs, and designated circumstances under which OSHA will inspect federal facilities. In the operation of their internal OSHA programs, agency heads have to meet requirements of basic program elements issued by the Department of Labor and comply with OSHA standards for the private sector unless they can justify alternatives.

9.2 State Employees

The OSH Act excludes the employees of state governments. Virtually all states with their own OSHA programs, about half, however, cover their state and local employees. Some labor unions believe this exclusion of state workers is one of the most serious gaps in the OSH Act, and several congressional bills have sought in vain to remedy the perceived omission. In light of recent Supreme Court decisions, however, such bills even if enacted might not be constitutional.

10.0 State OSHA Programs

The federal OSHA program was intended by many legislators and businesses only to fill the gaps where state programs were lacking. The states were to be the primary

---

[76] Executive Order 12196, signed 26 February 1980, 45 *FR* 12769, superseding E.O. 11807 of 28 September 1974.

regulatory control. It has not happened that way, of course, but approximately two dozen state programs are still important.[77]

10.1 Concept

The OSH Act requires OSHA to encourage the states to develop and operate their own job safety and health programs, which must be "at least as effective as" the federal program.[78] Until effective state programs were approved, federal enforcement of standards promulgated by OSHA preempted state enforcement,[79] and continue to do so where state laws have major gaps. Conversely, state laws remain in effect when no federal standard exists.

Before approving a submitted state plan, OSHA must make certain that the state can meet criteria established in the act.[80] Once a plan is in effect, the secretary may exercise "authority…until he determines, on the basis of actual operations under the State plan, that the criteria set forth are being applied."[81] But he cannot make such a determination for three years after the plan's approval. OSHA may continue to evaluate the state's performance in carrying out the program even after a state plan has been approved. If a state fails to comply, the approval can be withdrawn, but only after the agency has given due notice and opportunity for a hearing.

10.2 Critiques

The program has not developed as anticipated into an essentially state-oriented system, although almost half the states have their own system.

Organized labor has never liked the state concept, both because of its poor experience with state enforcement in the past and because it realized that its strength could more easily be exercised in one location—Washington, D.C.—than in all fifty states and the territorial capitals, many of which are traditionally hostile to labor unions. This has meant, ironically, that some of the better state programs, in areas where unions had the most influence, were among the first rejected by state legislators under strong union pressure.

---

[77] There are 23 states and territories with complete state plans for the private and public sectors, and two that cover only public employees. Eight others have withdrawn their programs.
[78] OSH Act §§ 2(b)(11) and 18(c)(2).
[79] OSH Act § 18(a).
[80] OSH Act § 18(c)(1)–(c)(8).
[81] OSH Act § 18(c).

Industry has cooled to the local concept, which requires multistate companies to contend with a variety of state laws and regulations instead of a uniform federal plan. Furthermore, state OSHAs are often considerably larger than the local federal force, so there can be more inspections.

It was therefore never clear what incentive a state had to maintain its own program, since a governor could always terminate his state's plan and save the budgetary expenses, knowing that the federal government would take up the slack. California's Governor Dukmejian, for example, came to this conclusion in 1987 and terminated the state Cal-OSHA. However, the idea did not stick; California's state program was soon reestablished and, surprisingly, the notion did not spread.

Organized labor and industry are not alone in their criticism of the state programs. Health research organizations, OSHA's own national advisory committee (NACOSH), and some of the states themselves have also voiced disapproval of the state program policy. Ineffective operations at the state level, disparity in federal funding, and the lack of the necessary research capability are just a few of the criticisms lodged.[82]

There is some defense of state control, however. "To the extent that local control increases the responsiveness of programs to the specific needs of people in that area, this [a state plan] is a potentially good policy."[83] But reevaluation and revision will be necessary in the next several years if OSHA's policy for state programs is to be accepted by all the factions involved.

11.0 Consultation

Employers subject to OSHA regulation, particularly small employers, would benefit from onsite consultation to determine what must be done to bring their workplaces into compliance with the requirements of the OSH Act. This was particularly true during the agency's formative years. Although OSHA's manpower and resources are limited, this assistance, where rendered, should be free from citations or penalties.

As in so many other areas of OSHA regulation, there has been a great deal of controversy surrounding the consultation process. Union leaders have always feared that OSHA could become merely an educational institution rather than one with effective enforcement. But Section 21(c) of the act does mandate consultation with

---

[82] Robert Hayden, "Federal and State Rules" in *Proceedings of the Occupational Health and Safety Regulation Seminar* (Washington, D.C.: Government Institutes, 1978), pp. 9–10.

[83] Nicholas A. Ashford, *Crisis in the Workplace: Occupational Disease and Inquiry* (Boston: MIT Press, 1976), p. 231.

employers and employees "as to effective means of preventing occupational injuries and illnesses."[84]

Along with the consultation provisions, the statute provides for "programs for the education and training of employers and employees in the recognition, avoidance, and prevention of unsafe or unhealthful working conditions in employments covered" by the act.[85] OSHA produces brochures and films to educate employees about possible hazards in their workplaces. But, there are problems at every stage of the information process, from generation to utilization.

In 1979, OSHA experimented with a New Directions Training and Education Program, which made available millions in grants to support the development and strengthening of occupational safety and health competence in business, employee, and educational organizations. This program supported a broad range of activities, such as training in hazard identification and control; workplace risk assessment; medical screening and recordkeeping; and liaison work with OSHA, the National Institute for Occupational Safety and Health, and other agencies. "The goal of the program was to allow unions and other groups to become financially self-sufficient in supporting comprehensive health and safety programs."[86] This program, criticized by some as a payoff to constituent groups, especially labor unions, was a natural target of the budget cutters during the Reagan administration, but the concept of increased consultation has been given even greater emphasis.

There is also a provision that state plans may include onsite consultation with employers and employees to encourage voluntary compliance.[87] The personnel engaged in these activities must be separate from the inspection personnel, and their existence must not detract from the federal enforcement effort. These consultants not only point out violations, but also give abatement advice.

12.0 Overlapping Jurisdiction

There are other agencies involved with statutory responsibilities that affect occupational safety and health. These agencies indirectly regulate safety and health matters in their attempt to protect public safety.

One example of an overlapping agency is the Department of Transportation and its constituent agencies, such as the Federal Railroad Administration and the

---

[84] OSH Act § 21(c)(2).
[85] OSH Act § 21(c)(1).
[86] U.S. Department of Labor, "OSHA News," 12 April 1978.
[87] 29 C.F.R. § 1902.4(c)(2)(xiii).

Federal Aviation Administration. These agencies promulgate rules concerned with the safety of transportation crews and maintenance personnel, as well as the traveling public, and consequently overlap similar responsibilities of OSHA.

Section 4(b)(1) of the OSH Act states that when other federal agencies "exercise statutory authority to prescribe or enforce standards or regulations affecting occupational safety or health," the OSH Act will not apply to the working conditions addressed by those standards. Memorandums of understanding (MOUs) between these agencies and OSHA have eliminated much of the earlier conflict.

The Environmental Protection Agency is the organization that overlaps most frequently with OSHA. When a toxic substance regulation is passed by EPA, OSHA is affected if that substance is one that appears in the workplace. For instance, both agencies are concerned with pesticides, EPA with the general environmental issues surrounding the pesticides and OSHA with some aspects of the agricultural workers who use them. During the early 1970s, there was a heated interagency conflict over field reentry standards for pesticides (*see* the chapter on Pesticides), a struggle which spilled over into the courts and eventually had to be settled by the White House in EPA's favor.[88] The OSHA-EPA MOU of 1990 will hopefully prevent a repetition of such problems.

Thus, although the health regulatory agencies generally function in a well-defined area, overlap does occur. As another example, there are toxic regulations under Section 307 of the Federal Water Pollution Control Act, Section 112 of the Clean Air Act, and under statutes of the FDA and CPSC. These regulatory agencies realized the need for coordination, particularly when dealing with something as pervasive as toxic substances, and under the Carter Administration combined their efforts into an interagency working group called the Interagency Regulatory Liaison Group (IRLG). Although the IRLG was abolished at the beginning of the Reagan administration, the concept of interagency working groups is a good one. The federal agencies involved in regulation should rid themselves of the antagonism and rivalry of the past and cooperate with one another to meet the needs of the public.

---

[88] *Florida Peach Growers Assn. v. Dept. of Labor*, 489 F.2d 120 (5th Cir., 1974). To avoid this type of confrontation, in 1976 Congress provided in Section 9 of the Toxic Substances Control Act for detailed coordination procedures to be followed when jurisdictional overlap occurs.

13.0 Occupational Safety and Health Review Commission

The OSH Act established the Occupational Safety and Health Review Commission (OSHRC) as "an independent quasi-judicial review board"[89] consisting of three members appointed by the President to six-year terms. Any enforcement actions of OSHA that are challenged must be reviewed and ruled upon by the Commission.[90]

13.1 OSHRC Appeal Process

Any failure to challenge a citation within fifteen days of issuance automatically results in an action of the Review Commission to uphold the citation. This decision by default is not subject to review by any court or agency. When an employer challenges a citation, the abatement period, or the penalty proposed, the Commission then designates a hearing examiner, an administrative law judge, who hears the case; makes a determination to affirm, modify or vacate the citation or penalty; and reports his finding to the Commission.[91] This report becomes final within thirty days unless a Commission member requests that the Commission itself review it.

The employer or agency may then seek a review of the decision in a federal appeals court.

13.2 Limitations of the Commission

One of the major problems with the Review Commission is the question of its jurisdiction. "The question has arisen of the extent to which the Commission should conduct itself as though it were a court rather than a more traditional administrative agency."[92] The Commission cannot look to other independent agencies in the government for a resolution of this problem "because its duties and its legislative history have little in common with the others."[93] It cannot conduct investigations, initiate suits, or prosecute; therefore, it is best understood as an administrative agency with the limited duty of "adjudicating those cases brought

---

[89] Ashford, *Crisis*, p. 145.
[90] OSH Act § 12(a)–(b).
[91] OSH Act § 12(j).
[92] Ashford, *Crisis*, p. 145.
[93] *Id.*, pp. 281-82.

before it by employers and employees who seek review of the enforcement actions taken by OSHA and the Secretary of Labor."[94]

Another problem inherent in the organization of the Commission is the separation from the President's administration. There has been a question of where the authority of the administration ends and the authority of the Commission begins. Because of the autonomous nature of the Review Commission, it cannot always count on the support of the Executive agencies. In fact, OSHA has generally ignored Review Commission decisions, and few inspectors are even aware of the Commission interpretations on various regulations.

14.0 National Institute of Occupational Safety and Health

Under the OSH Act, the Bureau of Safety and Health Services in the Health Services and Mental Health Administration was restructured to become the National Institute for Occupational Safety and Health (NIOSH), so as to carry out HEW's responsibilities under the act.[95] (HEW—the Department of Health, Education, and Welfare—has since become the Department of Health and Human Services, HHS.) For the past two decades NIOSH has reported, illogically, to the Centers for Disease Control (CDC), and the two organizations have headquarters in Atlanta.

Since mid-1971, NIOSH has claimed the training and research functions of the act, along with its primary function of recommending standards. For this latter task, NIOSH provides recommended standards to OSHA in the form of criteria documents for particular hazards. These are compilations and evaluations of all available relevant information from scientific, medical, and (occasionally) engineering research.

The order of hazards selected for criteria development is determined several years in advance by a NIOSH priority system based on severity of response, population at risk, existence of a current standard, and advice from federal agencies (including OSHA) as well as involved professional groups.[96]

The criteria documents may actually have some value apart from the role in standards-making. Even though they do not have the force of law, they are widely

---

[94] *Id.*

[95] OSH Act § 22(a).

[96] John F. Finklea, "The Role of NIOSH in the Standards Process," in *Proceedings of the Occupational Health and Safety Regulation Seminar*, p. 38.

distributed to industry, organized labor, universities, and private research groups as a basis to control hazards. The criteria documents also serve as a "basis for setting international permissible limits for occupational exposures."[97]

To the extent that certain criteria documents may be deficient, as discussed earlier, this expansive role for them among laymen poses a real problem. This problem may unfortunately become worse if NIOSH declines in both funds and morale. Nevertheless, there is some benefit in having the two organizations separate. NIOSH has on occasion criticized OSHA for regulatory decisions which the former believed were scientifically untenable.

15.0 Hazard Communication Regulations

OSHA's output of health standards has never been impressive. In recent years, it has tried three new approaches to get around this bottleneck. The first was the "federal" cancer policy designed to create a template for dealing in an expedited fashion with a number of hazardous chemicals. The second was the wholesale review initiated in 1988 of all the Z-1 list consensus standards—an effort struck down by the courts.

The third, characterized by one OSHA official as the agency's most important rulemaking ever, is the hazard communication (hazcom) regulation issued in November 1983.[98]

15.1 Reason for the Regulation

This standard, sometimes known as the *worker right-to-know rule*, provides that hazardous chemicals must be labeled, material safety data sheets (MSDS) on hazards be prepared, and workers and customers should be informed of potential chemical risks.

How could a rule with such far-reaching consequences be issued from an administration that so stressed deregulation and deliberately avoided issuing other protective regulations? The answer lies in an almost unprecedented grassroots movement at the state and municipal level to enact their own "worker right-to-know" laws which, many businessmen felt, could be a considerable burden on interstate commerce. They therefore lent their support to OSHA in its confrontation with the Office of Management and Budget (OMB) at the White House. A

---

[97] *Id.*, p. 39.
[98] 49 FR 52380, 25 November 1983.

federal regulation on this subject would arguably preempt the multiplicity of local laws.

The rule was originally presumed to apply to only a few hundred, perhaps a thousand, particularly hazardous chemicals. The individual employers would evaluate the risk and then decide for themselves which products merited coverage. Most employers were unable or unwilling to make such scientific determinations. Within a year or two, this limited program expanded into universal coverage.

15.2 Scope and Components

Published on 25 November 1983, OSHA's Hazard Communication or "Right-to-Know" Standard[99] went into effect two years later, in November 1985, for chemical manufacturers, distributors, and importers, and in May 1994 for manufacturers that use chemicals. It required that employees be provided with information concerning hazardous chemicals through labels, material safety data sheets, training and education, and lists of hazardous chemicals in each work area. Originally it covered only manufacturing industries classified in SIC codes 20–39, but by court order in 1987 it was extended to virtually all employers.[100]

Every employer must assess the toxicity of chemicals it makes, distributes, or uses based on guidelines set forth in the rule. Then it must provide this material downstream to those who purchase the chemicals through MSDSs.[101] The employers are then required to assemble a list of the hazardous materials in the workplace, label all chemicals, provide employees with access to the MSDSs, and provide training and education. While all chemicals must be evaluated, the "communication" provisions apply—in theory—only to those chemicals known to be present in the workplace in such a way as to potentially expose employees to physical or health hazards.

Special provisions apply to the listing of mixtures that constitute health hazards. Each component which is itself hazardous to health and which comprises one percent or more of a mixture must be listed. Carcinogens must be listed if present in quantities of 0.1 percent or greater.

---

[99] 48 FR 53280; 29 C.F.R. § 1200.

[100] 52 FR 31852, 24 August 1987, in response to *United Steelworkers of America, AFL-CIO v. Pendergrass*, 819 F.2d 1263 (3rd Cir., 1987).

[101] There is some legal question whether OSHA, which has jurisdiction over employer-employee health and safety relations, has authority over the relationship between a company and its downstream customers.

The Hazard Communication Standard is a performance-oriented rule. While it states the objectives to be achieved, the specific methods to achieve those objectives are at the discretion of the employer. Thus, in theory, employers have considerable flexibility to design programs suitable for their own workplaces. However, this may mean the employers will have questions about how to comply with the standard.

The purpose of labeling is to give employees an immediate warning of hazardous chemicals and a reminder that more detailed information is available. Containers must be labeled with identity, appropriate hazard warnings, and the name and address of the manufacturer. The hazard warnings must be specific, even as to the endangered body organs. For example, if inhalation of a chemical causes lung cancer, the label must specify that and cannot simply say "harmful if inhaled" or even "causes cancer." Pipes and piping systems are exempt from labeling, as are those substances required to be labeled by another federal agency.

Material safety data sheets (MSDSs), used in combination with labels, are the primary tools for transmitting detailed information on hazardous chemicals. A MSDS is a technical document which summarizes the known information about a chemical. Chemical manufacturers and importers must develop a MSDS for each hazardous chemical produced or imported and pass it on to the purchaser at the time of the first shipment. The employer must keep these sheets where employees will have access to them at all times.

The purpose of employee information and training programs is to inform employees of the labels and MSDSs and to make them aware of the actions required to avoid or minimize exposure to hazardous chemicals. The format of these programs is left to the discretion of the individual employer. Training programs must be provided at the time of initial assignment and whenever a new hazard is introduced into the workplace.

15.3 Hazard Evaluation

Chemical manufacturers are required to evaluate all chemicals they sell for potential health and physical hazards to exposed workers. Purchasers of these chemicals may rely on the supplier's determination or may perform their own evaluations.

There are really no specific procedures to follow in determining a hazard. Testing of chemicals is not required, and the extent of the evaluation is left to the manufacturers and importers of hazardous chemicals. However, all available scientific evidence must be identified and considered. A chemical is considered hazardous if it is found so by even a single valid study.

Chemicals found on the following *master* lists are automatically deemed hazardous under the standard:

- the International Agency for Research on Cancer (IARC) monograph;
- the *Annual Report on Carcinogens* published by the National Toxicology Program (NTP);
- OSHA's "Subpart Z" list, found in Title 29 of the Code of Federal Regulations, Part 1910; or
- Threshold Limit Values for Chemical Substances and Physical Agents in the Work Environment, published by the American Conference of Governmental Industrial Hygienists.

If a substance meets any of the health definitions in Appendix A of the standard, it is also to be considered hazardous. The definitions given are for a carcinogen, a corrosive, a chemical which is highly toxic, an irritant, a sensitizer, a chemical which is toxic, and target organ effects.

Appendix B of the standard gives the principal criteria to be applied in complying with the hazard determination requirement. First, animal as well as human data must be evaluated. Second, if a scientific study finds a chemical to be hazardous, the effects must be reported whether or not the manufacturers or importers agree with the findings.

Appendix C of the standard gives a lengthy list of sources that may assist in the evaluation process. The list includes company data from testing and reports on hazards, supplier data, MSDSs or product safety bulletins, scholarly text books, and government health publications.

In practice, as noted above, companies have begun requiring MSDSs from manufacturers for all chemicals they purchase, so the evaluation aspect of the standard has become unimportant.

15.4 Trade Secrets

Although there is agreement that there must be a delicate balance between the employee's right to be free of exposure to unknown chemicals and the employer's right to maintain reasonable trade secrets, the exact method of protection has been considerably disputed.

Under the standard, a trade secret is considered to be defined as in the *Restatement of Torts*; *i.e.*, something that is not known or used by a competitor. However, OSHA had to revise its definition to conform with a court ruling which said that a trade secret may not include information that is readily discoverable through reverse engineering.

Although the trade secret identity may be omitted from the MSDS, the manufacturer must still disclose the health effects and other properties about the chemical. A chemical's identity must immediately be disclosed to a treating physician or nurse who determines that a medical emergency exists.

In nonemergency situations, any employee can request disclosure of the chemical's identity if he demonstrates through a written statement a *need to know* the precise chemical name and signs a confidentiality agreement. The standard specifies all purposes which OSHA considers demonstrate a need to know a specific chemical identity.

The standard initially limited this access to health professionals, but in 1985, the U.S. Court of Appeals for the Third Circuit ruled that trade secrets protections must be narrowed greatly, allowing not only health professionals, but also workers and their designated representatives the same access as long as they follow the required procedures.[102] In response, OSHA issued a final rule on trade secrets in September 1986[103] which narrows the definition of *trade secret*. It denies protection to chemical identity information that is readily discoverable through reverse engineering. It also permits employees, their collective bargaining representatives, and occupational nurses access to trade secret information.

Upon request, the employer must either disclose the information or provide written denial to the requestor within 30 days. If the request is denied, the matter may be referred to OSHA, whereupon evidence to support the claim of trade secret, and alternative information that will satisfy the claimed need.

15.5 Federal Preemption Controversy

Several states and labor groups have filed suits challenging state laws that are more protective. New Jersey, for example, has enacted the toughest labeling law in the nation, requiring industry to label all its chemical substances, whether they are hazardous or not, and supply the information to community groups and health officials, as well as to workers.

They were also concerned that, because the original OSHA standard only covered the manufacturing sector, more than 50 percent of the workers (such as those workers in the agricultural and construction fields) would be unprotected, and OSHA did not cover (and still does not) such groups as state employees and

---

[102] *United Steelworkers of America, AFL-CIO-CLC v. Auchter, et al.*, 763 F.2d 728; 12 OSHC 1337 (3rd Cir., 1985).
[103] 51 FR 34590.

consumers. Moreover, they argued that OSHA would be incapable of enforcing worker protection because of the staff cuts made by the Reagan administration.

The chemical industry, on the other hand, favored a uniform federal regulation because they believed it would be less costly and easier to comply with one federal rule as opposed to several state and local rules that would often conflict or be confusing.

In October 1985, the U.S. Court of Appeals for the Third Circuit ruled that the federal Hazard Communication Standard does not preempt all sections of New Jersey's right-to-know laws designed to protect workers and the public from chemical exposure—only those which apply to groups the agency's rules covered, which were then only in the manufacturing sector.[104] Thus, while some parts of a state law may be preempted, other provisions may not be.

In September 1986, the Third Circuit also found that the federal Hazard Communication Standard did not entirely preempt requirements under Pennsylvania's right-to-know act pertaining to worker protection in the manufacturing industry where the state rules relate to public safety generally and for protection of local government officials with police and fire departments. However, five days later, also in September 1986, the U.S. Court of Appeals for the Sixth Circuit ruled that a right-to-know ordinance enacted by the city of Akron, Ohio, is preempted by the federal standard in manufacturing sector workplaces.

In 1992, the Supreme Court came down strongly on the side of preemption. The *Gade v. National Solid Waste Management Association* case, although it involved OSHA's so-called HAZWOPER regulations[105] rather than hazard communication, involved a state law requiring more additional training for heavy equipment operators on hazardous waste sites. The high court found that the OSHA regulations preempted the state despite arguments that the federal rules only set a minimum which the state could exceed—the situation in most environmental laws—and the more transparent claim that the state laws had a dual purpose in protecting the public as well as workers.[106]

---

[104] *New Jersey State Chamber of Commerce v. Hughey*, 774 F.2d 587, 12 OSHC 1589 (3rd Cir., 1985).

[105] Hazardous Waste Operations and Emergency Response (HAZWOPER) regulations in 29 C.F.R. § 1910.120.

[106] *Gade v. National Solid Wastes Management Ass'n*, 112 S. Ct. 2374 (1992).

16.0 Ergonomics Issues

For two decades OSHA officials have worked towards developing a standard on ergonomics. The original impetus was a series of reports from Midwest poultry and meatpacking plants that workers were developing "carpal tunnel syndrome" (CTS). This is a condition that develops from repetitive motion of the hand and wrist which irritates the nerve running through a bone channel near the thumb. Because similar conditions can develop from repetitive motion or strain in other parts of the body, such as with "tennis elbow," the malady was relabeled as "cumulative trauma syndrome," also conveniently abbreviated CTS, then changed to "repetitive motion syndrome," and so on until the more sweeping term "ergonomics" was adopted.

Along the way, OSHA was hitting offending companies with fines in the millions of dollars, some of the biggest in the agency's history. All of this had to be done under OSHA's "general duty clause," the famous Section 5(a)(1) of the Act, because there was no specific standard that addressed this particular condition. With the congressional rejection of OSHA's Ergonomic Standard, OSHA may again fall back to the general duty clause to deal with clear cases of abuse.

16.1 Overview

Ergonomics was not a new word or a new concept. It had long been used in Europe to denote arrangements of workers and tools that maximized productivity with a minimum of wasted effort. This was based on, ironically, the American-developed "time and motion studies" almost a century ago. The "erg" in ergonomics, after all, is from the Greek word meaning "work." The term also came to be used in furniture and office design with the connotation of comfortable and well laid out.

The workplace collision came when the concept of mass production—with each worker repeating a number of simple steps all day—clashed with the physical irritation that might cause to certain parts of the body. The better companies sought to deal with the problem, though most were concerned with the boredom and carelessness aspects of endless repetition rather than with possible deleterious effects on the body. The remedies, however, tended to be very specific to each worksite or even each job. So how could a general standard be developed?

A very different problem was raised for OSHA. Considering the host of unregulated chemicals, life-threatening workplace hazards, and a pathetically slow agency pace for dealing with them, is this where OSHA should be putting its priorities for at least a decade?

Congress did not think so, and for a number of years put a "rider" on OSHA's appropriation bills that such an omnibus ergonomic standard should not be developed. Due to a congressional slip-up, however, and the confusion in the year right before the 2000 presidential elections, OSHA was able to slide out a proposal in November 2000, due to take effect just days before the new president and Congress took office and could do anything about it. Once in effect, it was legally much more difficult to overturn it, except for denying appropriations for enforcement, and a closely-divided House and Senate had more pressing issues.[107]

16.2 Scope of the Standard

The new standard is designed to reduce the incidence of *musculoskeletal disorders* (MSDs) by requiring that companies establish programs to prevent them. In other words, the standard is not prescriptive but procedural.

The standard applies to all general industry workplaces under OSHA but not, for technical legal reasons, to the construction, maritime, agricultural, or most railroad operations. They will have their own standards once the legal steps are completed. Being subject to the standard, however, does not mean that it automatically applies in its entirety. Some actions have to be taken, and others need occur only after an *Action Trigger*. The trigger is, in Western parlance, a hair-trigger which goes off very easily. Therefore, most workplaces can expect to fall under its provision sooner rather than later.

16.3 Initial Actions and Action Triggers

Certain specified *initial actions* must be taken by every employer: Every employee must be given, first, basic information about MSDs, including symptoms and reporting obligations; second, a summary of the requirements of the act; and third, a written notice in a conspicuous place or by electronic communication.

Point one, above, more specifically, requires providing information to workers on:

i. Common musculoskeletal disorders (MSDs) and their signs and symptoms;

ii. The importance of reporting MSDs and their signs and symptoms early and the consequences of failing to report them early;

iii. How to report MSDs and their signs and symptoms in your workplace;

iv. The kinds of risk factors, jobs and work activities associated with MSD hazards; and

---

[107] OSHA ergonomics program standard final rule, 65 FR 68261 (14 November 2000), taking effect 60 days later.

v. A short description of the requirements of OSHA's ergonomics program standard.

An Action Trigger occurs when an employee reports a work-related MSD which rises above a certain threshold, namely when (1) it requires days away from work, restricted work, or medical treatment beyond first aid; or (2) when the symptoms last for more than seven consecutive days. The trigger is then met if the employee's job "routinely involves, on one or more days a week, exposure to one or more relevant risk factors at the levels described in the *Basic Screening Tool* in Table W-1." In making this determination an employer could seek assistance from a *Health Care Professional* (HCP), who plays a key role in implementation of the subsequent program.

16.4 The Required Ergonomics Program

A full description of the requirements of an acceptable ergonomics program is beyond the scope of this chapter. To summarize, in this author's opinion, the standard is a real contradiction: it sets forth a string of generalities spelled out in numbing detail.

There is a partial way out—a "quick fix" system instead of setting up a full ergonomics program. This may be done only if there has been no more than one MSD incident in that job, and no more than two MSD incidents in the whole establishment in the preceding 18 months. Given the hair-trigger nature of the ergonomics program, however, this is unlikely to provide much respite for most companies.

17.0 Legislation

The OSH Act has remained virtually untouched since its passage in 1970. With the inauguration of a Democratic president, William Clinton, in 1993 and Democratic control of both houses of Congress, the expectation was that the labor unions would secure the passage of the first significant revisions in the law.

Under the circumstances, the proposed legislation was surprisingly innocuous. It included verbose and often unnecessary sections on enforcement, refusal to work, and other issues. Among them was a seemingly innocuous section providing for labor-management safety committees in the workplace. Both employers and employees have found these committees quite useful, but some manufacturers' organizations criticized the language as forcing a much greater role for labor unions.

TABLE 1—COMPARISON OF PROPOSED BASIC OBLIGATIONS WITH FINAL GRANDFATHER CLAUSE PROGRAM ELEMENT CORE ELEMENTS AND SUBELEMENTS—Continued

Proposed basic obligation	Corresponding core elements and subelements of the final grandfather clause
Proposed Employee Participation Obligation: You must set up a way for employees to report MSD signs and symptoms and to get prompt responses. You must evaluate employee reports of MSD signs and symptoms to determine whether a covered MSD has occurred. You must periodically provide information to employees that explains how to identify and report MSD signs and symptoms.	(c)(1)(ii) Employee participation, as demonstrated by the early reporting of MSDs and active involvement by employees and their representatives in the implementation, evaluation, and future development of your program; [See also paragraph (c)(1)(iv).]
Proposed Job Hazard Analysis and Control Obligation: You must analyze the problem job to identify the ergonomic risk factors that result in MSD hazards. You must eliminate the MSD hazards, reduce them to the extent feasible, or materially reduce them using the incremental abatement process in this standard. If you show that the MSD hazards only pose a risk to the employee with the covered MSD, you may limit the job hazard analysis and control to that individual employee's job.	Final § 1910.900(c)(1)(iii): [Your program must contain the following elements:] Job hazard analysis and control, as demonstrated by a process that identifies, analyzes, and uses feasible engineering and administrative controls to control MSD hazards or to reduce MSD hazards to the levels specified in Appendix D or to the extent feasible, and evaluates controls to assure that they are effective. **Note to Paragraph (c)(1)(iii):** Personal protective equipment (PPE) may be used to supplement engineering and administrative controls, but you may only use PPE alone where other controls are not feasible. Where PPE is used you must provide it at no cost to employees.
Proposed Training Obligation: You must provide training to employees so they know about MSD hazards and your ergonomics program and measures for eliminating or materially reducing the hazards. You must provide training initially, periodically, and at least every 3 years at no cost to employees.	Final § 1910.900(c)(1)(iv): [Your program must contain the following elements:] Training of managers, supervisors, and employees (at no cost to these employees) in your ergonomics program and their role in it; the recognition of MSD signs and symptoms; the importance of early reporting; the identification of MSD hazards in jobs in your workplace; and the methods you are taking to control them.
Proposed MSD Management Obligation: You must make MSD management available promptly whenever a covered MSD occurs. You must provide MSD management at no cost to employees. You must provide employees with the temporary "work restrictions" and "work restriction protection (WRP)" this standard requires.	Final § 1910.900(c)(2): [Your program must contain the following elements:] By January 16, 2002, you must have implemented a policy that provides MSD management as specified in paragraphs (p), (q), (r) and (s) of this section.
Proposed Program Evaluation Obligation: You must evaluate your ergonomics program periodically, and at least every 3 years, to ensure that it is in compliance with this standard.	Final § 1910.900(c)(1)(v): [Your program must contain the following elements:] Program evaluation, as demonstrated by regular reviews of the elements of the program; regular reviews of the effectiveness of the program as a whole, using such measures as reductions in the number and severity of MSDs, increases in the number of jobs in which ergonomic hazards have been controlled, or reductions in the number of jobs posing MSD hazards to employees; and the correction of identified deficiencies in the program. At least one review of the elements and effectiveness of the program must have taken place prior to [insert date 60 days after the publication date of this standard].

Figure 1.5

With the Republican election victories in the House and Senate in late 1994, these Democratic legislative plans not only collapsed but the victors prepared their own onslaught on the OSH Act. To the surprise of many, the draconian Republican plans to curtail or even eliminate OSHA got no further than the previous Democratic plans. "Organized labor counted its victories in this year's Congress by the number of bills defeated rather than enacted."[108] However, the Republicans'

---

[108] "GOP Labor Bills Make Little Headway," AP, 28 October 1998.

hostile scrutiny of OSHA paralyzed the agency leadership and led to a sharp decline in both enforcement and standard setting.

Congressional oversight has been most intense regarding OSHA's proposed ergonomics standard. As discussed above, Congress used the appropriations process to order the agency not to issue the standard, while the Clinton Administration eventually refused to sign the legislation that included these prohibitions on the eve of the 2000 presidential elections. However, legislation was also introduced to direct OSHA to encourage safer medical needles, give small businesses more input into agency regulatory proceedings, bar home office inspections, and (signed into law) expand federal compensation for radiation-related exposure.[109]

---

[109] In the 106th Congress, H.R. 987, H.R. 4577, and S.1070 restricting OSHA on issuing an ergonomics standard; H.R. 5178 on needlestick prevention, S. 1156 amending the Small Business Regulatory Enforcement Fairness Act (SBREFA) of 1996; H.R. 4098 barring home office inspections; and S. 1515 signed into law by President Clinton on 10 July 2000 as P.L. 106-245.

Chapter 2

The Rulemaking Process

Lawrence P. Halprin, Esq.
Keller and Heckman, LLP
Washington, D.C.

1.0 Overview

The Occupational Safety and Health Act (OSH Act)[1] authorizes the Secretary of Labor to issue and enforce two types of rules that regulate the business operations of employers in the area of workplace health and safety. The Secretary of Labor has delegated this rulemaking authority to OSHA. The first type of rule is known as an "occupational safety and health standard." This type of rule requires that employers implement measures designed to control or eliminate particular workplace hazards.[2] The second type of rule is known as a "regulation." OSHA regulations establish rules governing OSHA inspections, OSHA administrative proceedings, various employer recordkeeping and reporting obligations, and other requirements of general applicability that do not mandate measures designed to control particular hazards.

The OSH Act contains provisions that specify the types of requirements that OSHA may, or in some areas, must adopt as standards or as regulations.[3] The OSH Act also contains provisions that specify most of the procedural requirements applicable to the adoption of standards. The balance of the procedural requirements applicable to OSHA standards as well as all of the procedural requirements applica-

---

[1] OSH Act § 2; 29 U.S.C. § 651, *et seq.*

[2] OSH Act § 3(8); 29 U.S.C. § 652(8) defines an occupational safety and health standard as a standard that "requires conditions or the adoption or use of one or more practices, means, methods, operations, or processes reasonably necessary or appropriate to provide safe and healthful employment and places of employment."

[3] OSH Act § 6; 29 U.S.C. § 655. *Industrial Union Dep't v. Hodgson*, 499 F.2d 467 (D.C. Cir. 1974).

ble to the adoption of OSHA regulations are provided by the Administrative Procedures Act (APA).[4]

1.1 The Decision to Initiate a Rulemaking

Under the OSH Act, any person may petition for the promulgation, modification, or revocation of a rule, or the Secretary may act on his / her own initiative or at the recommendation of the National Institute for Occupational Safety and Health (NIOSH). In the absence of a Congressional mandate to adopt or modify a particular rule,[5] OSHA generally applies the following criteria, adopted as part of its Priority Planning Process, to determine whether to proceed with rulemaking in a given area:

- the seriousness of the hazard,
- the number of workers exposed or the magnitude of the risk,
- the quality of available risk information,
- the potential for risk reduction,
- administrative efficiency or feasibility,
- legal feasibility; and
- other public policy considerations (such as intensity of public concern and public perception of the hazard).

1.2 The Traditional Rulemaking Process

1.2.1 Advisory Committees

Once the Secretary has determined that a rule should be promulgated, an *ad hoc* advisory committee may be appointed to study the matter more closely and assess the potential hazard at issue.[6] The advisory committee may even develop a proposed standard. The committee members must consist of representatives of management, labor, and state agencies, as well as one or more representatives of the Secretary of

---

[4] 5 U.S.C. § 501.

[5] Congress has mandated that OSHA issue or revise the following standards: Bloodborne Pathogens (29 C.F.R. § 1910.1030), Hazardous Waste and Emergency Response Operations (29 C.F.R. § 1910.120), and Process Safety Management (29 C.F.R. § 1910.119).

[6] OSH Act § 6(b)(1); 29 U.S.C. § 655(b)(1).

Health and Human Services (HHS). If an advisory committee is appointed to assist in the study of a particular issue, it is supposed to submit its recommendations within 90 days of its appointment, unless the Secretary grants an extension of up to 270 days from the appointment.

In addition to the reference *ad hoc* committees, there are two standing committees of stakeholders established by the Department of Labor to address OSHA issues: 1) The National Advisory Committee on Occupational Safety and Health (NACOSH); and 2) The Advisory Committee on Construction Safety and Health (ACCOSH). NACOSH may advise the Secretary of Labor on virtually any general enforcement, rulemaking or policy matter regarding the administration of the OSH Act, but generally defers on matters within the purview of ACCOSH. ACCOSH may advise the Secretary of Labor on virtually any general enforcement, rulemaking, or policy matter regarding the administration of the OSH Act with respect to the construction industry.

1.2.2 Request for Information / Advanced Notice of Proposed Rulemaking

OSHA may determine that it is necessary to solicit information from the public and other stakeholders prior to drafting a proposed rule. In such an instance, the Agency will issue a Request For Information (RFI) or an Advanced Notice of Proposed Rulemaking (ANPR) in the *Federal Register*, and provide a specific period of time for the public to offer comment. Comments received may then become the basis for the proposed rule.

1.2.3 Notice of Proposed Rulemaking

If the Secretary determines that it is necessary and appropriate to propose a rule, the process must proceed with the publication of a Notice of Proposed Rulemaking (NPRM) in the *Federal Register*.[7] The NPRM has two primary components: 1) the regulatory text of the rule OSHA proposes to issue; and 2) the Preamble. The Preamble explains why OSHA believes a rule is necessary and why OSHA believes the proposed rule would achieve OSHA's objective in a manner consistent with its statutory mandate and authority under the OSH Act, as well as all other applicable laws. The nature of the Preamble discussion (*e.g.*, scope, detail, and discussion of legal authority, relevant science and justification for the rule) generally reflects the complexity of the issues, the availability of relevant data, the potential costs of compliance, and the anticipated level of both support and opposition to the initiative. In the present climate, where virtually every OSHA standard is challenged

---

[7] OSH Act § 6(b)(2); 29 U.S.C. § 655(b)(2).

in court, and many are highly controversial, the NPRM is not only a required legal document but an important public relations document increasingly scrutinized by both the press and the Congress.

OSHA must provide at least 30 days for the public to comment on the proposal, and typically provides substantially longer. OSHA is required to hold a public hearing on a proposed standard if an interested person files written objections to the proposal and requests a public hearing in a timely fashion.[8] Recognizing that a request for a hearing is virtually inevitable, OSHA generally announces the dates of the public hearings and the deadlines for submitting hearing testimony in the NPRM.

Under OSHA's procedural rules, those interested parties who file a timely "Notice of Intent to Appear" at the public hearing have a right to testify at the hearing,[9] question other witnesses, and file post-hearing comments and briefs. The act of filing the Notice of Intent to Appear is viewed by OSHA as an indication that the filing party has a continuing interest and intent to participate in the rulemaking to the extent of its choosing. In other words, the person who files such a notice may choose not to testify or ask questions at the public hearing but retains the right to file a post-hearing comment or brief.

1.2.4 The Hearing

OSHA rulemakings are conducted under what is referred to as the informal rulemaking process. In other words, although the public hearing is presided over by an Administrative Law Judge (ALJ) from the Department of Labor, it is conducted under informal procedures more akin to a legislative hearing than a court trial. A transcript is created but never sent to the testifying witnesses to verify its accuracy. If a witness believes there is an error in the transcript of his or her testimony, it is up to the witness to file a notice of correction in the docket and it is unclear whether that notice will ever be formally linked to the transcript for the benefit of a subsequent reader of the transcript. Witnesses are not sworn in. They are not subject to a contempt action for declining to answer questions. Furthermore, the opportunity for cross-examination is restricted.

The apparently limited right to cross-examine witnesses is created by an OSHA rule applicable to these proceedings, rather than the OSH Act. The rule says that

---

[8] OSH Act § 6(b)(3); 29 U.S.C. § 655(b)(3).

[9] If a party will require a certain amount of time to present its testimony or has any scheduling limitations, it is incumbent on the party to make OSHA aware of these special needs at the time it files its notice.

"fairness may require an opportunity for cross-examination on crucial issues" and in those circumstances "the presiding officer shall provide an opportunity for cross-examination on crucial issues." However, the OSHA rule also says "the essential intent is to provide an opportunity for effective oral presentation by interested persons that can be carried out expeditiously and in the absence of rigid procedures which might unduly impede or protract the rulemaking process." It goes on to provide as follows:

> Upon reasonable notice to, interested persons, the Assistant Secretary may in any particular proceeding prescribe additional or alternative procedural requirements: (a) In order to expedite the conduct of the proceeding; (b) in order to provide greater procedural protection to interested persons whenever it is found necessary or appropriate to do so; or (c) for any other good cause which may be consistent with the applicable laws.

Relying on that provision, former Assistant Secretary Charles Jeffress issued a special procedural rule applicable to the rulemaking that led to OSHA's Ergonomics Program Standard (subsequently rescinded by P.L. 107-5). That special rule generally limited cross-examination of witnesses to a period of ten minutes per questioner for each round of questions. This process was to be continued through successive rounds of questions until either all questions were answered or the time for questioning of that panel of witnesses had expired.

At the close of the hearing, the ALJ must certify the complete record (*e.g.*, exhibits, written comments, transcripts) to the Secretary of Labor for a final decision. Normally, at the close of the public hearings, the hearing officer will announce the time periods during which those who filed a formal "Notice of Intent to Appear" will be allowed to submit post-hearing comments and post-hearing briefs. The purpose of these filings may be to rebut opposing testimony, to expand on points made during the hearing, and to make legal arguments.

1.2.5 The Final Rule

Possibly reflecting the thoughts (or dreams) of a different and simpler time, the OSH Act specifies that OSHA must publish the final rule in the *Federal Register* within 60 days of the end of the comment period or any later public hearing. No standard has been issued within that time frame in the 30 year history of the OSH Act. OSHA did issue the final Ergonomics Program Standard approximately 3-1/2 months after the deadline for filing post-hearing briefs. However, that result was achieved by an intensive and totally unprecedented commitment of OSHA's rulemaking resources that preempted every other rulemaking initiative and seems unlikely to be repeated. It also required the strong support of the White House to ensure that OMB approved the final rule package on an ongoing, expedited basis as OSHA developed it. Furthermore, the "rush to judgment," as reflected in OSHA's actions, was one of the primary reasons cited by members of Congress in voting to rescind that rule.

The Preamble to the *Federal Register* notice issuing the final rule must also contain a statement of reasons for OSHA's actions. While the nature of the rulemaking process is to solicit public comment with the expectation that they will result in appropriate changes to the regulatory text of the final rule, it is a well-established legal principle that the final rule must be a logical outgrowth of the proposal.[10] Otherwise, OSHA must re-issue the final rule as a proposal to allow for adequate public comment. The fairly universal position of the business community is that the final Ergonomics Program Standard issued on November 23, 2000 was not a logical outgrowth of the proposal where the proposal expressly stated that OSHA did not have the quantitative dose-response data needed to establish permissible exposure limits and similar objective standards, but the final standard contained numerical threshold triggers and "safe harbor" exposures for various so-called ergonomic risk factors.

In theory, if OSHA determines that a rule should not be issued, that decision must also be published in the *Federal Register*. A decision not to proceed is rarely made contemporaneously with the end of the comment period or public hearing, but is more often a reflection of changed priorities, political circumstances, or a stale record.

The OSH Act further specifies that the effective date for a new or modified rule may be delayed by no more than 90 days following publication to allow interested parties to become familiar with its requirements. OSHA frequently issues rules with a publication date, a delayed effective date and further delayed (phased in) compliance dates based on a determination that compliance within 90 days of publication would not be feasible.

In some cases, decisions as to effective dates have major consequences. OSHA's subsequently rescinded Ergonomics Program Standard initially went into effect four days before President George W. Bush's inauguration, and thus escaped the reach of the subsequent January 20, 2001 White House order to review and delay for 60 days the effect of all published final Executive Branch rules that had not yet gone into effect. OSHA apparently determined that it could not put the revised Injury and Illness Recording and Reporting Rule into effect before the inauguration while delaying its compliance date until January 1, 2002. As a result, that rule is covered by the January 20, 2001 White House order, and remained under review at the time this chapter was written.

---

[10] *United Steelworkers of America v. Marshall*, 647 F.2d 1189, 1221 (D.C. Cir. 1980).

1.3 Negotiated Rulemaking

Negotiated rulemaking is a consensual approach to developing regulations. The Negotiated Rulemaking Act (NRA) defines consensus as unanimous concurrence among the interests represented on a negotiated rulemaking committee, unless the committee itself unanimously agrees to use a different definition of consensus. Given that requirement, it is clear that negotiated rulemaking is only suited to a limited number of areas. Under this process, those with a stake in the potential rulemaking—including OSHA—jointly develop a proposed rule rather than simply responding to the details of an OSHA draft proposal. Basically, an *ad hoc* advisory committee is established to represent the various interests affected by the proposed action. This allows for an interactive exchange of information among interested parties rather than the restricted series of individual communications between individual parties and the agency associated with traditional rulemaking. To facilitate the communication process, the committee is chaired by a trained, impartial mediator rather than by the agency representative, who simply serves as a member of the committee.[11]

2.0 Safety and Health Standards

2.1 Basic Framework

Having discussed the OSHA rulemaking process in general terms, this chapter now examines its use in issuing or modifying occupational safety and health standards. The overarching goal to which all OSHA standards aspire is found in section 2(b) of the Act: "to assure so far as possible every working man and woman in the Nation safe and healthful working conditions."

A standard is a rule that "requires conditions or the adoption or use of one or more practices, means, methods, operations, or processes reasonably necessary or appropriate to provide safe and healthful employment and places of employment."[12] An occupational safety and health standard must possess two qualities: 1) it must

---

[11] The only OSHA standard issued through the negotiated rulemaking process was the recently issued Construction Safety Standards for Steel Erection, 66 *Fed. Reg.* 5195 (January 18, 2001). Participating stakeholders included representatives from labor, industry, government, professional construction safety experts and equipment specialists. At the time this chapter was written, even that rule was the subject of a relatively narrow legal challenge. For more information on the Department of Labor's negotiated rulemaking process, *see* Marshall J. Breger, "Framework for the Use of Negotiated Rulemaking in the Department of Labor." Office of the Solicitor, (1992).

[12] OSH Act § 3(8); 29 U.S.C. § 652(8).

address a particular workplace hazard; and 2) it must require workplace measures that will eliminate or substantially control exposure to that hazard.[13]

Safety standards are developed to regulate conditions (safety hazards) which are likely to cause immediate serious and potentially fatal injury. Health standards, on the other hand, regulate chronic workplace exposures to toxic materials or harmful physical agents (health hazards) that, over an extended period of time, are likely to cause a material impairment of health or functional capacity. Health hazards, generally speaking, are more difficult to define than safety standards because the dangers from health hazards are harder to identify and quantify, more subjective, and more subject to varying individual susceptibilities.

The OSH Act initially gave OSHA the authority to adopt three types of workplace safety and health standards.[14] First, for a two-year period from 1970 to 1972, section 6(a) gave OSHA the authority to adopt any appropriate "national consensus standard"[15] or "established Federal standard" as an occupational safety and health standard without using notice and comment rulemaking. The Congressional objective was to provide OSHA, on an interim basis, with what might be viewed as a baseline group of enforceable occupational safety and health standards. It was assumed that OSHA would review and revise these Section 6(a) standards as time permitted. Instead, OSHA determined that it generally would be more appropriate to allocate its resources to developing new standards rather than updating the old ones. Second, Section 6(b) permits OSHA to develop and adopt permanent standards in accordance with the applicable substantive and procedural requirements. Third, Section 6(c) allows OSHA to adopt Emergency Temporary Standards in extraordinary circumstances.

2.2 OSHA's Statutory Authority to Issue Standards

Section 3(8) of the Occupational Safety and Health Act (OSH Act) defines the term "occupational safety and health standard" as follows:

> The term "occupational safety and health standard" means a standard which requires conditions, or the adoption or use of one or more practices, means,

---

[13] *Chamber of Commerce v. Herman,* 174 F.3d 206 (D.C. Cir. 1999); *Louisiana Chemical Ass'n v. Bingham,* D.C. La.1980, 496 F.Supp. 1188, *reversed on other grounds* 657 F.2d. 777, *on remand* 550 F.Supp. 1136.

[14] OSH Act § 6; 29 U.S.C. § 655.

[15] The theory behind permitting OSHA to adopt national consensus standards was that these standards were developed and adopted through a system of due process that was generally open to the public and allowed all interested persons to participate and vote on the proposed standard as it moved through the process.

methods, operations, or processes, *reasonably necessary or appropriate* to provide safe or healthful employment and places of employment. [Emphasis added.]

Section 6(b)(5) of the OSH Act establishes the following additional requirements applicable to an occupational safety and health standard governing "toxic materials or harmful physical agents":

5. The Secretary, in promulgating standards dealing with toxic materials or harmful physical agents under this subsection, shall set the standard which most adequately assures, to the extent feasible, on the basis of the best available evidence, that no employee will suffer material impairment of health or functional capacity even if such employee has regular exposure to the hazard dealt with by such standard for the period of his working life. Development of standards under this subsection shall be based upon research, demonstrations, experiments, and such other information as may be appropriate. In addition to the attainment of the highest degree of health and safety protection for the employee, other considerations shall be the latest available scientific data in the field, the feasibility of the standards, and experience gained under this and other health and safety laws. Whenever practicable, the standard promulgated shall be expressed in terms of objective criteria and of the performance desired.

2.3 Other Controlling Statutes and Executive Orders

In addition to the OSH Act, a number of other federal laws and Executive Orders shape OSHA's authority and decision-making procedures. Four of the more significant ones are highlighted here. There are several other relevant statutory enactments and executive orders that may affect OSHA's rulemaking efforts. For example, the National Environmental Policy Act (NEPA) requires the Agency to consider the potential impact of a rulemaking on the environment, and if a significant impact is likely, an environmental impact statement must be prepared.[16]

2.3.1 Paperwork Reduction Act of 1995

The Paperwork Reduction Act (PRA) was enacted in 1995 to minimize the paperwork burden imposed by the federal government on the private sector and state and local governments.[17] At the same time, the PRA is intended to maximize the usefulness of the information that is collected. The Act applies to all rulemak-

---

[16] 42 U.S.C. § 4321 (1992).
[17] 44 U.S.C. § 3501.

ings with information collection requirements and requires an agency proposing a rule which includes an information collection to conduct a paperwork burden analysis and justify the information it intends to collect. If the required action is to keep records, to report, or to monitor, the agency must prepare an information collection request (ICR) which is submitted to the Office of Management and Budget (OMB). Unless the ICR is approved, the underlying paperwork requirement is unenforceable. The ICR is approved by OMB for a fixed period of time (generally three years) and then must be re-approved or the authorization expires.

As part of the approval process, the responsible OSHA official must submit a certification to OMB that covers the following points with respect to the proposed paperwork requirement:

a. It is necessary for the proper performance of agency functions;

b. It avoids unnecessary duplication;

c. It reduces burden on small entities;

d. It uses plain, coherent, and unambiguous language that is understandable to respondents;

e. Its implementation will be consistent and compatible with current reporting and recordkeeping practices;

f. It indicates the retention periods for recordkeeping requirements;

g. It informs respondents of the information called for under 5 C.F.R. § 1320.8 (b)(3) about:

   i. Why the information is being collected;

   ii. Use of information;

   iii. Burden estimate;

   iv. Nature of response (voluntary, required for a benefit, or mandatory);

   v. Nature and extent of confidentiality; and

   vi. Need to display currently valid OMB control number;

h. It was developed by an office that has planned and allocated resources for the efficient and effective management and use of the information to be collected;

i. It uses effective and efficient statistical survey methodology (if applicable); and

j. It makes appropriate use of information technology.

2.3.2 Regulatory Flexibility Act

The purpose of the Regulatory Flexibility Act (RFA), as amended by the Small Business Regulatory Enforcement Fairness Act (SBREFA), is to minimize the regulatory burden on small businesses by requiring agencies to consider the impact of any rulemaking on small entities which, based on their size, may not have the ability to absorb the costs of regulations in the same manner as larger companies which benefit from economies of scale.[18] Although the RFA does not require preferential treatment or special exemptions for small entities, the Act does require that each agency analyze how its regulations will affect the ability of small entities to invent, produce, and compete.

If the agency believes that the proposed rule may have a significant economic impact on a substantial number of small entities, an initial regulatory impact analysis (RIA) must be prepared and published in the *Federal Register* with the proposed rule. The analysis must describe the objectives of the proposed rule, discuss the proposed rule's direct and indirect effects, and explain why the agency chose that specific regulatory approach versus any other alternatives. When the notice and comment period is complete and the agency prepares to issue the final rule, it must prepare a final regulatory flexibility analysis or certify that the rule will not have a significant economic impact on a substantial number of small entities. The final RIA must address any comments received on the initial analysis, the alternatives considered, and the rationale for the final rule. A summary or a full copy of the final RIA must be published in the *Federal Register* with the final rule. Typically, OSHA reprints the Executive Summary from the final RIA in the Preamble to the final rule.

2.3.3 Small Business Regulatory Enforcement Fairness Act

The Small Business Regulatory Enforcement Fairness Act (SBREFA) was enacted in March 1996.[19] SBREFA amends the RFA, and expands judicial review to include RFA requirements, thereby providing small entities with a greater ability to force OSHA agency compliance with the goals only partially realized under the RFA. SBREFA also broadens the authority of the Chief Counsel for the Office of

---

[18] 5 U.S.C. § 601 (1992). Incorporating the definition established by the Small Business Act (15 U.S.C. § 632), small entities are defined as independently-owned and operated businesses that are not dominant in their field of operation. The SBA has further established size standards, based on the number of employees or dollar volume of business, which vary by industry. *See* 13 C.F.R. Part 121 (1996).

[19] SBREFA was enacted as part of the "Contract with America Advancement Act" (Public Law 104-121), and amends portions of 5 U.S.C. Chapter 6.

Advocacy of the U.S. Small Business Administration to file *amicus curiae* (friend of the court) briefs on cases under judicial review.

SBREFA adopts a special procedure applicable to rulemakings conducted by OSHA and the Environmental Protection Agency (EPA). It requires these agencies to notify the Small Business Administration (SBA) before issuing a *proposed* rule affecting small entities.[20] The SBA, in turn, convenes a Small Business Advocacy Review Panel consisting of representatives of OSHA (or EPA, as applicable), the Office of Advocacy, and OMB's Office of Information and Regulatory Affairs (OIRA). To date, the Panel has been chaired by an OSHA representative. The Panel names a group of approximately 10 to 15 small business entity representatives (SERs) to participate in the small business review process. The SERs typically receive the following documents: an explanation of the objectives of the draft rule, the draft regulatory text, a very preliminary economic impact analyses, a list of questions for their consideration, and any other pertinent information that OSHA chooses to provide. They are provided with a relatively short period of time to review all of that information and then participate in a conference call with the Panel and either all or a subgroup of the SERs to discuss their findings, concerns and recommendations. Following the conference call, the SERs are provided with another fairly short period of time to submit written comments to the Panel. An official Panel report is developed and submitted to OSHA, which must then explain how it responded to the report. As one might expect, the nature of the report and the interaction between the three agencies in seeking to develop a consensus document (OSHA, OMB, SBA) is significantly affected by the political environment.

2.3.4 Executive Order 12866—Regulatory Impact Analyses

Executive Order 12866[21] reinforces and expands on the principles in the Regulatory Flexibility Act. It sets out principles and requirements which are viewed as part of the inherent authority of the Executive Branch to the extent not foreclosed by either the language of the OSH Act or other applicable laws. E.O. 12866 focuses on the need to properly weigh the costs and benefits of available regulatory alternatives before choosing a course of action. Section 1(b) of E.O. 12866 states that federal agencies engaged in rulemaking must, to the extent permitted by its authorizing statute:

1. identify and assess available alternatives to direct regulation, including providing economic incentives to encourage the desired behavior, such as user

---

[20] SBREFA § 244.
[21] 58 *Fed. Reg.* 51735 (Oct. 4, 1993).

fees or marketable permits, or providing information upon which choices can be made by the public;

2. consider ... the degree and nature of the risks posed by various substances or activities within its jurisdiction;

3. design its regulations in the most cost-effective manner to achieve the regulatory objective;

4. assess both the costs and the benefits of the intended regulation and, recognizing that some costs and benefits are difficult to quantify, propose or adopt a regulation only upon a reasoned determination that the benefits of the intended regulation justify the costs;

5. base its decisions on the best reasonably obtainable scientific, technical, economic, and other information concerning the need for, and consequences of, the intended regulation;

6. identify and assess alternative forms of regulation and...to the extent feasible, specify performance objectives, rather than specifying the behavior or manner of compliance that regulated entities must adopt;

7. wherever feasible...seek views of appropriate State, local, and tribal officials before imposing regulatory requirements that might significantly or uniquely affect those governmental entities;

8. avoid regulations that are inconsistent, incompatible, or duplicative with its other regulations or those of other Federal agencies;

9. tailor its regulations to impose the least burden on society ... consistent with obtaining the regulatory objectives; and

10. draft its regulations to be simple and easy to understand, with the goal of minimizing the potential for uncertainty and litigation arising from such uncertainty.

Only those rules deemed a "significant regulatory action" are subject to OMB review under E.O. 12866. "Significant" means having: (1) an annual impact on the economy of $100 million or more; (2) a serious inconsistency or interference with action taken or planned by another agency; (3) an impact on entitlement, grants, user fees, or loan programs; or (4) novel legal or policy issues. The agency must document its determination of whether the action is "significant" in a Regulatory Impact Analysis (RIA). It is then OMB's duty to review those actions determined to be "significant" and assure that the agency has assessed all cost and benefits of available regulatory alternatives. Unlike the RFA and SBREFA, there is some uncertainty, with apparently conflicting court decisions, as to whether private parties have a right to enforce the requirements of E.O. 12866 through judicial review or otherwise.

2.4 Characterization of the Safety and Health Standards

Based on the application of Sections 3(8) and 6(b)(5) of the OSH Act, OSHA standards issued under Section 6(b) of the OSH Act have been characterized as either "safety standards" as defined by Section 3(8) of the OSH Act or "health standards" as defined by Sections 3(8) and 6(b)(5) of the OSH Act. Much of the language of the OSH Act reflects an ambiguous political compromise by Congress. The courts have been left to clarify those provisions in a manner which furthers the Congressional mandate while recognizing that OSHA may only act through a proper delegation of Congressional authority.

2.4.1 Generally Applicable Principles for OSHA Standards

Generally, to sustain either a safety standard or a health standard on judicial review as being "reasonably necessary and appropriate,"[22] OSHA must demonstrate the following:[23]

   a. Current workplace exposure levels to the identified hazards pose a significant risk of harm to the workers who would be covered by the standard;[24]

   b. The requirements of the rule would significantly or materially reduce the workplace risk to workers exposed to those identified hazards;

---

[22] According to OSHA: "A standard is reasonably necessary or appropriate within the meaning of Section 652(8) if it substantially reduces or eliminates significant risk, and is economically feasible, technologically feasible, and cost effective, and is consistent with prior Agency action or is a justified departure, is supported by substantial evidence, and is better able to effectuate the Act's purposes than any national consensus standard it supersedes. See 58 *Fed. Reg.* 16612-16616 (March 30, 1993)."

[23] *See* Control of Hazardous Energy Sources, Supplemental Statement of Reasons, 58 *Fed. Reg.* 16612, 16614, cols. 2 and 3 (March 30, 1993), upheld in *International Union, UAW v. Occupational Safety and Health Administration, U.S. Department of Labor*, 37 F.3d 665 (D.C. Cir. 1994) (Lockout / Tagout II).

[24] *Industrial Union Department, AFL-CIO v. American Petroleum Institute*, 448 U.S. 607, 615 (1980) (*Benzene*)(vacating the benzene standard). OSHA has generally considered, at minimum, a fatality risk of 1/1000 over a 45-year working *lifetime* to be a significant health risk. *See* the *Benzene* decision *Industrial Union Dep't v. American Petroleum Institute*, 448 U.S. 607, 646 (1980); the *Asbestos* decision *Building and Constr. Trades Dep't, AFL-CIO v. Brock*, 838 F.2d 1258, 1265 (D.C. Cir. 1988); the *Formaldehyde* decision *International Union, UAW v. Pendergrass*, 878 F.2d 389, 392 (D.C. Cir. 1989). More recently, OSHA apparently determined that an *annual* risk of 1/1000 of experiencing a lost work day musculoskeletal disorder was a significant risk. *See* 65 *Fed. Reg.* 68556.

c. The requirements of the rule are technically feasible;[25]

d. The requirements of the rule are economically feasible[26]

e. The requirements of the rule are the most cost-effective approach for achieving the reduction in risk posed by those identified hazards;[27]

f. The requirements of the rule, to the extent practical and feasible, specify performance objectives (based on objective criteria) rather than specifying the behavior or manner of compliance.[28]

g. OSHA's determinations are supported by substantial evidence in the record taken as a whole;[29]

h. OSHA's findings and decisions are based on the basis of the best reasonably available evidence.[30]

i. Interested parties have been provided with adequate notice and an opportunity for participation, the rulemaking has otherwise been conducted in accordance with applicable legal procedures, and the final rule is a logical outgrowth of the proposal.[31]

j. The final rule does not supersede or in any manner affect any workmen's compensation law or enlarge, diminish or affect in any other manner the common law or statutory rights, duties, or liabilities of employers and em-

---

[25] *American Textile Mfrs. Inst., Inc. v. Donovan*, 452 U.S. 490, 513 (1981) (*"ATMI"* or *"Cotton Dust"*). A standard is technologically feasible if the protective measures it requires already exist, can be brought into existence with available technology, or can be created with technology that can reasonably be expected to be developed. *Id.* AISI v. OSHA, 939 F.2d 975, 980 (D.C. Cir. 1991)(*"AISI"*).

[26] *ATMI.* A standard is economically feasible if industry can absorb or pass on the costs of compliance without threatening its long-term profitability or competitive structure. *See ATMI*, 452 U.S. at 530 n. 55; *AISI*, 939 F.2d at 980. There is also case law indicating that a requirement must not only be feasible but within the bounds of what is reasonable for each affected industry. *American Dental Association v. Secretary of Labor*, 984 F.2d 823 (7th Cir. 1993), *cert. denied*, 510 U.S. 859.

[27] A standard is cost effective if the protective measures it requires are the least costly of the available alternatives that achieve the same level of protection. ATMI, 453 U.S. at 514 n. 32; *International Union, UAW v. OSHA*, 37 F.3d 665, 668 (D.C. Cir. 1994) ("LOTO III").

[28] OSH Act § 6(b)(5); 29 U.S.C. § 655(b)(5); Executive Order 12866, 1(b)(8).

[29] OSH Act § 6(f); 29 U.S.C. § 655(f).

[30] OSH Act §§ 6(b)(5) and 6(f); 29 U.S.C. § 655(b) and § 655(f); Executive Order 12866, § 1(b)(7).

[31] OSH Act § 6(b); 29 U.S.C. § 655(b), Administrative Procedures Act § 553.

ployees under any law with respect to injuries, diseases, or death of employees arising out of, or in the course of, employment.[32]

k. The language of the final rule must be written in language that can reasonably be understood by those who must comply with it.[33]

2.4.2 Safety Standards v. Health Standards

For safety standards (the category that does not apply to "toxic materials or harmful physical agents"), the costs of compliance must be reasonably related to the benefits of compliance and there is some support for the proposition that they must yield a positive result under a cost-benefit analysis.[34] There is some question, however, as to whether the costs and benefits to be considered should be limited to the direct impact on employers, or whether it should extend to the impacts on employees as well as society at large.[35]

For health standards governing "toxic materials and harmful physical agents" (in contrast to safety standards governing safety hazards), it is not enough that the standard significantly reduce the workplace risk. Rather, a health standard must, to the extent feasible and within reasonable bounds for each industrial sector, reduce workplace exposures to a level below that which presents a significant risk of material impairment of health or functional capacity to employees. In other words, when adopting a health standard, OSHA is required to perform the described feasibility analysis but may not base its decision-making on a cost-benefit analysis.[36] In addition, whenever practicable, the standard must be expressed in terms of objective criteria and of the performance desired.[37]

2.4.3 The Content of Permanent Standards

Under the OSH Act, the final rule may include, as appropriate, in addition to maximum permitted exposure levels, requirements for labels, medical examinations, protective equipment, and engineering controls, and such provisions as are necessary and feasible to reduce to acceptable levels the risk of injury or illness of exposed

---

[32] OSH Act § 4(b)(4); 29 U.S.C. § 655(4(b)(4).
[33] U. S. Const., Amend. V (requiring due process of law).
[34] *National Grain and Feed Association, Inc. v. OSHA*, 866 F.2d 717, 733 (5th Cir. 1989).
[35] *International Union v. OSHA*, 938 F.2d 1310, 1320 (D.C. Cir. 1991).
[36] *ATMI* 453 U.S. at 514.
[37] OSH Act § 6(b)(5); 29 U.S.C. § 655(b)(5).

employees.[38] The major focus of contention in the development of standards frequently involves: (a) the nature or significance of the alleged risk, and (b) the degree to which the risk can be reduced in a technologically and economically feasible way.

2.4.4 Specification Standards versus Performance Standards

Occupational safety and health standards may either: (1) set a performance standard that allows employers the flexibility to develop the most cost-effective approach for each facility; or (2) require that specific action be taken by employers to abate a specific hazard. One example of a performance standard is represented by the permissible exposure limits (PELs) found in OSHA's Air Contaminants Standard.[39] The standard lists the exposure limit (*e.g.*, 10 ppm 8-hour TWA), and leaves the selection of engineering controls up to the employer.[40] Another example of OSHA's use of performance standards is its Construction Safety Standard for Scaffolds. For many situations, instead of prescribing the specific means by which employers must protect their workers from such hazards as falls, falling objects and structural instability, the standard identifies compliance options or establishes performance-based criteria and provides the employer with the flexibility to choose the proper protection system for its workplace.

Specification standards, on the other hand, prescribe the particular means an employer must use to abate a hazard. Again, using the construction industry as an example, OSHA standards requiring specific kinds of guardrails to protect workers from open-sided floors, wall openings, and floor openings. The standards not only describe the length, height and strength of the railings but also specifies that those protective devices include a top rail, intermediate rail, toeboard and posts, *etc.*[41]

2.4.5 Horizontal versus Vertical Standards

Occupational safety and health standards are also characterized as "horizontal standards " or "vertical standards." "Horizontal standards" regulate occupational safety and health issues across industry lines and are generally applicable to all employers. An example of a "horizontal standard" is the Hazard Communication

---

[38] OSH Act § 6(b)(7); 29 U.S.C. § 655(b)(7).

[39] 29 C.F.R. § 1910.1000.

[40] The rule does contain a hierarchy of controls (engineering, work practices and PPE) and from that perspective is not strictly a performance standard. If it were, employers could elect to use respiratory protection instead of engineering controls.

[41] *See* 29 C.F.R. § 1926.500 (d) and (f) (1).

Standard, which applies to all employers covered by the OSH Act.[42] "Vertical standards," on the other hand, apply to a specific industry (*e.g.,* telecommunications, logging, pulp and paper, construction). In some situations, an industry-specific or "vertical standard" may address the same hazards covered by a general industry or "horizontal standard". In those situations, the standard specifically applicable to a condition, practice, or chemical process will prevail over any general standard which may otherwise be applicable.[43] However, employers that are covered by a specific industry standard must also comply with any general industry standard if the general industry standard is not in direct conflict with the industry-specific standard.

2.5 Emergency Temporary Standards

If OSHA determines that employees are in grave danger due to a newly identified hazard, such as exposure to a toxic substance or harmful physical agent determined to be highly toxic or physically harmful, OSHA is authorized to issue an emergency temporary standard.[44] The emergency standard takes effect immediately upon publication in the *Federal Register* (*i.e.,* without having to meet notice and comment requirements). The *Federal Register* notice also serves as formal notice of a proposed permanent standard subject to normal notice and comment procedures. However, an emergency temporary standard (ETS) has a limited effective life of six months. Therefore, as a practical matter, OSHA must issue a final rule within six months of issuing the ETS or there will be a gap between expiration of the ETS and adoption of a permanent standard. In theory, OSHA could conclude, based on further review, that no standard is needed.

Only nine emergency temporary standards have been issued since 1970. Three were not challenged and went into effect—vinyl chloride monomer, asbestos, and dibromochloropropane (DBCP).[45] Six were successfully challenged in court. The ETS for benzene[46] was invalidated on procedural grounds. The other five were overturned by the courts on substantive grounds—asbestos,[47] 14 carcinogens,[48] field reentry (pesticides),[49] acrylonitrile, and commercial diving.[50]

---

[42] See 29 C.F.R. § 1910.1200.

[43] See, e.g., *Western Water Proofing Co.,* 7 O.S.H. Cas. (BNA) 1499, 1502 (Rev. Comm'n 1979); *Lloyd C. Lockrem Inc.,* 3 O.S.H. Cas. (BNA) 2045 (Rev. Comm'n 1976). 29 C.F.R. § 1910.5 (c) (1).

[44] 29 U.S.C. § 655 (c).

[45] 48 Fed. Reg. 51086 (November 4, 1983) (asbestos). 43 Fed. Reg. 45536 (March 17, 1978) (DBCP).

[46] *Industrial Union Dept., AFL-CIO v. Bingham,* 570 F.2d 965 (D.C. Cir. 1977).

[47] *Asbestos Information Association / North America v. OSHA,* 727 F.2d 415 (5th Cir. 1984).

OSHA appears to have largely abandoned the use of emergency temporary standards as a regulatory vehicle primarily because of its inability to satisfy the stringent statutory criteria, as interpreted by the courts, under which OSHA is authorized to take this unilateral *ex parte* action. Secondarily, as discussed earlier, an ETS sets in motion a timetable for publishing a permanent standard that OSHA has extraordinary difficulty in meeting. As a result, emergency temporary standards are not viewed as a viable alternative to the normal rulemaking process, and the General Duty Clause (discussed in Chapter 4) has been used in its stead to address pressing issues.

3.0 State Jurisdiction and Plans

At the time the OSH Act was adopted, Congress recognized that many states already administered or would like the opportunity to administer effective worker protection programs. Thus, Congress included provisions in the OSH Act[51] permitting states:

1. to exercise jurisdiction over any occupational safety and health issue where no federal standard exists;
2. to assume responsibility for development and enforcement of state rules at least as effective as the corresponding federal standards;
3. to operate an occupational health and safety program comparable in funding, authority, and staffing to the federal program; and
4. to enforce more stringent standards applicable to products in interstate commerce where they are compelled by local conditions and do not unduly burden interstate commerce.

A state must submit a proposed state program plan to OSHA if the state desires to assume responsibility for the development and enforcement of standards "relating to" any occupational safety or health issue with respect to which a federal standard has been issued. Congress provided for funding of up to 50 percent of the cost of administering and enforcing programs in the states, and up to 90 percent to improve state capabilities in this area.

---

[48] *Dry Color Manufacturers' Association v. Department of Labor*, 486 F.2d 98 (3rd Cir. 1973).
[49] *Florida Peach Growers Ass'n v. U.S. DOL*, 489 F.2d 120 (5th Cir. 1974).
[50] *Taylor Diving & Salvage Co., Inc. v. U.S. Department of Labor*, 537 F.2d 819 (5th Cir. 1976).
[51] OSH Act § 18; 29 U.S.C. § 667.

Obtaining "final approval" of a state plan is a multi-step process. First, the state must submit a "developmental plan" in which the state assures OSHA that within 3 years it will have a structural framework in place for the effective administration of a safety and health program. Once the "developmental plan" is completed, the plan is eligible for certification. Certification attests to the completeness of the plan. At any time after this initial plan approval, OSHA and the state may enter into an "operational status agreement" whereby OSHA suspends federal enforcement with respect to any workplace activities covered by the state plan. Ultimately, if the state plan receives "final approval," the health and safety program is operated by the state and OSHA relinquishes total authority. Approximately twenty-four states and 2 territories have approved plans, with a few being limited in scope to state employees.[52]

States with approved state plans promulgate and enforce their own standards. Generally, the states have six months after the promulgation of a new federal standard to adopt an equivalent standard. The majority of states simply adopt the federal standard. The states can also act in advance of the federal government because of perceived needs that have not been addressed. In several cases, the states adopted standards which were significantly different from the federal rule. For example, although OSHA's effort to update the PELs in its Air Contaminants Standard through a generic rulemaking was vacated, several states have proceeded to update their PELs using that approach. California adopted a rule known as Proposition 65 to expand the requirements of the Hazard Communication Standard for California manufacturers, and more recently adopted an ergonomics standard addressing repetitive stress injuries.[53] A number of states have adopted variations of a generic safety and health program standard (*e.g.,* California, Nevada). One of the unsettled state plan issues is OSHA's position that state plans can adopt a standard and begin enforcement before OSHA approves that standard, although OSHA approval may be delayed for years.

---

[52] Information on the current status of state programs can be obtained from OSHA's Office of State Programs in Washington, D.C. or OSHA's website at <http://www.osha.gov/oshdir/states.html>. Although the law provides for withdrawal of approval by the Secretary after notice and hearing, no state plans have been involuntarily withdrawn. In 1991, as a result of allegations of ineffective operation of the state's plan, OSHA proposed to withdraw approval of North Carolina's program. The state vigorously opposed OSHA's action, and succeeded in retaining jurisdiction after making significant changes in funding, resources, and enforcement.

[53] 8 C.F.R. § 5110.

4.0 Challenging OSHA Standards

OSHA standards can be challenged within 59 days of promulgation (sometimes called pre-enforcement review) by any person who may be adversely affected by the standard.[54] The challenge is made by filing a petition for review with the U.S. Court of Appeals for the circuit in which the objector resides or has its principal place of business.[55] In a pre-enforcement review, the burden is on OSHA to demonstrate that the standard at issue is "reasonably necessary and appropriate" (as that phrase is defined above) based on "substantial evidence in [the] record considered as a whole"[56] and otherwise in compliance with the applicable laws discussed in Section 2.0 of this chapter.

The validity of a standard may also be challenged during enforcement proceedings.[57] An employer cited for failure to comply with a standard may, for instance, claim that the standard is invalid because compliance is technologically or economically infeasible,[58] or the underlying rulemaking was procedurally defective.[59] In an enforcement proceeding, the standard is presumed to be valid and the burden of proof is on the employer.

5.0 Variances

If an employer can adequately demonstrate that it will need additional time to come into compliance with a standard, the employer may be able to obtain a temporary variance from the obligation to comply with the standard. If an employer can show by a preponderance of the evidence that its alternative practices and control measures provide equivalent employee protection, the employer may be able to obtain a permanent variance from the obligation to comply with the standard. When the problem (*e.g.*, inability to comply, the prescribed measure is not cost-

---

[54] OSH Act § 6(f); 29 U.S.C. § 655(f). Promulgation generally occurs upon publication in the *Federal Register*, but may occur earlier if publicly announced by the Secretary.

[55] If challenging a state standard, then judicial review takes place in state court.

[56] OSH Act § 6(f); 29 U.S.C. § 655(f).

[57] Cf. In *RSR Corp. v. Donovan*, the Court barred the challenge because the plaintiff had an opportunity to raise the issue, during his participation in the rulemaking and pre-enforcement review proceedings. 747 F.2d 294 (5th Cir. 1984).

[58] *See* OSH Act § 6(b)(5); 29 U.S.C. § 655(b)(5); *Loomis Cabinet Co. v. OSHRC*, 20 F.3d 938 (9th Cir. 1994).

[59] *See, e.g., Deering Milliken, Inc. v. OSHRC*, 630 F.2d 1094 (5th Cir. 1980); *Marshall v. Union Oil Co.*, 616 F.2d 1113 (9th Cir. 1980); contra, *National Industrial Constructors, Inc. v. OSHRC*, 583 F.2d 1048 (8th Cir. 1978), which suggests that some procedural attacks are limited to pre-enforcement challenges.

effective and an effective alternative exists) is broader in scope, an employer may be able to obtain a letter of interpretation from OSHA's Office of Compliance Programs stating that the employer's arrangement provides equivalent protection and constitutes, at worst, a *de minimis* violation for which no enforcement action will be taken by OSHA.

5.1 Temporary Variances

A temporary variance from the OSHA standard may be obtained when the employer can show that it: 1) is unable to comply with the deadlines of a standard; 2) is taking steps to safeguard employees; and 3) has developed a plan to come into compliance and will implement it as soon as possible.[60] Before a temporary variance can be granted, affected employees must be notified and provided with an opportunity for a hearing. The temporary variance is initially effective for one year and may be renewed for up to two more years. A renewal application is supposed to be filed at least 90 days in advance of expiration of the current variance.

5.2 Permanent Variances

Permanent variances address situations where compliance with the specific control method required by the standard is obviated by the use of an equivalent alternative means of protection.[61] They are obtained through a process which is essentially a mini-rulemaking for the individual employer. Before a permanent variance can be granted, affected employees must be notified and provided with an opportunity for a hearing. The critical and very difficult showing that must be made by the employer is that the employer's alternative control measures are as effective, in terms of employee protection, as those specified in the standard. Relatively few permanent variances have been issued in OSHA's 30-year history, and most have been issued for situations related to traditional safety problems as opposed to health standards. A permanent variance is subject to modification or revocation at any time after it has been in effect for six months.

5.3 Other Variances

OSHA has also provided an opportunity for employers to obtain an experimental variance from OSHA if the employer is participating in an experiment to demon-

---

[60] OSH Act § 6(b)(6)(A); 29 U.S.C. § 655(b)(6)(A); 29 C.F.R. § 1905.10.
[61] OSH Act § 6(d); 29 U.S.C. § 655(d).

strate or validate new job safety and health techniques.[62] Additionally, the Secretary may find that certain variances are justified for purposes of national defense.[63]

5.4 Interim Order

An employer may apply to OSHA for an interim order authorizing the employer to continue to operate under existing conditions while the variance request is being considered.[64] The application must explain why the order should be granted, including how employee protection is being maintained in the interim, and may be filed at the same time or after the application for a variance is submitted. If the request is denied, OSHA must specify the reasons why the application was rejected.

---

[62] OSH Act § 6(b)(6)(C); 29 U.S.C. § 655(B)(6)(C).
[63] OSH Act §16; 29 C.F.R. § 1905.12.
[64] 29 C.F.R. § 1905.11(c).

Chapter 3

The Duty to Comply

Arthur G. Sapper, Esq.
McDermott, Will & Emery
Washington, D.C.

1.0 Overview

Section 5(a)(2) of the OSH Act, 29 U.S.C. § 654(a)(2), states that "each employer...shall comply with occupational safety and health standards promulgated under this Act." Other sections of the Act impose an implicit duty to comply with OSHA's regulations.[1] Although the duty to comply with standards and regulations seems unqualified, the courts and the Occupational Safety and Health Review Commission have held that the duty is qualified in various ways.

2.0 Applicability of OSHA Standards

2.1 The General Principle of Preemption

The OSHA standards themselves state a general principle—the more specific standard prevails over the more general.[2] For this reason, decisions speak of the defense of preemption – *i.e.*, a citation will be vacated if the cited condition is regulated by a more specifically applicable standard.[3] While many factors are relevant to such an inquiry,[4] the basic question is whether application of the more

---

[1] Section 9(a), 29 U.S.C. § 658(a) (permitting citation to be issued for violation of regulation); Sections 17(a)-(c) & (e), 666(a)-(c) & (e) (permitting civil and criminal penalties to be imposed for violating regulations).

[2] 29 C.F.R. § 1910.5(c).

[3] *E.g., McNally Constr. & Tunneling Co.*, 16 BNA OSHC 1886 (Rev. Comm'n 1994), *aff'd and approved*, 71 F.3d 208, 17 BNA OSHC 1412 (6th Cir. 1995); *Bratton Corp.*, 14 BNA OSHC 1893 (Rev. Comm'n 1990) (steel erection standards do not preempt general fall protection standards) (acceding to view of several circuits); *New England Telephone & Telegraph Co.*, 8 BNA OSHC 1478 (Rev. Comm'n 1980).

[4] See, for example, the various factors examined in the cases in note 3.

generally applicable standard would defeat a rulemaking decision implicit in the more specifically applicable standard.[5]

In accordance with this principle, an employer must first determine whether his industry is specially regulated by one of the several industry-specific "parts" in Title 29 of the Code of Federal Regulations. These industry-specific parts are: Part 1913, which applies to shipyards; Part 1917, which applies to marine terminals; Part 1918, which applies to longshoring; Part 1926, which applies to construction; and Part 1928, which applies to agriculture.

If no industry-specific part applies, then an employer must look to Part 1910, which is entitled "General Industry Standards" and which applies to all employers engaged in businesses affecting commerce. The employer must then determine whether a special, industry-specific section within Subpart R of Part 1910, or an industry-specific subpart within Part 1910, regulates both his industry and the particular condition cited. For example, § 1910.261, the first section in Subpart R, regulates the paper industry, while Subpart T of Part 1910 covers commercial diving. If no industry-specific standard in Subpart R of Part 1910 applies, then the employer is regulated by the many generally applicable sub-parts in Part 1910. For example, Subpart O regulates the guarding of machinery generally.

The preemption principle – *i.e.,* the principle that the specific standard prevails over the general standard – applies even to standards within an industry-specific part (*e.g.,* within Part 1926, the construction part). Thus, the general provisions in Part 1926 governing the use of non-sparking electrical equipment in flammable gas concentrations were held to have been preempted by the specific provisions on the use of such equipment in flammable gas concentrations in tunnels under construction.[6]

2.2 Special Applicability Problems

May a standard in Part 1910 be applied to work regulated by an industry-specific part (*e.g.,* construction work)? It has been held that, if there is no applicable construction standard in Part 1926, OSHA may cite an employer for a violation of a Part 1910 standard.[7] In addition, some Part 1910 standards expressly state that they apply to construction work.[8]

---

[5] *Lowe Constr. Co.,* 13 BNA OSHC 2182 (Rev. Comm'n 1989).
[6] *McNally Constr. & Tunneling Co.,* 16 BNA OSHC 1886 (Rev. Comm'n 1994), *aff'd and approved,* 71 F.3d 208, 17 BNA OSHC 1412 (6th Cir. 1995).
[7] *Western Waterproofing Inc.,* 7 BNA OSHC 1499, 1501-02 (Rev. Comm'n 1979).
[8] *E.g.,* § 1910.134 (introductory provision).

Nevertheless, some Part 1910 standards do not apply to construction work. Some expressly state they do not so apply,[9] and the preamble to at least one other set of standards states that those standards were not intended to so apply.[10] In at least one case, OSHA has refrained from applying a standard to construction work because its proposed version did not give that industry notice and opportunity to comment on its applicability.[11]

3.0 General Principles of the Duty to Comply

Although OSHA must show that a condition violative of a standard existed, OSHA need not always show that the cited employer *himself* violated the standard, *i.e.*, that the cited employer created the violative condition.[12]

Unless a standard explicitly or implicitly incorporates hazardousness as an element of a violation, OSHA need not show that a failure to comply with a standard creates a hazard.[13]

3.1 The Exposure Rule

With respect to many[14] standards, employers must comply only if there is, or reasonably predictably will be, exposure of employees to the violative condition. This principle is reflected in the allocation to OSHA of the burden of proving either:

  a. that employees are or were in the zone of danger created by a violative condition; or

  b. that it is reasonably predictable that employees, by "operational necessity" or otherwise (including inadvertence) in the course of their work or associ-

---

[9] *E.g.*, § 1910.147(a)(1)(ii)(A) (lockout standard).

[10] *E.g.*, the electrical standards in Subpart S of Part 1910. See 46 *Fed. Reg.* 4034, 4039 ("the electrical standards of Part 1910 do not apply to construction activities").

[11] Thus, OSHA has stated that the Bloodborne Pathogens Standard, § 1910.1030, does not apply to construction work because the construction industry did not receive public notice of, and an advisory committee was not consulted about, such an application when the standard was proposed. Letter from Secretary of Labor Lynn Martin to Robert Georgine, "Construction Activities and Operations and the Bloodborne Pathogens Standard" (Dec. 23, 1992), <http://www.osha-slc.gov/OshDoc/Interp_data/I19921223.html>.

[12] See § 3.1 of this chapter.

[13] *Kaspar Electroplating Corp.*, 16 BNA OSHC 1517 (Rev. Comm'n 1993); *Bunge Corp. v. Secretary of Labor,* 638 F.2d 831, 834 (5th Cir. 1981).

[14] See text accompanying note 24 below.

ated activities (*e.g.*, going to rest rooms) will be in the zone of danger created by the cited condition.[15]

The term "zone of danger" refers to "that area surrounding the violative condition that presents the danger to employees which the standard is intended to prevent."[16]

The Commission adopted this "reasonably predictable exposure" test after the courts rejected or suggested disapproval of the Commission's early requirement that OSHA prove actual exposure – *i.e.*, that an employee had actually been endangered by a violation.[17] Nevertheless, the mere possibility of exposure is insufficient.[18] That the employer is expected in the future to create a violative condition, but has not yet done so, is insufficient.[19] On the other hand, exposure of just a single employee is sufficient to trigger the employer's duty and to satisfy OSHA's burden of proof.[20]

OSHA need not show that a compliance officer personally witnessed facts supporting an exposure finding.[21] Nor need OSHA show that an employee is, for example, teetering on the edge of a unguarded floor.[22] Brevity of exposure is immaterial.[23] For some standards and regulations – particularly those requiring recordkeeping (*e.g.*, 29 C.F.R. § 1904.2(a) (requiring injury log)—no showing of exposure need be made.[24]

---

[15] *Fabricated Metal Prods.*, 18 BNA OSHC 1072, 1074 (Rev. Comm'n 1997) (surveying cases); *Gilles & Cotting, Inc.*, 3 BNA OSHC 2002 (Rev. Comm'n 1976).

[16] *RGM Constr. Co.*, 17 BNA OSHC 1229, 1234 (Rev. Comm'n 1995).

[17] *Brennan v. Gilles & Cotting, Inc.*, 504 F.2d 1255, 1263-66, 2 BNA OSHC 1243 (4th Cir. 1974) (remanding for reconsideration of actual exposure test); *Brennan v. OSHRC (Underhill Construction Corp.)*, 513 F.2d 1032, 2 BNA OSHC 1641 (2d Cir. 1975) (rejecting actual exposure test). See also *Adams Steel Erec., Inc.*, 766 F.2d 804, 812, 12 BNA OSHC 1393, 1398 (3d Cir. 1985) (same).

[18] *Fabricated Metal Prods.*, 18 BNA OSHC 1072, 1074 & n.8 (Rev. Comm'n 1997) (rejecting "physically possible" test); *Rockwell Intl. Corp.*, 9 BNA OSHC 1092 (Rev. Comm'n 1980).

[19] *Sharon Steel Corp.*, 12 BNA OSHC 1539 (Rev. Comm'n 1985).

[20] *E.g., Mineral Industries v. OSHRC*, 639 F.2d 1289, 1294-95, 9 BNA OSHC 1387 (5th Cir. 1981).

[21] *E.g., Brennan v. OSHRC (Underhill Construction Corp.)*, 513 F.2d 1032, 1038, 2 BNA OSHC 1641 (2d Cir. 1975); *see North Berry Concrete Corp.*, 13 BNA OSHC 2055, 2055-56 (Rev. Comm'n 1989).

[22] *E.g., Underhill Construction*, 513 F.2d at 1038.

[23] *E.g., Morgan & Culpepper, Inc. v. OSHRC*, 676 F.2d 1065, 1069, 10 BNA OSHC 1629, 1632 (5th Cir. 1982) (short duration of exposure no defense); *Brock v. L.R. Wilson & Sons*, 773 F.2d 1377, 1386, 12 BNA OSHC 1499 (D.C. Cir. 1985); *Walker Towing Corp.*, 14 BNA OSHC 2072, 2074 (Rev. Comm'n 1991); *Frank Swidzinski Co.*, 9 BNA OSHC 1230, 1232 (Rev. Comm'n 1981); *see Flint Engineering & Construction Co.*, 15 BNA OSHC 2052, 2056 (Rev. Comm'n 1992).

[24] *Thermal Reduction Corp*, 12 BNA OSHC 1264, 1268 (Rev. Comm'n 1985).

3.2 To Whose Employee Does the Duty Run?

This question was most vexing in the early years of the Act and, in some respects, the answer is only somewhat clearer today. The question first arose on multi-employer worksites, such as construction sites, where employees of Employer A (usually, a subcontractor) may be exposed to a violative condition created or controlled by Employer B (usually, the general contractor or another sub-contractor). In its early days, the Review Commission followed a simple rule: The employer of the employees exposed to a violative condition (Employer A) could be cited, regardless of whether the condition had been created by another employer (Employer B) and regardless of whether abatement of the condition was controlled by that employer.[25] Moreover, *only* the employer of the exposed employee (Employer A) could be cited;[26] the employer who created or controlled the violative condition (Employer B) could not be cited unless one of his own employees was exposed to it as well. Complicating the matter was that Employer A may lack the expertise to even recognize that the condition is violative.

Sub-contractors complained that this policy was highly unfair to them, and they and OSHA complained that it allowed the most guilty to escape liability. Eventually, beginning with a decision by the Second Circuit,[27] the rules of liability changed in two ways: A new, expanded liability rule was developed (Section 3.2.1 below); and a new series of affirmative defenses was established (Section 3.2.2 below). OSHA recently issued a directive to its enforcement personnel attempting to explain these rules in great detail.[28]

3.2.1 The Multi-Employer Worksite Liability Rules

Although OSHA may satisfy its burden of proving exposure by proving an employment relationship between the exposed employees and the cited employer (*i.e.,* showing that the exposed employees are those of the cited employer),[29] this is not necessary. OSHA may instead show exposure of an employee of *some* other em-

---

[25] *R.H. Bishop Co.,* 1 BNA OSHC 1767, 1769 (Rev. Comm'n 1974).
[26] *Martin Iron Works, Inc.,* 2 BNA OSHC 1063 (Rev. Comm'n 1974). *See also Hawkins Constr. Co.,* 1 BNA OSHC 1761 (Rev. Comm'n 1974); *Gilles & Cotting, Inc.,* 1 BNA OSHC 1388 (Rev. Comm'n 1973), *aff'd in relevant part,* 504 F.2d 1255 (4th Cir. 1974).
[27] *Brennan v. OSHRC (Underhill Construction Corp.),* 513 F.2d 1032, 2 BNA OSHC 1641 (2d Cir. 1975).
[28] CPL 2-0.124, "Multi-Employer Citation Policy" (Dec. 10, 1999), available at <http://www.osha-slc.gov/OshDoc/Directive_data/CPL_2-0_124.html>.
[29] *E.g., Van Buren-Madawaska Corp.,* 13 BNA OSHC 2157 (Rev. Comm'n 1989); *MLB Industries, Inc.,* 12 BNA OSHC 1525, 1528 (Rev. Comm'n 1985).

ployer *and* that the cited employer controlled or created the violative condition.[30] The current general principle on which multi-employer liability is based is that "an employer who either creates or controls the cited hazard has a duty...to protect not only its own employees, but those of other employers 'engaged in the common undertaking.'"[31]

3.2.1.1 General Construction Contractors

Hence, OSHA may cite general contractors for violations to which employees of subcontractors are exposed or which subcontractors created.[32] The general contractor is deemed to "have sufficient control over its subcontractors to require them to comply with the safety standards and to abate violations."[33] General contractors must take whatever measures are "commensurate with its degree of supervisory capacity,"[34] which includes some oversight over the work of subcontractors.[35] Hence, a general contractor was expected to detect a problem with a ground fault circuit interrupter installed by a subcontractor even though the condition was by nature latent and hidden from view.[36] On the other hand, a general contractor is responsible for only those violations that "it could reasonably be expected to prevent or detect."[37]

---

[30] *See* note 31-38 below.

[31] *McDevitt Street Bovis, Inc.*, 19 BNA OSHC 1108 (Rev. Comm'n 2000).

[32] *E.g., Huber, Hunt & Nichols, Inc.*, 4 BNA OSHC 1406, 1407-08 (Rev. Comm'n 1976).

[33] *Gil Haugan d/b/a Haugan Construction Company*, 7 BNA OSHC 2004, 2006 (Rev. Comm'n 1979). *Also see Lewis & Lambert Metal Contract, Inc.*, 12 BNA OSHC 1026, 1030 (Rev. Comm'n 1984).

[34] *Marshall v. Knutson*, 566 F.2d 596 (8th Cir. 1977).

[35] *McDevitt Street Bovis, Inc.*, 19 BNA OSHC 1108 (Rev. Comm'n 2000); *Centex-Rooney Construction Co.*, 16 BNA OSHC 2127 (Rev. Comm'n 1994).

[36] *Blount International Ltd.*, 15 BNA OSHC 1897, 1899-1900 (Rev. Comm'n 1992). *But see Knutson Construction Co.*, 4 BNA OSHC 1759, 1761 (Rev. Comm'n 1976), aff'd 566 F.2d 596 (8th Cir. 1977) (general contractor not liable for failing to detect a one-inch crack on the underside of a scaffolding platform; unreasonable to expect general contractor to detect such a crack).

[37] *David Weekley Homes*, 19 BNA OSHC 1116, 1119 (Rev. Comm'n 2000); *Centex-Rooney Constr. Co.*, 16 BNA OSHC 2127, 2130 (Rev. Comm'n 1994); *Blount Intl. Ltd.*, 15 BNA OSHC 1897, 1899 (Rev. Comm'n 1992).

3.2.1.2 Legal Status of the Multi-Employer Liability Rules

Several courts have, to one degree or another, upheld this extra-employment liability theory in the construction context.[38] The Fifth Circuit disagreed with the idea of extra-employment liability in a maritime industry case.[39] The D.C. Circuit has twice reserved ruling on whether the imposition of extra-employment liability in the construction context is inconsistent with 29 C.F.R. § 1910.12, which regulates the application of the construction standards in Part 1926.[40] That provision requires a construction employer to protect "*his* employees" (emphasis added) by complying with the construction standards.

3.2.1.3 Non-Construction Applications of the Multi-Employer Liability Rules

Outside the construction industry, OSHA has occasionally attempted to hold boat owners[41] and factory owners[42] liable for violations committed by their independent contractors and to which only the contractors' employees are exposed. In a recent attempt, the D.C. Circuit held that a factory owner is not liable for a contractor's lockout violation that affected only its own employees when it had no authority over the contractor's employees and its only "control" stemmed from its ownership of the property or its contract with the contractor.[43] See the additional discussion of this point in § 5.3.2 of Chapter 11.

---

[38] *Universal Construction Corp.*, 182 F.3d 726, 728-31, 18 BNA OSHC 1769 (10th Cir. 1999); *United States v. Pitt-Des Moines, Inc.*, 168 F.3d 976, 18 BNA OSHC 1609 (7th Cir. 1999); *R.P. Carbone Constr. Co. v. OSHRC*, 166 F.3d 815, 18 BNA OSHC 1551 (6th Cir. 1998); *Beatty Equip. Leasing, Inc. v. Secretary of Labor*, 577 F.2d 534, 6 BNA OSHC 1699 (9th Cir. 1978); *Marshall v. Knutson Constr. Co.*, 566 F.2d 596, 6 BNA OSHC 1077 (8th Cir. 1977); *Brennan v. OSHRC (Underhill Construction Corp.)*, 513 F.2d 1032, 1038, 2 BNA OSHC 1641 (2d Cir. 1975).

[39] *Melerine v. Avondale Shipyards, Inc.*, 659 F.2d 706, 10 BNA OSHC 1075 (5th Cir. 1981).

[40] *IBP, Inc. v. Herman*, 144 F.3d 861, 865 & n.3, 18 BNA OSHC 1353 (D.C. Cir. 1998); *Anthony Crane Rental, Inc. v. Reich*, 70 F.3d 1298, 1306-07, 17 BNA OSHC 1447 (D.C. Cir. 1995) (noting "tension" between wording of § 1910.12 and liability doctrine).

[41] *Harvey Workover, Inc.*, 7 BNA OSHC 1687, 1689 (Rev. Comm'n 1979); *Camden Drilling Co.*, 6 BNA OSHC 1560, 1561 (Rev. Comm'n 1978) (barge owner responsible for compelling subcontractor to have its employees stop using their own defective fan and either repair it or remove it).

[42] *E.g., IBP, Inc. v. Herman*, 144 F.3d 861, 18 BNA OSHC 1353 (D.C. Cir. 1998), *rev'g* 17 BNA OSHC 2073 (Rev. Comm'n 1997).

[43] *IBP, Inc. v. Herman*, 144 F.3d 861, 18 BNA OSHC 1353 (D.C. Cir. 1998), *rev'g* 17 BNA OSHC 2073 (Rev. Comm'n 1997).

3.2.2 Multi-Employer Worksite Defense Rules

As noted above, the liability rule followed in the early days of the Act was that exposure of one's own employee to a violative condition meant that one was liable, even if the cited employer did not create or control the violative condition. This rule has been partially reversed by the creation of a series of affirmative defenses by the Commission,[44] which has been accepted by several courts of appeals.[45] Today, a citation will be vacated if the cited employer on a multi-employer worksite:

a. Did not create or control the allegedly violative condition (such that he could not realistically correct the condition); and

b. Either:

1. Took reasonable alternative protective measures; or

2. Did not know, nor with the exercise of reasonable diligence, could have known of the *hazardousness* of the cited condition.[46]

Element (a) may be established by, for example, showing that the employer was prevented from abating a hazardous working condition due to union jurisdictional rules.[47]

Although these defenses originally arose in the context of construction sites, where there are frequently a number of different employers working at the same time, the Commission later applied them to all multi-employer worksites.[48]

4.0 Actual or Constructive Knowledge

OSHA must prove that the cited employer actually knew, or could have known, with the exercise of reasonable diligence, of the physical circumstances that com-

---

[44] *See generally Anning-Johnson Co.*, 4 BNA OSHC 1193 (Rev. Comm'n 1976); *Grossman Steel & Aluminum Corp.*, 4 BNA OSHC 1185 (Rev. Comm'n 1976).

[45] *Dun-Par Engineered Form Co. v. Marshall*, 676 F.2d 1333, 10 BNA OSHC 1561 (10th Cir. 1982); *Electric Smith, Inc. v. Secretary of Labor*, 666 F.2d 1267, 10 BNA OSHC 1329 (9th Cir. 1982); *DeTrae Enterprises, Inc. v. Secretary of Labor*, 645 F.2d 103, 9 BNA OSHC 1425 (2d Cir. 1980); *Bratton Corp. v. OSHRC*, 590 F.2d 273, 7 BNA OSHC 1004 (8th Cir. 1979).

[46] *E.g., LeeRoy Westbrook Construction Co.*, 13 BNA OSHC 2101, 2103 (Rev. Comm'n 1989); *Lewis & Lambert Metal Contractors*, 12 BNA OSHC 1026 (Rev. Comm'n 1984).

[47] *See McLean-Behm Steel Erectors, Inc.*, 6 BNA OSHC 1712, 1715 (Rev. Comm'n 1978).

[48] *Rockwell International Corp.*, 17 BNA OSHC 1801 n. 11 (Rev. Comm'n 1996).

prise the violative condition.[49] This element must also be proved in General Duty Clause cases.[50] The element pertains to the physical circumstances that comprise the violative condition, not the violativeness or hazardousness of the condition.[51]

An employer is not reasonably diligent if he neither makes an attempt to become aware of the physical conditions facing his employees, nor trains his employees to recognize hazards arising from them.[52] Reasonable diligence includes "the obligation to inspect the work area, to anticipate hazards to which employees may be exposed, and to take measures to prevent the occurrence."[53] A citation will be vacated on this ground if an employer reasonably relies on the expertise of the independent contractor who created the condition to which his employees were exposed.[54]

In general, if a compliance officer can see a physical condition during a normal inspection, it will be inferred that the employer could, with reasonable diligence, have seen it too.[55] However, OSHA must show that the cited condition was present for a sufficient amount of time such that, with the exercise of reasonable diligence, the employer could have discovered its existence.[56]

5.0 Additional Elements That OSHA Must Sometimes Prove

Sometimes a standard is so vague as to deprive employers of fair notice of its requirements, contrary to the Due Process clause of the Fifth Amendment to the

---

[49] *E.g., Ragnar Benson, Inc.,* 18 BNA OSHC 1937, 1939 (Rev. Comm'n 1999); *Continental Electric Co.,* 13 BNA OSHC 2153, 2154 (Rev. Comm'n 1989); *Prestressed Systems, Inc.,* 9 BNA OSHC 1864, 1870 (Rev. Comm'n 1981).

[50] *See U.S. Steel Corp.,* 12 BNA OSHC 1692, 1699 (Rev. Comm'n 1986).

[51] *Ormet Corp.,* 14 BNA OSHC 2134, 2138 (Rev. Comm'n 1991); *Southwestern Acoustics & Specialty, Inc.,* 5 BNA OSHC 1091 (Rev. Comm'n 1977) (employer need be shown only to have had knowledge of "physical conditions which constitute a violation," not that condition was prohibited by law).

[52] *Beaver Plant Operations, Inc.,* 18 BNA OSHC 1972, 1976 (Rev. Comm'n 1999).

[53] *Frank Swidzinski* Co., 9 BNA OSHC 1230, 1233 (Rev. Comm'n 1981).

[54] *E.g., Sasser Elec. & Mfg. Co.,* 11 BNA OSHC 2133 (Rev. Comm'n 1984), *aff'd,* 12 BNA OSHC 1445 (4th Cir. 1985) (not officially published).

[55] *See Green Construction Co.,* 4 BNA OSHC 1808, 1810 (Rev. Comm'n 1976) (Barnako, Chairman, concurring).

[56] *David Weekley Homes,* 19 BNA OSHC 1116, 1120-21 (Rev. Comm'n 2000).

Constitution.⁵⁷ To cure this vagueness, OSHA may be required to prove additional elements. For example, in *Granite City Terminals Corp.*,⁵⁸ the Commission held that, if a standard is vague, OSHA must prove that a reasonable person would have recognized a hazard warranting protective measures, and that the sought measures are feasible. A showing that a "reasonable person" would have recognized the violativeness of the cited condition has been required, and held or implied to be sufficient, in a number of circuits and by the Commission; they have not expressly required, or have suggested to be unnecessary, a showing that the employer or its industry follow the practice that OSHA seeks to impose.⁵⁹ At least for the generally-worded personal protection standards at 29 C.F.R. 1910.132(a) and 1926.28(a), however, the Fifth Circuit has determined that the employer's own conduct or "industry custom and practice will generally establish the conduct of the reasonably prudent employer ...."⁶⁰

6.0 The Employer's Substantive Affirmative Defenses

This section discusses defenses that an employer may raise to a citation. The burden of proving these defenses is on the employer. Additional discussion of some of these defenses is in § 5.3.2 of Chapter 11.

6.1 Infeasibility

The Commission has created a limited affirmative defense for the employer who finds that compliance is infeasible. A citation may be vacated if the employer proves that:

---

⁵⁷ *E.g., Kropp Forge Co. v. Secretary of Labor*, 657 F.2d 119, 9 BNA OSHC 2133 (7th Cir. 1981); *Georgia Pacific Corp. v. OSHRC*, 25 F.3d 399, 16 BNA OSHC 1895 (11th Cir. 1994) (standard vague as interpreted by OSHA).

⁵⁸ 12 BNA OSHC 1741 (Rev. Comm'n 1986).

⁵⁹ *E.g., Voegele Company, Inc. v. OSHRC*, 625 F.2d 1075, 1078, 8 BNA OSHC 1631 (3d Cir. 1980); *American Airlines, Inc. v. Secretary of Labor*, 578 F.2d 38, 41, 6 BNA OSHC 1691, 1692-93 (2d Cir. 1978); *Ray Evers Welding Co. v. OSHRC*, 625 F.2d 726, 731-32, 8 BNA OSHC 1271 (6th Cir. 1980); *Bristol Steel & Iron Works, Inc. v. OSHRC*, 601 F.2d 717, 722-23, 7 BNA OSHC 1462 (4th Cir. 1979).

⁶⁰ *Cotter & Co. v. OSHRC*, 598 F.2d 911, 913, 7 BNA OSHC 1510 (5th Cir. 1979); *S & H Riggers & Erectors, Inc. v. OSHRC*, 659 F.2d 1273, 1285, 10 BNA OSHC 1057 (5th Cir. 1981); *Owens-Corning Fiberglas Corp. v. Donovan*, 659 F.2d 1285, 1288, 10 BNA OSHC 1070 (5th Cir. 1981); *B & B Insulation, Inc. v. OSHRC*, 583 F.2d 1364, 1370, 6 BNA OSHC 2067 (5th Cir. 1978); *Power Plant Div., Brown & Root, Inc. v. OSHRC*, 590 F.2d 1363, 1365, 7 BNA OSHC 1137 (5th Cir. 1979).

"(1) [The Infeasibility Element:] the means of compliance prescribed by the applicable standard would have been infeasible under the circumstances in that either "(a) its implementation would have been technologically or economically infeasible or "(b) necessary work operations would have been technologically or economically infeasible after its implementation; and

"(2) [The Alternative Measures Element:] either "(a) an alternative method of protection was used or "(b) there was no feasible alternative means of protection."[61]

Element (2) effectively compels an employer to show that, although strict compliance was necessary, he took whatever steps were feasible. See § 6.1.2 below for more detail on this point. An employer need not show that a variance application was inappropriate,[62] which is an element of the defense of greater hazard. See § 6.2 below.

6.1.1 The Infeasibility Element of the Defense

In the early days of the Act, this defense was known as "impossibility."[63] In 1986, in *Dun-Par Engineered Form*,[64] the Commission changed the name of the defense and its first element to "infeasibility" in part because "[s]trict application of an 'impossibility' defense does not accommodate considerations of reasonableness or common sense, or reflect the strong sense of the practical implicit in the standards adopted under § 6(a)" and the feasibility element in § 6(b)(5) of the Act.[65]

The defense has often proved difficult to establish. An employer must at least attempt to adapt existing technology and use some creativity to solve the infeasibility problem.[66] An inability to comply because the appropriate equipment was not on site is insufficient, for "[i]t is the duty of an employer to use equipment that

---

[61] *Beaver Plant Operations, Inc.*, 18 BNA OSHC 1972, 1977 (Rev. Comm'n 1999), *citing Gregory & Cook, Inc.*, 17 BNA OSHC 1189, 1190 (Rev. Comm'n 1995); *Seibel Modern Manufacturing & Welding Corp.*, 15 BNA OSHC 1219, 1228 (Rev. Comm'n 1991); *Mosser Constr. Co.*, 15 BNA OSHC 1408, 1416 (Rev. Comm'n 1991); *Dun-Par Engineered Form Co.*, 12 BNA OSHC 1949 (Rev. Comm'n 1986), *rev'd on another ground*, 843 F.2d 1135, 13 BNA OSHC 1652 (8th Cir. 1988).

[62] *Dun-Par Engineered Form*, 12 BNA OSHC at 1956.

[63] E.g., *M.J. Lee Construction Co.*, 7 BNA OSHC 1140, 1144 (Rev. Comm'n 1979).

[64] 12 BNA OSHC at 1953-56.

[65] 12 BNA OSHC at 1955.

[66] See *Pitt-Des Moines Inc.*, 16 BNA OSHC 1429, 1433-34 (Rev. Comm'n 1993); *Gregory & Cook Inc*, 17 BNA OSHC 1189, 1191 (Rev. Comm'n 1995) (employer should attempt to acquire more suitable guard; Commission "expect[s] employers to exercise some creativity in seeking to achieve compliance").

permits him to comply with the Secretary's standard."[67] The defense may also be rejected if it was feasible to preclude employee access to the zone of danger.[68]

Courts that have considered the infeasibility defense have concluded that it encompasses both technological and economic factors.[69] At one time, the Commission took the position that the economic effect of compliance was irrelevant.[70] However, in *State Sheet Metal Co.*,[71] the Commission stated that "evidence as to the unreasonable economic impact of compliance with a standard may be relevant to the infeasibility defense."[72] In *Peterson Bros. Steel Erec. Co.*,[73] the Commission stated that it would look to the effect that compliance would have on the company's "financial position as a whole" to determine whether the company would be "adversely affected." It is not sufficient that an employer who has failed to use safety measures would be at a competitive disadvantage with others that did not use the measures, for an "employer cannot be excused from compliance with the Act on the basis that everyone else will ignore the law."[74]

---

[67] *Williams Enterprises Inc.*, 13 BNA OSHC 1249 (Rev. Comm'n 1987).

[68] *Walker Towing Corp.*, 14 BNA OSHC 2072, 2075–76 (Rev. Comm'n 1991).

[69] *Quality Stamping Products v. OSHRC*, 709 F.2d 1093, 1099, 11 BNA OSHC 1550 (6th Cir. 1983) (employer must show not "economically practicable" because "prohibitively expensive"); *Donovan v. Williams Enterprises, Inc.*, 744 F.2d 170, 178, 11 BNA OSHC 2241 (D.C. Cir. 1984); *Faultless Div., Bliss & Laughlin Indus. v. Secretary*, 674 F.2d 1177, 1189, 10 BNA OSHC 1481 (7th Cir. 1982); *Southern Colo. Prestress Co. v. OSHRC*, 586 F.2d 1342, 1351, 6 BNA OSHC 2032 (10th Cir. 1978); *Atlantic & Gulf Stevedores, Inc. v. OSHRC*, 534 F.2d 541, 4 BNA OSHC 1061 (3d Cir. 1976).

[70] *See, e.g., Stan-Best, Inc.*, 11 BNA OSHC 1222, 1231 (Rev. Comm'n 1983); *Research Cottrell, Inc.*, 9 BNA OSHC 1489, 1498 (Rev. Comm'n 1981).

[71] 16 BNA OSHC 1155 (Rev. Comm'n 1993).

[72] In *State Sheet Metal*, the Commission stated that in *Dun Par Engd. Form Co.*, 12 BNA OSHC 1962, 1966 (Rev. Comm'n 1986), it first implied that an infeasibility defense may include economic factors. There, it found that the employer had not demonstrated that the costs were unreasonable in light of the protection afforded and had not shown what effect, if any, the added costs would have on its contract or on its business as a whole. *See also Walker Towing Corp.*, 14 BNA OSHC 2072, 2077 (Rev. Comm'n 1991).

[73] 16 BNA OSHC 1196, 1203 (Rev. Comm'n 1993), *aff'd*, 26 F.3d 573 (5th Cir. 1994).

[74] *Gregory & Cook, Inc.*, 17 BNA OSHC 1189, 1192 (Rev. Comm'n 1995). *Accord, Peterson Bros. Steel Erection Co. v. Secretary of Labor*, 26 F.3d 573, 16 BNA OSHC 1900 (5th Cir. 1994) (employer's claim that it would be disadvantaged as against competitors that did not comply is not relevant because "[a]n employer cannot be excused from non-compliance on the assumption that everyone else will ignore the law"). *See also Peterson Bros.*, 16 BNA OSHC 1196, 1203 (Rev. Comm'n 1993) (evidence that costs, while substantial, could be absorbed on the project negated the employer's claim of economic infeasibility); *State Sheet Metal*, 16 BNA OSHC 1155, 1159, 1160–61 (Rev. Comm'n 1993).

6.1.2 The Alternative Measures Element of the Infeasibility Defense

This element—which also appears in the great hazard and multi-employer defenses—reflects the view that, even if full compliance is not feasible, "an employer [must] comply to the extent feasible...."[75] "[B]efore an employer will be excused from ignoring a standard's requirements and leaving its employees unprotected, it must show that it has explored all possible alternate forms of protection."[76] At one time, the Commission in *Dun-Par Engineered Form*[77] shifted the burden of persuasion on this element to OSHA, but the Eighth Circuit held otherwise,[78] and the Commission later followed that holding.[79]

6.2 The Greater Hazard Defense

The Commission has also held that employers need not strictly comply with a standard to the extent that compliance would create greater hazards than non-compliance would. It created an affirmative defense based on the idea that "industry is so diverse that any rule is bound to be counterproductive now and again."[80] The defense has three elements:

1. Compliance with the standard would create greater hazards than non-compliance;

2. Alternative protective measures were taken or were not available; and

3. A variance application is inappropriate.[81]

The defense does not apply to the General Duty Clause because the usefulness of a proposed abatement method is part of the Secretary's burden in General Duty Clause cases.[82]

---

[75] *Donley's Inc.*, 17 BNA OSHC 1227 (Rev. Comm'n 1995). *But see Spancrete Northeast Inc. v. OSHRC*, 905 F.2d 589, 14 BNA OSHC 1585 (2d Cir. 1990), which suggests that the defense does not have a second element. The court held that if compliance with the cited standard is infeasible, the Secretary must plead in the alternative and prove a failure to comply with another standard.

[76] *State Sheet Metal Co.*, 16 BNA OSHC 1155 (Rev. Comm'n 1993).

[77] 12 BNA OSHC at 1956-59.

[78] *Brock v. Dun-Par Engineered Form Co.*, 843 F.2d 1135 (8th Cir. 1988), *rev'g* 12 BNA OSHC 1949 (Rev. Comm'n 1986).

[79] *Seibel*, 15 BNA OSHC at 1227-28.

[80] *Caterpillar Inc. v. Herman*, 131 F.3d 666, 18 BNA OSHC 1104 (7th Cir. 1997).

[81] *Russ Kaller Inc.*, 4 BNA OSHC 1758 (Rev. Comm'n 1976). *See also PBR, Inc. v. Secretary of Labor and OSHRC*, 643 F.2d 890, 9 BNA OSHC 1357 (1st Cir. 1981); *John H. Quinlan*, 17 BNA OSHC 1194 (Rev. Comm'n 1995).

The first element of the defense requires a showing that compliance would create *greater* hazards than non-compliance—not new or different hazards.[83] It also requires a showing that *all* alternative ways of protection are more dangerous than non-compliance, not just the means of protection mentioned in the standard.[84]

The reason for the third element of the defense—the inappropriateness of a variance—is that "some employers will believe *incorrectly* that their working conditions are safer than those prescribed in the standards....[R]emoving this incentive to seek variances [by eliminating the element]...would be allowing an employer to take chances not only with his money, but with the lives and limbs of his employees."[85] The third element does not apply to regulations because a variance cannot be sought from a regulation.[86]

6.3 Unpreventable Employee Misconduct

This defense has been stated in various ways, but it basically requires an employer to show that he required his employees to take protective measures that comply with the standard and that he enforced that requirement.[87] The Commission has distilled its decisions as requiring four elements of proof—that:

1. the employer has established work rules designed to prevent the violation;
2. it has adequately communicated those rules to its employees;
3. it has taken steps to discover violations; and
4. it has effectively enforced the rules when violations have been discovered.[88]

---

[82] *Royal Logging Co.*, 7 BNA OSHC 1744, 1751 (Rev. Comm'n 1979), *aff'd*, 645 F.2d 822, 9 BNA OSHC 1755 (9th Cir. 1981).

[83] *See Dun-Par*, 12 BNA OSHC at 1967; *Williams Enterprises Inc.*, 13 BNA OSHC 1249 (Rev. Comm'n 1987).

[84] *John H. Quinlan*, 17 BNA OSHC 1194 (Rev. Comm'n 1995).

[85] *General Electric Co. v. Secretary*, 576 F.2d 558, 561, 6 BNA OSHC 1541 (3d Cir. 1978) (emphasis in the original). *See also Reich v. Trinity Industries, Inc.*, 16 F.3d 1149, 1154, 16 BNA OSHC 1670 (11th Cir. 1994); *Modern Drop Forge Co. v. Secretary of Labor*, 683 F.2d 1105, 1116, 10 BNA OSHC 1852 (7th Cir. 1982); *Dole v. Williams Enterprises, Inc.*, 876 F.2d 186, 190 n.7, 14 BNA OSHC 1001 (D.C. Cir. 1989).

[86] *Caterpillar Inc. v. Herman*, 131 F.3d 666, 18 BNA OSHC 1104, 1106 (7th Cir. 1997).

[87] *E.g., Secretary of Labor v. L.E. Myers Co.*, 818 F.2d 1270, 13 BNA OSHC 1289 (6th Cir.), *cert. denied*, 484 U.S. 989 (1987); *Texland Drilling Corp.*, 9 BNA OSHC 1023 (Rev. Comm'n 1980).

[88] *E.g., Capform, Inc.*, 16 BNA OSHC 2040, 2043 (Rev. Comm'n 1994).

Effective enforcement generally must be progressive, *i.e.,* it must become increasingly severe as an employee commits additional infractions. Thus, an employer was held to have failed to establish the defense when an employer who broke a safety rule for the second time was given only an oral warning instead of a written reprimand.[89]

Although there is a similar doctrine of supervisory misconduct,[90] some cases characterize it as not an affirmative defense but as a rebuttal of the imputation to the employer of the supervisor's knowledge.[91] The Commission has stated that involvement by a supervisor in a violation is "strong evidence that the employer's safety program was lax."[92] "Where a supervisory employee is involved, the proof of unpreventable employee misconduct is more rigorous and the defense is more difficult to establish since it is the supervisor's duty to protect the safety of employees under his supervision."[93]

The objection has been made that the overlap of this defense with the knowledge element of OSHA's case is confusing, for OSHA must prove knowledge while the employer must prove the defense.[94]

6.4 Invalidity of the Standard

A standard is invalid if it was not adopted in accordance with a statutory procedural requirement. *See generally* Chapter 2. Two examples of invalidity resulting from violations of such requirements are:

1. Making a substantive change in a national consensus or established Federal standard adopted under § 6(a) of the OSH Act, 29 U.S.C. § 655(a).[95] This is a special case of failing to give the public notice or an opportunity for comment on the adoption or amendment of a standard.[96]

---

[89] *E.g., Gem Industrial, Inc.,* 17 BNA OSHC 1861 (Rev. Comm'n 1996), and particularly note 8 of the lead opinion and note 12 of the dissenting opinion there.

[90] *Daniel Construction Co.,* 10 BNA OSHC 1549, 1552 (Rev. Comm'n 1982).

[91] *E.g., Consolidated Freightways Corp.,* 15 BNA OSHC 1317, 1321 (Rev. Comm'n 1991).

[92] *Daniel Construction,* 10 BNA OSHC at 1552.

[93] *Seyforth Roofing Co.,* 16 BNA OSHC 2031 (Rev. Comm'n 1994).

[94] *See, e.g., New York State Gas & Elec. v. Secretary of Labor,* 88 F.3d 98, 17 BNA OSHC 1650, 1655-1657 (2d Cir. 1996) (reviewing confusing state of law and criticizing Commission for inconsistency); *L.E. Myers v. Brock,* 484 U.S. 989 (1987) (White, J., dissenting from denial of certiorari) ("confusing patchwork of conflicting approaches").

[95] *E.g., Usery v. Kennecott Copper Corp.,* 577 F.2d 1113, 6 BNA OSHC 1197 (10th Cir. 1977).

[96] *See* 5 U.S.C. § 553(b); OSH Act § 6(b)(2); 29 U.S.C. § 655(b)(2); *Kooritzky v. Reich,* 17 F.3d 1509 (D.C. Cir. 1994) (non-OSHA case).

2. Failing to consult the Advisory Committee on Construction Safety and Health before proposing a standard regulating construction work.[97]

6.5 De Minimis

Section 9(a) of the OSH Act, 29 U.S.C. § 658(a), states: "The Secretary may prescribe procedures for the issuance of a notice in lieu of a citation with respect to *de minimis* violations which have no direct or immediate relationship to safety or health." The consequence of characterizing a violation as *de minimis* is that the violation carries neither an abatement requirement nor a monetary penalty.[98] The Commission has long asserted that it may characterize a violation as *de minimis*.[99] There is a split in the circuits as to whether the Commission has this authority. The First, Third, Fifth, and Ninth Circuits have held that it does,[100] while the Seventh Circuit disagrees.[101]

As to what a *de minimis* violation is, the Commission has formulated the test in various ways, including asking whether the violation is "trifling."[102] In another case, it stated: "A *de minimis* violation is one in which there is technical noncompliance with a standard but the departure from the standard bears such a negligible relationship to employee safety and health as to render inappropriate the assessment of a penalty or the entry of an abatement order."[103] One circuit has held that a violation is *de minimis* if the employer's safety measures are as safe as those required by a standard.[104] The Commission has in effect held that the employer bears the burden of proof on the *de minimis* issue.[105]

---

[97] *National Constructors Ass'n v. Marshall*, 581 F.2d 960, 970-71, 6 BNA OSHC 1721 (D.C. Cir. 1978).

[98] E.g., *Keco Indus., Inc.*, 11 BNA OSHC 1832, 1834 (Rev. Comm'n 1984).

[99] E.g., *General Electric Co.*, 3 BNA OSHC 1031, 1040 (Rev. Comm'n 1975).

[100] *Donovan v. Daniel Constr. Co.*, 692 F.2d 818, 10 BNA OSHC 2188 (1st Cir. 1982); *Secretary of Labor v. OSHRC (Erie Coke Corp.)*, 998 F.2d 134, 16 BNA OSHC 1241 (3d Cir. 1993); *Phoenix Roofing, Inc. v. Dole*, 874 F.2d 1027, 14 BNA OSHC 1036 (5th Cir. 1989); *Chao v. Symms Fruit Ranch, Inc.*, 242 F.3d 894, 19 BNA OSHC 1337 (9th Cir. 2001).

[101] *Caterpillar Inc. v. Herman*, 131 F.3d 666, 18 BNA OSHC 1104 (5th Cir. 1989).

[102] *El Paso Crane & Rigging Co.*, 16 BNA OSHC 1419, 1429 (Rev. Comm'n 1993) (failure to attest and sign OSHA injury log "trifling").

[103] *Keco Indus., Inc.*, 11 BNA OSHC 1832, 1834 (Rev. Comm'n 1984).

[104] *Phoenix Roofing, Inc. v. Dole*, 874 F.2d 1027, 14 BNA OSHC 1036 (5th Cir. 1989).

[105] See *Holly Springs Brick & Tile Co.*, 16 BNA OSHC 1856 (Rev. Comm'n 1994) (rejecting *de minimis* argument for lack of evidence).

Chapter 4

The General Duty Clause

Peter L. de la Cruz, Esq.
Keller and Heckman, LLP
Washington, DC

1.0 Overview

The general duty clause is the common name for section 5(a) of the Occupational Safety and Health Act.[1] In drafting the Act, Congress recognized that it would be virtually impossible for the Occupational Safety and Health Administration (OSHA) to anticipate every possible hazardous condition that might exist in every place of employment and promulgate appropriate standards. To fill that gap, Congress imposed a general duty on employers to address hazards to which no specific standard applied. Thus, each employer must provide employment which is "free from recognized hazards that are causing or likely to cause death or serious physical harm to his employees."[2] In simple terms, to prove a general duty clause violation, OSHA must show that a hazard: (1) foreseeably exists in the workplace, (2) is 'recognized' by the employer or the industry, (3) is likely to cause death or serious physical injury, and (4) can be feasibly abated.

To set the framework for understanding the scope of the general duty clause, we begin with an overview of the legislative context and then review the facts or elements that OSHA must prove. Because many companies tend to focus their resources on compliance with all of the applicable OSHA regulations, the general duty clause occasionally takes a secondary place. But unlike most OSHA standards, the general duty clause applies to all employers, in all industries and at all times.

---

[1] OSH Act § 5(a)(1); 29 U.S.C. § 654(a)(1).
[2] OSH Act § 5(a)(1); 29 U.S.C. § 654(a)(1).

Two aspects of the general duty clause highlight its unique importance. First, the general duty clause imposes an obligation on all employers to be familiar with voluntary standards, trade association guidelines, and practices common to their industry. Second, OSHA has used the general duty clause as the basis for enforcement actions involving new or controversial areas. For example, OSHA used the general duty clause initially to address ergonomics issues and bloodborne pathogen concerns. Thus, the general duty clause presents its own compliance challenges for employers.

2.0 Guidance Provided by Congress

In the OSH Act, Congress adopted a lofty goal in seeking to "assure, so far as possible, every working man and woman in the Nation safe and healthful working conditions...."[3] The legislature sought to achieve this through the twin duties of compliance with specific OSHA standards and the general duty clause.

As explained by the Committee which reported OSH legislation to the Senate in October 1970:

> The committee recognizes that precise standards to cover every conceivable situation will not always exist. This legislation would be seriously deficient if any employee were killed or seriously injured on the job simply because there was no specific standard applicable to a recognized hazard which could result in such a misfortune. Therefore, to cover such circumstances the committee has included a requirement to the effect that employers are to furnish employment and places of employment which are free from recognized hazards to the health and safety of their employees.[4]

The General Duty Clause states that: "Each employer shall furnish to each of his employees employment and a place of employment which are free from recognized hazards that are causing or are likely to cause death or serious physical harm

---

[3] OSH Act § 2(b).

[4] S. Rep. No. 1282, 91st Congress, 2d Session 9 (1970), *reprinted* in Senate Committee on Labor and Public Welfare, *Legislative History of the Occupational Safety and Health Act of 1970*, at 149, 92d Cong., 1st Sess. (Committee Print)(1971)(hereinafter *Legislative History*). Substantially similar reasoning was contained in the report of the House of Representatives committee that reported a somewhat similar bill. H.R. Rep. No. 1291, 91st Cong., 2d Sess., at 22 (1970), *reprinted* in *Legislative History, supra,* at p. 852 (the provision's purpose is "to provide for the protection of employees who are working under such *unique* circumstances that no standard has yet been enacted to cover this situation").

to his employees."⁵ Both the House and the Senate Committees relied upon, and quoted in their Reports, the following statement of Governor Howard Pyle, then-president of the National Safety Council, on the desirability of a general duty provision:

> If national policy finally declares that all employees are entitled to safe and healthful working conditions, then all employers would be obligated to provide a safe and healthful workplace rather than only complying with a set of promulgated standards. The absence of such a general obligation provision would mean the absence of authority to cope with a hazardous condition which is obvious and admitted by all concerned for which no standard has been promulgated.⁶

Yet, there was a considerable amount of controversy in the legislative development of the General Duty Clause. Senate Bill S. 2193 contained a General Duty Clause that would have required employers to furnish workplaces "free from recognized hazards so as to provide safe and healthful working conditions."⁷

In the House, Congressman Daniels introduced H.R. 16785 which provided that "[e]ach employer shall furnish to each of his employees employment and a place of employment which is safe and healthful."⁸ Opponents of the Daniels' bill referred to the clause as a "sweeping general requirement" couched in "language...so broad, general and vague as to defy practical interpretation let alone responsible enforcement."⁹ The Republican alternative, H.R. 19200,¹⁰ offered by Congressmen Steiger and Sikes, forbade "any hazards which are readily apparent and are causing or are likely to cause death or serious physical harm."¹¹ Opponents of the Steiger / Sikes bill contended that the cited language would also cause problems in interpretation, because it was possible that a hazard readily apparent to a safety expert might not be readily apparent to a layman or ordinary worker. Nevertheless, the "readily apparent" language, the version reported by the Education & Labor Committee to

---

⁵ OSHA §5(a)(1); 29 U.S.C. §654(a)(1)(1982).
⁶ *Legislative History, supra,* at 150 and 851.
⁷ S. 2193, 91st Cong., 2d Sess. (1970), *reprinted in Legislative History* at 204-295.
⁸ H.R. 16785, 91st Cong., 2d Sess. (1970), *reprinted in Legislative History* at 721-762.
⁹ 116 Cong. Rec. H. 10616 (daily ed. Nov. 23, 1970) (remarks of Congressman Anderson), *reprinted in Legislative History* at 982.
¹⁰ H.R. 19200, 91st Cong., 2d Sess. (1970), *reprinted in Legislative History* at 763-830.
¹¹ *Legislative History* at 769.

the floor of the House, was thought to be more specific than the H.R. 16785 version, and the "readily apparent" version was adopted by the House.[12]

A conference committee, appointed to resolve the differences in the language of the bills adopted by the House and Senate, essentially settled upon the Steiger-Sikes version but substituted the phrase 'recognized hazards' for 'hazards which are readily apparent.' The compromise produced the present General Duty Clause language in §5(a)(1) of the OSH Act. There is no discussion in the conference report of the intent of Congress in adopting the "recognized hazards" terminology.

Ultimately, the legislative history, coupled with the sweeping language used in the provision, left many unanswered questions in interpreting and applying the General Duty Clause which have been the subject of both administrative and judicial decisions.

3.0 Basic Framework

A 1973 decision, just three years after the OSH Act became law, has set the tone for general duty clause analysis over the last 20 years. In *National Realty and Construction Co. v. Occupational Safety and Health Review Commission*,[13] the United States Court of Appeals for the District of Columbia Circuit formulated a test for determining whether the general duty clause of the OSH Act has been violated. Under *National Realty* and its progeny, to prove a General Duty Clause violation, OSHA must show that:

1. the employer failed to keep the workplace "free" of a hazard to which the employees were exposed,

2. the hazard at issue is "recognized,"

3. the hazard was "causing or likely to cause death or serious physical harm";[14] and

4. "demonstrably feasible measures would have materially reduced the likelihood that such misconduct would have occurred."[15]

---

[12] *Legislative History* at 1091-1117.

[13] 489 F.2d 1257 (D.C. Cir. 1973).

[14] *Id.* at 1265.

[15] 489 F.2d at 1267. In *National Realty*, OSHA issued a general duty citation for a construction site fatality, when a foreman rode as a passenger on the running board of a front end loader that rolled over. The company contested the citation by showing that the company had a work rule prohibiting its employees from riding as passengers on moving equipment, and that this

(continued on the following page)

This formulation of the burden of proof has often been restated with approval by the Occupational Safety and Health Review Commission[16] (OSHRC) and other courts of appeals.[17]

The Court's decision in *National Realty* established a number of important criteria governing General Duty Clause enforcement. For example, the court held that the General Duty Clause does not impose absolute or strict liability.[18] Further, employers' general duty clause obligations are not governed by the legal concept of 'reasonableness,' but a general and common duty to "bring no adverse effects to the life and health of their employees..."[19]

Congress intended that the General Duty Clause obligation be achievable and that the hazards creating a general duty violation must not only be "recognized," but they must also be preventable.[20] To sustain a violation, OSHA must show that "demonstrably feasible measures would have materially reduced the likelihood that such misconduct would have occurred."[21] This aspect of preventability, in the court's view, required the informed judgement of safety experts.[22]

Although the court vacated the citation in the *National Realty* case, it noted that an accident or hazardous conduct need not have occurred to sustain a gen-

---

was not a common or accepted practice at the worksite. The Administrative Law Judge (ALJ) vacated the citation with a finding that foreman Smith had violated the company's safety policy and that this did not constitute a violation of §5(a)(1) of the Act by the company. The ALJ's decision was reversed by a two to one decision of the three-member Occupational Safety and Health Review Commission, in which each member gave a different reason for his opinion. The Court of Appeals reversed the Commission's ruling and held that no general duty clause violation occurred because the record OSHA presented did not carry its burden of proof for all the necessary elements of the case. For example, the court found that OSHA did not specify the steps the company could have taken to prevent unsafe equipment riding.

[16] *See Connecticut Light & Power Co.*, 13 O.S.H. Cas. (BNA) 2214 (Rev. Comm'n 1989); *Kastalon Inc.*, 12 O.S.H. Cas. 1928 (BNA) (Rev. Comm'n. 1987).

[17] *See e.g., Donovan v. Missouri Farmers Ass'n*, 674 F.2d 690 (8th Cir. 1982); *St. Joe Minerals Corp. v. OSHRC*, 647 F.2d 840 (8th Cir. 1981); *Magma Copper Co. v. Marshall*, 608 F.2d 373, 375 (9th Cir. 1979); *Georgia Electric Co. v. Marshall*, 595 F.2d 309, 321 (5th Cir. 1979); *Titanium Metals Corp. of Am. v. Usery*, 579 P.2d 536 (9th Cir. 1978); *Empire - Detroit Steel v. OSHRC*, 579 F.2d 378, 383 (6th Cir. 1978); *Getty Oil Co. v. OSHRC*, 530 F.2d 1143, 1145 (5th Cir. 1976).

[18] *National Realty*, 489 F.2d at 1265-1266.

[19] *National Realty*, 489 F.2d at 1266, n.34.

[20] *National Realty*, 489 F.2d at 1266.

[21] *National Realty*, 489 F.2d at 1267.

[22] *National Realty*, 489 F.2d at 1266.

eral duty clause violation.[23] At the same time, "actual occurrence of hazardous conduct is not, by itself, sufficient evidence of a violation, even when the conduct has led to injury."[24]

4.0 Recognized Hazards

In a general duty citation, OSHA must describe the hazard and then demonstrate that the hazard is "recognized" and likely to cause death or serious physical harm. A safety hazard has been defined as a "condition that creates or contributes to an increased risk that an event causing death or serious bodily harm to employees will occur."[25]

The issue of what is considered a "recognized hazard" under § 5(a)(1) is probably the most litigated aspect of the General Duty Clause. Recognition of a hazard can be established on the basis of: (1) an individual employer's recognition, (2) general recognition within the employer's industry, or (3) common sense recognition.

Recognized hazards include not only those hazards which are detectable by human senses, but also non-obvious hazards, including those that can only be detected through the aid of instrumentation. As one court observed, the Act's purpose would be defeated by a narrow construction of recognized hazards. "[T]o limit the general duty clause to dangers only detectable by the human senses seems to us to be a folly....Where hazards are recognized but not detectable by the senses, common sense and prudence demand that instrumentation be utilized."[26]

The policy of recognizing non-obvious hazards as being within the scope of the general duty requirement has consistently been followed by OSHA in its enforcement activity. For example, the now superseded edition of the OSHA *Field Manual* included the following:

A hazard is recognized if it is a condition that is (a) of a common knowledge or general recognition in the particular industry in which it occurs, and (b) detectable (1) by means of the senses (sight, smell, touch, and hearing), or (2) is of such wide, general recognition as a hazard in the industry that even if it is not detectable by

---

[23] *National Realty,* 489 F.2d at 1267.
[24] *National Realty,* 489 F.2d at 1267.
[25] *Baroid Div. v. OSHRC,* 660 F.2d 439, 444 (10th Cir. 1981) (substantial accumulation of gas near an oil rig constitutes a hazard).
[26] *American Smelting and Refining Co. v. OSHRC,* 501 F.2d 504 at 511 (8th Cir. 1974).

means of the senses, there are generally known and accepted tests for its existence which should make its presence known to the employer.[27]

4.1 Employer Recognition

One way in which a hazard can be established is through evidence that the employer had actual knowledge that the condition was hazardous. This issue was discussed in *Brennan v. OSHRC* (Vy Lactos Industries Inc.) where the court observed that an employer's personal knowledge of the existence of a hazard was sufficient to make the hazard recognized.[28] The Eighth Circuit stated that "[e]ven a cursory examination of the Act's legislative history clearly indicates that the term recognized was chosen by Congress not to exclude actual knowledge, but rather to reach beyond an employer's actual knowledge to include the generally recognized knowledge of the industry as well."[29] This view has been followed in subsequent decisions by the Review Commission[30] and the courts.[31]

OSHA need not show that prior accidents have occurred to prove that an employer had actual knowledge of a hazardous condition. Moreover, evidence of employer recognition can be based upon an employer's actual knowledge of a hazard, such as written or oral statements made by the employer or the employer's representatives before or during the OSHA inspection. Employer knowledge or constructive knowledge may be based upon other extrinsic evidence such as the requirements of state and local laws.

For instance, according to Chapter IV of the OSHA *Field Operations Manual*:[32]

1. Company memorandums, safety rules, operating manuals or operating procedures and collective bargaining agreements may reveal the employer's awareness of the hazard. In addition, accident, injury and illness reports

---

[27] OSHA *Field Manual*, 46 (CCH ed., 1979).

[28] 494 F.2d 460, 464 (8th Cir. 1974).

[29] 494 F.2d at 464.

[30] See e.g., *Secretary v. General Electric Co.*, 10 O.S.H. Cas. (BNA) 2034 (Rev. Comm'n 1982); *Secretary v. Copperweld Steel Co.*, 2 O.S.H. Cas. (BNA) 1602 (Rev. Comm'n 1975).

[31] See *Pratt & Whitney Aircraft Div. v. Secretary*, 649 F.2d 96 (2d Cir. 1981); *St. Joe Minerals v. OSHRC*, 647 F.2d 840 (8th Cir. *1981); Magma Copper Co. v. Marshall*, 608 F.2d 373 (9th Cir. 1979).

[32] OSHA *Field Operations Manual*, Chapter IV, p. IV-9 (CPL 2.45B, CH-4, 12/13/93).

prepared for OSHA, workmen's compensation, or other purposes may show this knowledge.[33]

2. Employee complaints or grievances to supervisory personnel may establish recognition of a hazard, but the evidence should show that the complaints were not merely infrequent offhand comments.[34]

3. The employer's own corrective action may serve as the basis for establishing employer recognition of the hazard if the employer did not adequately continue or maintain the corrective action or if the corrective action did not afford any significant protection to the employees.[35]

4.2 Industry Recognition

A recognized hazard may also be shown to exist if the hazard is considered common knowledge in the employer's industry. *In National Realty* the court stated in a footnote that "an activity may be a 'recognized hazard' even if the defendant employer is ignorant of the activity's existence or its potential for harm."[36] The court defined "recognized hazard" as a condition that is known to be hazardous, and is known, not necessarily by each and every individual employer, but "taking

---

[33] *See, e.g., Secretary v. General Dynamics Land Systems Div. Inc.*, 15 O.S.H. Cas. (BNA) 1275 (Rev. Comm'n. 1991) (evidence of employee injuries from chemical exposure and oxygen deficiencies put employer on notice that its employees were being exposed to a confined space hazard); *Ryder Truck Lines, Inc. v. Brennan*, 497 F.2d 230, 234 (5th Cir. 1974) (prevalence of employees' foot injuries should have made employer aware of need to comply with general protective equipment regulation); *American Airlines Inc.*, 6 O.S.H. Cas. (BNA) 1252, 1254 (Rev. Comm'n 1977); *But see, Donovan v. General Motors Corp.*, 764 F.2d 32, 36-37 (1st Cir. 1985) (history of foot injuries does not constitute actual knowledge); *Secretary v. Atlantic Sugar Assoc.*, 4 O.S.H. Cas. (BNA) 1355 (Rev. Comm'n 1976) (evidence of correspondence and highly publicized accidents that transporting field workers in standing position without rear barriers constitutes actual knowledge).

[34] *See, e.g., General Dynamics Land Systems*, 15 O.S.H. Cas. (BNA) 1285 (formal grievances concerning adverse consequences from chemical exposure were filed); *Koppers Inc.*, 14 O.S.H. Cas. (BNA) 1269 (OSHRC 1989) (hazard recognized where employees had observed "tilting" and "sagging" of platform and had reported those observations to management officials). This kind of decision indicates how judges may find that an employer had constructive knowledge of a hazard. The question of how much employee recognition should also constitute employer recognition of a hazard is, of course, a question of fact.

[35] *General Electric*, 10 O.S.H. Cas. (BNA) 2034 (Rev. Comm'n 1982); *Armstrong Cork Co.*, 8 O.S.H. Cas. (BNA) 1070, 1071 (installation of a spring-loaded switch as a safety measure before initiation of a citation).

[36] 489 F.2d at 1265 n.32.

into account the standard of knowledge in the industry."[37] The court determined that industry recognition would be measured by "the common knowledge of safety experts who are familiar with the circumstances of the industry or activity in question."[38] The Review Commission and other courts have followed *National Realty's* rationale of industry recognition.[39]

Information contained in a supplier's specification sheets or material safety data sheets (MSDSs) is another basis for showing industry recognition or imputing recognition to the employer. In *Young Sales Corp.*, for example, major manufacturers of corrugated sheeting provided written warnings to their customers that workers should not walk directly on the sheeting because it was very brittle.[40]

Other methods of proof that a particular hazard is recognized by an employer's industry include the use of industry publications,[41] advisory standards such as those by the American National Standards Institute (ANSI)[42] and state[43] and local laws.[44] Recognition of a hazard in one industry, however, does not prove that the hazard is recognized in the cited employer's industry.

---

[37] *Id.*

[38] *Id.*

[39] *Pratt & Whitney Aircraft Div.*, 649 F.2d 96 at 100 (2d Cir. 1981), 9 O.S.H. Cas. (BNA) at 1557 (indoor chemical storage of acids and cyanides is recognized hazard in the industry); *Secretary v. Beird-Poulan*, 7 O.S.H. Cas. (BNA) 1225, 1229 (Rev. Comm'n 1979) (production of sparks in the presence of magnesium dust was recognized hazard in industry); *Usery v. Marquette Cement Mfg. Co.*, 568 F.2d 902 at 910.

[40] *Young Sales Corp.*, 7 O.S.H. Cas. (BNA) 1297 (Rev. Comm'n 1979); *Great Southern Oil & Gas Co.*, 10 O.S.H. Cas. (BNA) 1996 (Rev. Comm'n, 1982) (failed to install wires and anchors in accordance with rig manufacturers specifications).

[41] *R. L. Sanders Roofing Co.*, 7 O.S.H. Cas. (BNA) 1566 (Rev. Comm'n 1979), *reversed* 620 F.2d 97 (5th Cir 1980). *Parker Drilling Co.*, 8 O.S.H. Cas. (BNA) 1717 (RCALJ 1980) (International Assoc. of Drilling Contractors guidelines); *Georgia Electric Co.* 5 O.S.H. Cas. (BNA) 1112 (Rev. Comm'n 1977) (industry safety manual).

[42] *St. Joe Minerals Corp. v. OSHRC*, 647 F.2d at 845 n. 8. *St. Joe. Lead Co.*, 9 O.S.H. Cas. (BNA) at 1649 (standards for elevators); *Betten Processing Corp.*, 2 O.S.H. Cas. (BNA) 1724 (Rev. Comm'n 1975) (standard for cranes).

[43] *Ford Motor Co.* 5 O.S.H. Cas. (BNA) 1765 (Rev. Comm'n 1977); *Sugar Cane Growers Coop of Fla.*, 4 O.S.H. Cas. (BNA) 1320, 1323 (Rev. Comm'n 1976) (Florida statute requiring securely attached seating for transportation of migrant farm workers); *M.A. Swatek & Co.*, 1 O.S.H. Cas. (BNA) 1191 (Rev. Comm'n 1973) (state safety criteria used to guard against cave-in hazards).

[44] *Williams Enterprises*, 4 O.S.H. Cas. (BNA) 1663 (Rev. Comm'n 1976).

4.3 'Common Sense' Recognition

If a reasonable person having knowledge of the alleged conditions would have recognized that a hazard existed, the courts and the Commission have inferred recognition.[45] For example, in *Donovan v. Missouri Farmers Association*, the court held that an unventilated, almost completely enclosed pit not having a ready means of exit in case of an emergency was an obvious hazard and therefore warranted a safety belt with a lifeline for workers entering the pit.[46] Another example is contained in *Marquette Cement*, when bricks were dumped using an unenclosed chute into an alleyway.[47] Employees were working 26 feet below and did not know of the hazard. The OSHA *Field Inspection Reference Manual* and the OSHA *Field Operations Manual* suggest using "common-sense recognition" only in flagrant cases.[48]

5.0 Unforeseeable Hazards

When a general duty citation was precipitated by an accident, as is often the case, proof of recognition includes a showing that the accident was foreseeable by the cited employer. Recognition is established if the employer admits that he had reason to believe that the accident would happen, or if OSHA produces competent evidence to show that the particular industry of which the cited employer was a part recognized that the conditions which caused the accident were hazardous.

But the mere occurrence of an accident does not establish its foreseeability. For example, in *Vineyard*, the court of appeals found no general duty violation when the employee-lineman was electrocuted while working on a utility pole because the lineman's conduct which caused him to be electrocuted was not reasonably foreseeable.[49] The task which was being performed by the deceased lineman did not make it necessary for the lineman to move into a dangerous position which would have required more protection than was used. The likelihood that serious physical harm to the lineman would result from being in a dangerous position without proper safety protection is insufficient to demonstrate the existence of a recognized hazard. It must also be shown that it was reasonably foreseeable that the lineman would, in

---

[45] *Eddy's Bakeries Co.*, 9 O.S.H. Cas. (BNA) 2147, 2150 (Rev. Comm'n 1981) (danger of gasoline vapors near open flame heaters is common knowledge).

[46] 674 F.2d 690 (8th Cir. 1982).

[47] *Usery v. Marquette Cement Mfg. Co.*, 568 F.2d 902 (2d Cir. 1977).

[48] OSHA *Field Inspection Reference Manual*, Chapter III, p. III-11 (CPL 2.103, 9/26/94). OSHA *Field Operations Manual*, Chapter IV S 185 (C.C.H. ed 1983).

[49] *Cape and Vineyard Division v. OSAHRC*, 512 F.2d 1148 (1st Cir. 1975).

fact, move into a dangerous position. In addition, the absence of any evidence of prior similar accidents over many years may demonstrate that the accident in question was not foreseeable.[50]

6.0 Employee Misconduct and the General Duty Clause

As with other alleged OSHA violations, employee misconduct may serve as a defense. The General Duty Clause does not impose strict liability "for the results of idiosyncratic, demented, or perhaps suicidal self-exposure of employees" to known hazardous conditions.[51] "An employer cannot be held to guard against hazards created by employee conduct which is not reasonably foreseeable."[52] Both the OSH Review Commission and the courts have held that the occurrence of employee conduct which results in an accident cannot be held to be reasonably foreseeable in those situations where the cited employer has established work rules designed to prevent the violation, has adequately communicated work rules to its employees, has taken steps to discover violations of the rules, and has effectively enforced the rule in the event of infractions.[53]

> The court in *National Realty* stated:
>
> Congress intended to require elimination only of preventable hazards. It follows, we think, that Congress did not intend unpreventable hazards to be considered 'recognized' under the clause. Though a generic form of hazardous conduct, such as equipment riding, may be 'recognized,' unpreventable in-

---

[50] *Secretary v. FMC Corp.,* 12 O.S.H. Cas. (BNA) 2008, 2010 (Rev. Comm'n 1986); *Secretary v. Cerro Metal Products,* 12 O.S.H. Cas. (BNA) 1821, 1824 (Rev. Comm'n 1986); *Secretary v. Rockwell International Corporation,* 9 O.S.H. Cas. (BNA) 1092, 1098 (Rev. Comm'n 1980); *Ray Evers Welding Co. v. OSAHRC,* 625 F.2d 726, 732 (6th Cir. 1980); *Brennan v. Smoke-Craft Inc.,* 530 F.2d 843 (9th Cir. 1976) (evidence of no injuries in 10 years of cutting sausages is probative); *Canrad Precision Indust.,* 3 O.S.H. Cas. (BNA) 1198 (Rev. Comm'n 1975); *Brennan v. OSHRC (Republic Creosoting Co.),* 501 F.2d 1196, 1201 n.9 (7th Cir. 1974). "This is particularly so where, as here, the employer is required to comply with a generally worded standard." *Secretary v. Granite city Terminals Corp.,* 12 O.S.H. Cas. (BNA) 1741, 1746 (Rev. Comm'n 1986).

[51] *Brennan v. OSAHRC (Hanovia Lamp Division),* 502 F.2d 946, 951 (3d Cir. 1974).

[52] *Cape and Vineyard Division v. OSAHRC,* 512 F.2d 1148, 1152 (lst Cir. 1975).

[53] *Jones & Laughlin Steel Corp. v. OSAHRC,* 10 O.S.H. Cas. (BNA) 1778, 1782 (Rev. Comm'n 1982); *Pennsylvania Power & Light Co. v. OSAHRC,* 737 F.2d 350, 358 (3d Cir. 1984).

stances of it are not, and thus the possibility of their occurrence at a workplace is not inconsistent with the workplace being 'free' of recognized hazards.[54]

To establish the existence of a recognized hazard, the allegedly noncomplying condition or practice must also be shown to be one over which the cited employer "can reasonably be expected to exercise control. "[T]he employer is only required to do that which it would be negligent for it not to do."[55] The cases have acknowledged that no system can guarantee that safe[56] procedures will always be followed, and that a hazard resulting from unexpected employee conduct cannot be totally eliminated.[57]

7.0 Demonstration of Serious Harm

To establish a General Duty Clause violation, the hazard must not only be recognized, it must also be likely to cause death or serious bodily harm.[58] The legislative history does not directly address the meaning of the statutory phrase "causing or are likely to cause death or serious physical harm." The courts have focused primarily on the probability of serious physical harm, should an accident occur. The likelihood of the accident occurring need not be considered. For instance, the court in *Titanium Metals Corp. v. Usery*, declared that "[i]n applying the 'likely to cause' element of the general duty clause, it is improper to apply mathematical tests relating to the probability of a serious mishap occurring."[59]

Given its factual and expert knowledge underpinnings, the courts show great deference to the Review Commission's determination of whether such an accident would result in death or serious physical harm. In *National Realty*, the D.C. Circuit

---

[54] *National Realty*, 489 F.2d at 1266.
[55] *Ray Evers Welding Co.*, 625 F.2d at 732 (6th Cir. 1980).
[56] *Secretary v. Northwest Airlines, Inc.*, 8 O.S.H. Cas. (BNA) 1982, 1992 (1980).
[57] *National Realty*, 489 F.2d at 1266.
[58] This requirement appears to mirror section 17(k) of the OSH Act, which states that a serious violation exists if:

> there is a substantial probability that death or serious physical harm could result from a condition which exists, or from one or more practices, means, methods, operations, or processes which have been adopted or are in use, in such place of employment unless the employer did not, and could not with the exercise of reasonable diligence, know of the presence of the violation.

OSHA §17(k); 29 U.S.C. §666(k)(1982).
[59] 579 F.2d 536, 543 (9th Cir. 1978).

explained in a footnote that "[i]f evidence is presented that a practice could eventuate in serious physical harm upon other than a freakish or utterly implausible concurrence of circumstances, the Commission's expert determination of likelihood should be accorded considerable deference by the courts."[60]

The hazard need not actually have caused an injury or death for the "causing or likely to cause harm" requirement to be satisfied. For example, in *Kelly Springfield*, the Fifth Circuit determined that, despite the fact that no injuries occurred, the employer violated the General Duty Clause when his dust collection system caused an explosion.[61]

8.0 Feasible Hazard Abatement

Once the existence of a recognized hazard which is likely to cause death or serious harm is established, OSHA must then prove that the cited hazard could have been substantially reduced or eliminated by a feasible means of abatement.[62] The term "feasible" has been held to mean economically and technologically capable of being done.[63]

A "feasible means of abatement," *i.e.*, the technology to abate the hazard, must be identified and shown to be something that would materially reduce or eliminate the "hazard" which caused OSHA to issue the citation. When, for example, OSHA claims that the "hazard" was the absence of an effective employee safety program, it must "prove that the employer's (existing) safety program was lacking and that specific alternative measures would be demonstrably safer."[64] Similarly, if OSHA has alleged that, *e.g.*, a toxic substance presents a hazard, the agency must establish that feasible technology exists to reduce or eliminate the hazard. The mere fact that something is technologically feasible does not mean that it necessarily meets the

---

[60] *National Realty*, 489 F.2d at 1265 n. 33.

[61] See *Kelly Springfield Tire Co. v. Donovan*, 729 F.2d at 325 (court deferred to Commission in weighing evidence).

[62] *Kastalon Inc.*, 12 O.S.H. Cas. (BNA) at 1931. When an employer contests a citation, OSHA is obligated to file a more complete statement of the citation's charges with the OSH Review Commission. That statement is known as the "Complaint." The Commission's procedural rules impose additional requirements on OSHA in general duty clause cases by providing that the Complaint must "identify the alleged hazard and specify the feasible means by which the employer could have eliminated or materially reduced the alleged hazard." Commission Rule 35(c), 29 C.F.R. §2200.35(c).

[63] *United Steelworkers v. Marshall*, 647 F.2d 1189, 1264 (D.C. Cir. 1980), *cert. denied*, 453 U.S. 913, 101 S. Ct. 3148, 69 L. Ed. 2d 997 (1981).

[64] *Secretary v. Bechtel Power Corp.*, 12 O.S.H. Cas. (BNA) 1509, 1511 (Rev. Comm'n 1985).

requirement of being economically feasible. Feasible can mean either what is achievable or what can be accomplished in a practical sense.[65]

The evidence necessary to prove the feasibility and likely utility of any abatement measure which OSHA may propose to correct the hazard must be specific.[66] In *Cerro Metal Products*, an employee was crushed while attempting to repair the loader of a press when his assistant energized the press after having initially de-energized it. The employer had a safety rule requiring de-energization of the press before repairs and that rule was explained in a safety manual given to all employees. Furthermore, there had been a recent safety meeting concerning the rule and no evidence was present indicating that supervisors were aware of any violations of the rule. Since OSHA failed to show that the existing safety program was lacking and what further feasible steps the employer could have employed to make the workplace safer, the citation was vacated.[67]

9.0 Replacement of the General Duty Clause with OSHA Standards

Because the General Duty Clause was intended to fill gaps that may exist in OSHA standards, it is well-established that an employer cannot be held in violation of §5(a)(1) when OSHA has adopted an occupational safety and health standard that addresses the very hazard that is the subject of an alleged violation. As explained in *Brisk Waterproofing*:

> [A]ny other interpretation...could lead to wholesale abandonment of the specific standards....[T]o do so would provide little advance warning of what specifically is required in order that employers could maintain a safe and healthful

---

[65] *Compare American Textile Mfgrs. Inst. v. Donovan*, 452 U.S. 490 (1981) (OSH Act does not require OSHA to engage in a cost-benefit analysis before promulgating a health standard under OSHA § 6(b)(5). Feasible means "capable of being done, executed, or effected.") *with Donovan v. Castle & Cooke Foods*, 692 F.2d 641, 647-649 (9th. Cir. 1982) (economic analysis may require that a cost / benefit analysis be performed for a Section 6(a) standard).

[66] *See, for example, Secretary v. FMC Corp.*, 12 O.S.H. Cas. (BNA) at 2013 (details as to how proposed measure would work at the cited workplace are necessary to prove its feasibility); *Pelron Corp.*, 12 O.S.H.Cas. (BNA) 1833 at 1836 (Secretary must show the specific measures that should have been taken that would have materially reduced the risk of harm); *Carlyle Compressor Co. v. OSHRC*, 683 F.2d 673, 676 (2d Cir. 1982) (Secretary must prove particular steps employer could have taken to prevent citation); *Champlin Petroleum Co. v. OSHRC*, 593 F.2d 637, 641 (it is incumbent on the Secretary to demonstrate exactly how the company should and could improve communications with employees so as to better reduce cited hazard).

[67] *Secretary v. Cerro Metal Products*, 12 O.S.H. Cas. (BNA) 1821, 1824 (Rev. Comm'n 1986).

workplace. It is our view that the purposes of the Act would be ill served by such a situation.[68]

To implement this principle, the OSHA *Field Operations Manual*[69] instructs OSHA inspectors that the General Duty Clause:

1. Shall not be used when a standard applies to a hazard.

2. Shall not normally be used to impose a stricter requirement than that required by the standard. For example, if data establishes that a 3 ppm level is a recognized hazard but the standard provides for a permissible exposure limit of 5 ppm, Section 5(a)(1) generally is not to be cited to require that the 3 ppm level be achieved.

3. Shall normally not be used to require an abatement method not set forth in a specific standard. A specific standard is one that refers to a particular toxic substance or deals with a specific operation, such as welding. If a toxic substance standard covers engineering control requirements but not requirements for medical surveillance, Section 5(a)(1) generally is not to be cited to require medical surveillance.

4. Shall not be used to enforce "should"[voluntary consensus] standards. If a standard or its predecessor, such as an ANSI standard, uses the word "should," rather than "shall" or "must," neither the standard nor Section 5(a)(1) shall ordinarily be cited with respect to the hazard addressed by the "should" portion of the standard.

5. Shall not normally be used to cover categories of hazards exempted by a standard. For example, in those cases where specific categories of hazards, types of machines, operations, or industries are exempted from coverage by a standard, they normally are not to be cited under Section 5(a)(1) if the reason for the exemption is the lack of a hazard.

In some cases, however, compliance with a specific OSHA standard does not absolve an employer from being found guilty of a General Duty Clause violation. This is particularly so in cases when the employer knows that mere compliance with the standard is inadequate. In *United Auto Workers v. General Dynamics*,[70] the court declared that if an employer knows that a particular OSHA standard is inadequate

---

[68] 1 O.S.H. Cas. (BNA) at 1264.
[69] CPL 2.45B, pages IV-12 to IV-14. *See also* pages III-12 and III-13 of OSHA's *Field Inspection Reference Manual*.
[70] 815 F.2d 1570, 1577 (D.C. Cir.), *cert. denied*, 484 U.S. 976 (1987).

to protect its workers against the specific hazard cited by OSHA, then the employer is required by the General Duty Clause to take whatever additional protective measures may be required. According to *General Dynamics*, employer knowledge is the key to a General Duty Clause violation. If an employer knows that compliance with a specific OSHA standard will not protect its workers from a hazard, the company will not satisfy the General Duty Clause obligation no matter how faithfully it observes that standard. The decision states that:

> By the same token, absent such knowledge, an employer may rely on his compliance with a safety standard to absolve him from liability for an injury actually suffered by employees as a consequence of a hazard the standard was intended to address, and he will be deemed to have met his obligation under the general duty clause with respect thereto. In other words, compliance with a safety standard will not relieve an employer of his duty under section 5(a)(1); rather, it satisfies that duty. It is in this sense that it may be said that an OSHA standard preempts obligations under the general duty clause.[71]

Although, in general, prosecution of §5(a)(1) violations should not be substituted for OSHA's obligation to promulgate standards, such General Duty Clause citations should not be precluded simply because of the existence of a standard if the employer knows that a dangerous hazard exists notwithstanding its compliance with the requirements of the standard.

10.0 Conclusion

The obligations employers face under the General Duty Clause should be considered in developing a health and safety compliance plan. Steps to ensure compliance might include: a hazard assessment of the workplace and ongoing activities; monitoring applicable voluntary standards issued by ANSI, ASTM and other recognized standards-making bodies; or, if there is a significant trade association in your industry, reviewing relevant guidelines issued by trade associations.

Congress showed its continuing endorsement of the General Duty Clause contained in the OSH Act, when it amended the Clean Air Act in 1990. At that time, Congress directly referenced §5(a)(1) as the basis for creating a new general duty under the Clean Air Act.[72]

---

[71] *Id.*

[72] Clean Air Act § 112(r)(1), 42 U.S.C. §7412(r)(1) reads in part: "Owners and operators of stationary sources producing, processing, handling or storing such [hazardous] substances have a general duty in the same manner and to the same extent as section 654 of Title 29

(continued on the following page)

[OSHA Act] to identify hazards which may result from such releases using appropriate hazard assessment techniques, to design and maintain a safe facility taking such steps as a re necessary to prevent releases, and to minimize the consequences of accidental releases which do occur...." This provision of the Clean Air Act Amendments of 1990 can be found at Pub. L. 101-549, 104 Stat. 2564.

Chapter 5

Recordkeeping

Martha E. Marrapese, Esq.
Keller & Heckman, LLP
Washington, D.C.

1.0 Overview

When the Occupational Safety and Health Act ("OSH Act" or "Act") was passed, Congress recognized that there was a need to collect statistical information concerning accidents, injuries and illnesses.[1] Congress also recognized that while thousands of new chemicals were being introduced every year little was known about the potential adverse health effects chemicals would pose for workers. Accordingly, Congress gave the Occupational Safety and Health Administration ("OSHA") broad powers to require employers to develop statistical information concerning health and safety matters and to require employers to create and maintain records documenting compliance.

When Congress passed the OSH Act, it provided authority for the Secretary of Labor, through OSHA, to promulgate rules regulating employer's conduct in operation of their businesses. Section 8 of the Act provides OSHA with its general rulemaking authority, and includes specifications on the content of the rules OSHA may impose. Congress specifically authorized OSHA to include provisions requiring employers to maintain records for various purposes. This chapter reviews the statutory basis for OSHA's authority and then identify the kinds of records that must be kept, the retention period, and the rules regarding access to the records by various parties.

2.0 Statutory Basis for OSHA Records

Section 8(c)(1) of the OSH Act authorizes OSHA to issue regulations requiring employers to "make, keep and preserve, and make available" records regarding their activities which relate to the Act, particularly as they relate to the "causes and prevention of occupational accidents and illnesses."

---

[1] 15 U.S.C. § 651 *et seq.*

Section 8(c)(2) required OSHA to publish recordkeeping regulations which require employers to report and record "work-related deaths, injuries and illnesses other than minor injuries requiring only first aid treatment and which do not involve medical treatment, loss of consciousness, restriction of work or motion, or transfer to another job."

Section 8(c)(3) gives OSHA the authority to issue regulations that require "employers to maintain accurate records of employee exposures to potentially toxic materials or harmful physical agents." OSHA's authority to require employers to conduct exposure monitoring is found in § 6(b)(7) of the Act. OSHA also is authorized in this section to require medical examinations and tests to be provided at the employer's cost.

OSHA does not have a generic recordkeeping regulation. Throughout the OSHA standards are requirements for the preparation and maintenance of records as a result of the agency's § 8 authority. An overview of the landscape includes OSHA's requirements in 29 C.F.R. Part 1904 to record work-related injuries and illnesses. This rule also contains provisions that require employers to report fatalities and accidents which resulted in five or more employees being hospitalized, and participate in Bureau of Labor Statistics ("BLS") and OSHA surveys of illness and injuries in the workplace. 29 C.F.R. § 1910.1020 contains rules pertaining to maintenance and access to employee medical records.

In addition, OSHA, when promulgating regulations under chemical-specific standards such as the lead, cotton dust, asbestos, and methylene chloride standards, normally requires employers to create and maintain exposure monitoring records as well as medical records. Safety standards, such as the process safety management, scaffolding, and confined spaces programs, will typically have recordkeeping requirements as well. Where OSHA's standards do not specifically direct that OSHA records be kept, employers may still find themselves in need of records simply to demonstrate compliance. In some cases, records requirements are inferred from the performance oriented nature of the standard. Finally, even where no chemical or safety standard exists, but there is an obligation under § 5, the so-called "General Duty Clause" of the OSH Act, to address a workplace hazard (i.e., musculoskeletal conditions) certain records may prove crucial to avoiding a citation.

3.0 Reasons for Requiring Records

As indicated above, recognizing that government must have information with which to be effective, Congress specifically authorized OSHA to require employees to keep records when it gave OSHA authority to regulate employer's actions with respect to the health and safety of employees.

A further reason for OSHA to require records is the need to document the history of employee's work environment for future research and studies, and as a basis for identifying the need for new regulations. One of the major reasons that Congress enacted the statute was the recognition that some kinds of work-related illnesses develop after exposure over long periods of time, and that little information was available in 1970 to define on the basis of objective science "safe" working conditions.

In practice, OSHA makes regulatory decision based on factors which seemingly have little to do with the causes of accidents. The ergonomics rule overturned by Congress is an example. The initial push for the standard came from employee unions, in the face of a declining statistical incidence of musculoskeletal conditions in the workplace. Notwithstanding the paucity of information concerning these conditions and the measures necessary to alleviate them, OSHA steamrolled the rulemaking through, forcing Congress to get into the act and void the rule. There are many other examples of outside forces constituting the driving forces to generate OSHA standards. The point, however, is that standards are rarely generated as a result of the statistical data Congress thought in 1970 should be the basis for OSHA standards.

There is a more basic reason for the requirement to keep records which also explains the tendency of OSHA to increase these requirements continually. That reason is the need for the Agency in its enforcement posture to document the existence of violations of the rules it issues. It is not possible for OSHA inspectors to be present on the job site every day, or even to inspect each workplace each year. It thus falls to the inspector to ascertain the compliance status of the employer through other means. Principal among these is the inspection of the records kept in the ordinary course of business. This is the classic approach of the government lawyer in searching the "paper trail" for evidence of wrongdoing.

Thus, recordkeeping can be a powerful tool to assist the agency in its enforcement efforts. Obviously, records provide OSHA inspectors with a ruler to measure the employer's compliance with the standard. Recordkeeping, or rather the alleged failure to keep records, is used by OSHA to levy huge fines under OSHA's "egregious" violations policy. The "egregious" violation policy permits OSHA to levy a fine for each alleged failure to record an injury or illness that OSHA believes should be recorded. Under the egregious policy, violations are alleged as willful violations. Willful violations are punishable by maximum penalties of up to $70,000 per willful violation with minimum penalties of $5,000 per violation. OSHA has used this policy to levy multimillion dollar fines, again, largely based upon alleged recordkeeping violations.

4.0 Types of Records

Records will generally fall into four classes. First, there are the usual kinds of communications between individuals both within the organization and with those outside the organization. These document the steps the responsible manager takes to carry out company policies and activities.

The second type of records kept are the written expressions of company policies, procedures and instructions. These document the specific actions that the company takes to implement both mandated and voluntary programs. In some case, the statutes or regulations require that a written program be developed; in others, it is simply good management practice to prepare such documents to consolidate the instructions to company personnel.

The third kind of records are those documents that describe the specific day-to-day activities of company employees in carrying out the mandates of company programs and policies. These include daily, weekday or monthly inspection reports, summary management reports, records of training and discipline and similar documents.

The last category of records are those mandated by government at all levels. In many instances, they are the same as those that are kept for other purposes. Often, they can and should be discarded after the record retention period specified in the statute or regulation expires. Government, on the other hand, wants the paper trail to remain, at least until the statute of limitations runs, in order to facilitate its enforcement functions.

5.0 Recordkeeping Requirements for Chemical-Specific Standards

Recall that §§ 6(b) and 8(c)(3) of the OSH Act permit OSHA to require exposure monitoring and issue recordkeeping rules for new standards that imposes exposure monitoring requirements for toxic materials or harmful physical agents. Typically, then, every standard of this kind which OSHA has promulgated since the asbestos rule was first promulgated in the early 1970's include a strong exposure monitoring as well as a medical surveillance component, records of which must be kept.

Standards in this category include the asbestos rule[2] and the thirty-two health-based standards promulgated with numbers higher than 1910.1001.[3] Employers

---

[2] 29 C.F.R. § 1910.1001.

who have reason to believe their employees are exposed to regulated substances should consult the OSHA standards to determine their compliance obligations.

Initial monitoring is required to determine whether the substance is present in quantities which exceed action levels. These standards set eight-hour time-weighted average (TWA) permissible exposure level ("PEL") and short term exposure level ("STEL") ceiling limits. Employers having employees exposed above the action level are usually required to perform periodic monitoring and to record the levels measured. The monitoring programs will specify monitoring frequencies, a confidence level to ensure the accuracy of the monitoring method that is used, and include monitoring for the effectiveness of any ventilation used to control exposures.

It is worth mentioning that employees or their representatives have a statutory right to observe employer monitoring activities.[4] In this case, employees were represented by a collective bargaining agent. The employer monitored for airborne levels of lead, and it monitored for excessive noise. OSHA's lead and noise[5] standards include provisions which permit the representative of affected employee or employees to observe the monitoring while it is being done. In this case, the employer did not afford the union an opportunity to observe, and the Secretary issued citations for violations of both regulations and for violating § 8(c)(3). The employer argued for an interpretation of the standard which would have, if adopted by OSHRC, excused its failures. The Commission rejected the interpretation, and found the employer in other than serious violation as charged. No monetary penalties were assessed in the case, but that failure should not be read as a signal that penalties will not be assessed in future cases.

The toxic substance and harmful physical agent standards will include medical surveillance, biological monitoring, medical removal, and medical recordkeeping requirements. Thus the asbestos standard, for example, requires exposed employees to be periodically examined by a physician. The physician is required, among other things, to take chest X-rays. The lead standard requires that blood lead values be obtained from employees. The records which are developed must also be maintained by the affected employer.

Other compliance activities are usually required, such as employee notification of exposure monitoring and medical surveillance results, the establishment of

---

[3] Notably, these include, among others, the standards for occupational exposure to inorganic arsenic (1910.1018), inorganic lead (1910.1025), coke oven emissions (1910.1029), cotton dust (1910.1043), 1,3-butadiene (1910.1051), and methylene chloride (1910.1052).

[4] *See Secretary v. American Sterilizer Co.*, 15 BNA OSHC 1476 (OSHRC 1992).

[5] 29 C.F.R. 1910.95.

regulated areas, and specifications for the selection, use, removal, cleaning, storage, and disposal of respirators and other personal protective equipment (PPE). The standards typically require employers to generate written compliance programs tailored to reduce employee exposure to or below the PEL, such as the one for cadmium.[6] The compliance hierarchy in these rules typically favors the use of engineering and work practice controls to comply with the PEL for the chemical. Requirements for hygiene areas and practices and hazard communication (including warnings on labels and material safety data sheets and employee training) are specified.

Each of OSHA's chemical and physical agent standards has its own recordkeeping provision. This provision identifies the exposure monitoring and medical surveillance records, other medical records, administrative memoranda, reports, and certifications that must be generated and maintained. These standards also identify the rules employers must follow for retention and access, such as the standard for vinyl chloride.[7] Table 1 lists the chemical and physical agent specific standards presently in effect.

Table 1: Chemical (and Physical Agent) Specific Standards as of 2001

Asbestos	1001	Coal Tar Pitch Volatiles	1002
4-Nitrobiphenyl (13 carcinogens)	1003	alpha-Naphthylamine	1004
Methyl chloromethyl ether	1006	3'-dichlorobenzidine (and its salts)	1007
bis-Chloromethyl ether	1008	beta-Naphthylamine	1009
Benzidine	1010	4-Aminodiphenyl	1011
Ethyleneimine	1012	beta-Propiolactone	1013
2-Acetylaminofluorene	1014	4-Dimethylaminoazobenzene	1015
N-Nitrosodimethylamine	1016	Vinyl Chloride	1017
Inorganic arsenic	1018	Lead	1025
Cadmium	1027	Benzene	1028
Coke Oven Emissions	1029	Bloodborne pathogens	1030
Cotton Dust	1043	1,2-dibromo-3-chloropropane	1044
Acrylonitrile	1045	Ethylene Oxide	1047
Formaldehyde	1048	Methylenedianiline	1050
1,2-Butadiene	1051	Methylene Chloride	1052
Ionizing Radiation	1096		

---

[6] 29 C.F.R. § 1910.1027(f)(2).

[7] 29 C.F.R. § 1910.1017(m).

As alluded to in the beginning of this chapter, however, the recordkeeping section of these standards is typically only the tip of the iceberg. To appreciate the number of records that should be kept to demonstrate compliance with these standards, review the lead standard, which includes the following provisions:

(d)(2)	Initial Determination
(d)(4)	Initial Monitoring
(d)(5)	Negative Initial Determination Record
(d)(7)	Additional Monitoring
(e)(1)	Infeasible Engineering Controls
(e)(3)	Written Compliance Program
(e)(5)	Measurements of Mechanical Ventilation
(e)(6)	Job Rotation Schedules
(f)(1)	Written Respirator Program
(f)(2)	Respirator Selection
(f)(3)	Respirator Fit Testing
(g)(2)(vi)	Notice to laundries
(g)(2)(vii)	Laundry Container Labels
(j)(2)	Blood lead monitoring
(j)(2)(iii)	Employee Blood Lead Notifications
(j)(3)	Medical Examinations
(j)(3)(iii)(B)	Second Opinion Notifications
(j)(3)(iv)	Information to Physicians
(j)(3)(v)	Written Medical Opinions
(l)(1)(ii)	Training Program
(l)(1)(iii)	Initial Training
(l)(1)(iv)	Annual Training
(l)(2)(i)	Employee Requests for Information
(l)(2)(ii)	DOL Requests for Information
(m)(2)(ii)	Sign Cleaning
(n)(1)	Exposure Monitoring Recordkeeping

(n)(2)   Medical Surveillance Recordkeeping

(n)(3)   Medical Removal Recordkeeping

(o)(2)(ii)(c) Observation of Monitoring

That is, there are 29 provisions in the lead standard in which some form of record is implied or required. This is, of course, an enormous burden, but more importantly, creates a significant problem for assuring compliance. Not only do the records have to be kept, they must be maintained. This is interpreted by OSHA to mean that the information in the records must be updated when it changes. Each of the standards specifies the period for record retention which typically is for either 40 years or the length of employment plus 20 or 30 years.

Do not forget to examine the appendices to these standards for any additional recordkeeping obligations they may generate. The appendices for 1,3-butadiene, for instance, have MSDS recommendations, medical screening, sampling and analytical methods, and a sample (non-mandatory) health questionnaire that may affect company recordkeeping.

Complementary to the chemical-specific standards are the recordkeeping requirements for respirators specified by § 1910.134. These requirements should be consulted in addition to any recordkeeping requirements specified by particular toxic substance standards.

Finally, be aware that there are a large number of chemical substances for which PEL levels are set but which do not specify exposure monitoring and recordkeeping requirements.[8] OSHA has for a number of years indicated that it will adopt a generic standard which will require exposure monitoring and recordkeeping for these substances, but that standard has never yet been issued.

6.0 Recordkeeping Requirements for Safety Standards

Only some of the more established safety standards for the general, construction and maritime industries have explicit recordkeeping provisions.[9] Each of these standards contain requirements for employers to inspect the devices and to make

---

[8] These substances are regulated under the provisions of the PEL standard at 29 C.F.R. § 1910.1000.

[9] Examples of standards that specify recordkeeping requirements are those which relate to fire extinguishers (1910.157, 1910.160), cranes and derricks (1910.179 through 181, and 1926.550) and mechanical power presses (1910.217).

records of those inspections. In the case of the power press requirements, the employer is also required to report point of operation injuries to OSHA within 30 days of the occurrence of the injury. These standards should be consulted to determine the precise extent of the recordkeeping requirements they impose. Other standards such as those for slings, powered platforms, telecommunications, and commercial diving specify recordkeeping requirements.

The more recent OSHA standards, like the lockout/tagout standard (1910.147) and the respirator standard (1910.134), have employers generating what amount to compliance plans tailored to their particular situations, and leave it largely to the employer to decide what documents to retain to demonstrate compliance. These standards usually require the employer to create and maintain employee training records as well. We can anticipate that all future standards will be equally designed to provide an adequate record for inspectors to use in evaluating compliance.

In general, all of the safety standards a facility is subject to should be reviewed with an eye toward the significant number of implied records that are required. For instance, the standard on Portable Wood Ladders (§1910.25) requires that "[l]adders shall be inspected frequently and those which have developed defects shall be withdrawn from service..." This provision does not explicitly require records. However, most employers would find prior purchase and repair receipts (or, in the exceptional case, a maintenance log) useful if an inspector found an employee using a ladder that had a defect.

The defense to such a citation would likely be one of "employee misconduct," that the employee failed to perform his job properly in selecting a defective ladder for use. The employer would have to show that he had a policy requiring employees to inspect the ladders "frequently," that the employee had been told of the policy, and that the policy was actively enforced. The contemporaneous records to make this demonstration to the inspector would show that (1) there was a policy, (2) the employee had been trained, (3) inspections had been conducted regularly by others, and (4) supervisors had enforced the requirement. This last element would likely be shown through evidence that employees had been disciplined for failure to perform the required acts.

This illustrates that the rule could be interpreted to require that at least four types of records would have to be kept to demonstrate employer compliance with the inspection requirement. Of course, not all employers will have such records, and in fact, many do not keep them. Nevertheless, because the burden is on the employer of demonstrating compliance with the standard when the *prima facie* case is made in the Agency's complaint, records of such kinds of activities are desirable.

Table 2 is an illustrative listing of OSHA safety standards that have either written program, training or inspection requirements.

Table 2: Illustrative List of Safety Standards and Types of Records

	Written Program	Specific Training	Routine Inspections	1910 Section
Portable Wood Ladders	N	N	Y	25
Emergency Response	Y	Y	N	38
Walking Working Surfaces	N	N	Y	68
Ventilation Systems	N	N	Y	94
Noise	N	Y	Y	95
Ionizing Radiation	N	Y	Y	96
Flammable / Combustible Liquids	N	N	N	106
Hazardous Waste	Y	Y	N	120
Personal Protective Equipment	Y*	Y	N	134
Lockout / Tagout	Y	Y	Y	147
Medical Services and First Aid	N	Y	N	151
Fire Brigades	Y	Y	Y	156
Fire Extinguisher	Y	Y	Y	157
Sprinklers / Hoses Standpipes	Y?	Y?	Y	158
Employee Alarms	Y	Y	Y	165
Servicing Truck Tires	N	Y	N	177
Powered Industrial Trucks	N	Y	Y	178
Cranes	N	N	Y	178,179
Derricks	N	Y	Y	181
Slings	N	N	Y	184
Machine Guarding	N	N	Y	217
Mechanical Power Presses	N	Y	Y	217
Portable Power Tools	N	N	Y?	241-244
Welding	N	Y	N	252
Electrical Systems	N	Y	N	301-399
Hazard Communication	Y	Y	Y	1200
Laboratories (Non-production)	Y	Y	N	1450
Confined Space Entry	Y	Y	Y	146
Process Safety Management	Y	Y	Y	119
Hazardous Waste Operations and Emergency Response	Y	Y	Y	120
Respiratory Protection	Y	Y	Y	134

7.0 Illness and Injury Recordkeeping Requirements

OSHA published its general injury and illness recordkeeping requirements in 29 C.F.R. Part 1904 early in the history of the agency. After being a work in progress for over ten years, amendments to the rule were rushed through the Department of Labor and Office of Management and Budget in the waning hours of the Clinton Administration and published in the *Federal Register* on the last publication date before President George W. Bush was inaugurated. The publication of this final rule completes OSHA's illness and injury recordkeeping trilogy. The rule authorizing OSHA's annual data collection survey[10] took effect on March 3, 1997. The revised requirements for reporting work-related fatalities and multiple hospitalizations[11] took effect on May 2, 1994, and were further modified in this rulemaking.

Prospects for change continue to exist. The National Association of Manufacturers (NAM) has a complaint filed in the U.S. District Court for the District of Columbia to ask the court to declare the rule invalid, enjoin OSHA from enforcing the rule, and directing OSHA to rescind it. The challenged provisions include those concerning work-relatedness and how that determination should be made. Moreover, the rule remains subject to final review and approval under the Paperwork Reduction Act (PRA). This will allow the business community another crack at both the substantive requirements of the rule and the associated new forms.[12] Given the required time frame likely to be required for settlement discussions and quite possibly a reopened rulemaking, the January 1, 2002 start-up date for the revised rule appears to be very much in question.

7.1 Those Subject to the Rule

This rule requires employers to record and report work-related fatalities, injuries, and illnesses. As part of its effort to simplify the existing scheme, OSHA has shifted to a "plain English" question and answer format. The "Blue Book" Guideline last amended by the Bureau of Labor Statistics in 1986 is largely incorporated into the latest changes in the standard. BLS retains a statistical function under this rule, and OSHA is responsible for determining whether the records are being kept accurately. The BLS Guideline continues to have historical value in providing employers with detailed information concerning what BLS believed constitutes recordable injuries and illnesses under the Act. Detailed information may also be found in the hundreds of interpretation letters and guidance memoranda issued by OSHA in response to specific compliance inquiries, and by combing through the background and explanatory materials in the 200 page preamble to the latest final rule.

The rule applies to all employers covered by the OSH Act. Section 1904.2 of the rule specifies that an employer must maintain a log and summary of all recordable occupational injuries and illnesses for each "establishment." Section 1904.12 of the rule defines an establishment as "a single physical location where business is conducted or where services or industrial operations are performed, e.g., a factory, store, warehouse, etc." OSHA recognizes that some firms and some operations are

---

[10] Now codified in § 1904.21.
[11] Now codified in § 1904.39.
[12] OSHA 300, 300A, 301, privacy concern case listing.

conducted over dispersed areas, e.g., construction work, sales activities, communications. The regulations provide that in such event records shall be maintained at the central reporting facility or establishment used by the workers.

7.2 Exemptions

Generally, all employers with 10 or fewer full or part time employees at any one time in the prior year are exempt from recording illnesses and injuries. Also exempt from this requirement are employers in a number of standard industrial classification ("SIC") codes which are considered to be low hazard industries. Examples of employers in this category include automotive dealers and gasoline service stations, retail stores such as apparel and accessory stores, furniture dealers, dining establishments, membership organizations, banks, real estate operators, insurance companies and the like.

Employers who are exempt from the recordkeeping requirements may be asked to participate in a given year in the statistical survey by the BLS or OSHA and, to that extent, would be required to keep these records.

Small and otherwise exempt employers are not exempt from complying with other requirements of this rule, such as the requirement to report hospitalizations of three or more employees and fatal accidents to OSHA within 8 hours of the occurrence.

Major changes to anticipate in this aspect of the rule involve the types of reportable fatalities. Employers will need to report a heart attack fatality at work, affording the local area office director an opportunity to investigate based on circumstances. While fatalities that occur on a commercial / public transport system or a motor vehicle accident on public roads will not need to be reported to OSHA, these will be potentially recordable incidents. Hospitalization or fatalities that occur 31 days after an incident will not require OSHA notification. Conversely, if the heart attack at work is not immediately fatal, but the employee dies within 30 days, OSHA will need to be notified.

Some states may impose additional recordkeeping requirements or other safety and health requirements which small employers are required to obey.

It should also be noted that employers who are exempt from illness and injury recordkeeping requirements are not thereby exempt from other safety and health obligations. Small employers are generally expected to comply with OSHA standards and, while not subject to programmed safety inspections, may be inspected as the result of an employee complaint or to determine compliance with OSHA health standards.

In addition, many employers who are subject to the safety and health requirements of other federal agencies are nonetheless required to make and maintain illness and injury records under OSHA with the exception of small employers. Thus, motor carriers are required to comply with OSHA recordkeeping requirements. Employers subject to the requirements of the 1977 Mine Safety and Health Act are not, however, subject to the OSHA recordkeeping rule. The agency having jurisdiction for the health and safety of miners, the Mine Safety and Health Administration (MSHA), has its own published recordkeeping requirements. Persons who are subject to the jurisdiction of other federal agencies should check recordkeeping regulations of that agency to determine whether they need to use OSHA forms or similar forms to comply with the Act.

7.3 Recordkeeping Forms

Employers are required by §§ 1904.2 through 1904.6 to make and maintain injury and illness records for each establishment they maintain. OSHA prints a "log and summary" which is useful for making the recordings. The log and summary is called the OSHA 200 Form, although it is the information required to be entered and not the form itself that is important. In other words, what OSHA requires is that the employer enter recordable injury and illness data on a summary form of some kind. The information is to be recorded within six days of the occurrence of the injury or illness, and the log and summary is to be maintained at the establishment wherein the injury or illness was incurred. Under the new rule, a new Form 300 will be used, and employers will be required to record each case on the Form 300 within seven calendar days of the occurrence, replacing the current six-business -day requirement.

In addition to the log and summary, § 1904.4 provides that each employer shall maintain a "supplementary record for each occupational injury or illness at that establishment." The information required on this form includes identity information for the injured or ill employee, information concerning where the injury or exposure to an occupational illness occurred, a description of the nature of the injury or illness, the date it occurred, and the part of the body affected. The supplementary information should also include information concerning hospitalization and the identity of the person making the report. OSHA provides a Form 101 for the supplementary information, but workmen's compensation or other insurance forms which contain the required information are acceptable. Under the new regulations, OSHA has developed a new Form 301 for recording supplementary information. Workmen's compensation or other insurance forms which contain the required information will continue to be accepted.

7.4 Posting

Employers who are required to make and maintain the aforementioned records are also required by § 1904.5, during the period February 1 to March 1, to post an annual summary of the injuries and illnesses reported for the establishment for the previous year. The information recorded on the OSHA 200 is used to compile the summary, and the person who makes or supervises the compilation is to certify that the compilation is "true and complete."

The summary is to be completed and posted even if there were no recordable injuries or illnesses in the prior year. The summary is to be posted in the place where employee notices, such as the OSHA poster, are posted.[13]

Under § 1904.32(b)(4) of the new rule, a company executive, such as a plant manager, will have to certify the accuracy of the Log. The final rule defines an executive as: 1) the company owner if the company is a sole proprietorship or partnership; 2) a corporate officer; 3) the highest ranking company official working at the establishment or 4) the immediate supervisor of the highest ranking company official working at the establishment. OSHA anticipates that this requirement will foster greater management involvement and support. Moreover, the new rule expands the year-end posting-period of the OSHA 300 summary from February 1 through April 30, a total of three months. Year 2001 summary information will need to be posted only from Feb. 1, 2002 to March 1, 2002.

7.5 Maintaining and Retaining These Records

Section 1904.6 requires an employer to not only keep, but actively maintain, illness and injury records for a period of five years and to provide access to the records when requested by OSHA. Employers have to ensure that the forms are kept up-to-date for the five year period. OSHA will as part of the opening of an inspection usually request access to the OSHA Log.

The regulations require that the records be kept at the establishment where the employees are employed. More than one establishment may be present at the same location. Thus a warehouse operated in conjunction with a factory may constitute two separate establishments for recordkeeping purposes. The essential purpose is that warehousing is classified apart from manufacturing, and it is important to gather statistical data for each affected industry.

---

[13] OSHA requires employers to post a poster informing employees of their rights under the Act. The poster should be at least 8 1/2 x 14 inches, and OSHA has issued citations when the poster is not of this size. It is to be posted in a place where employees' information is normally posted such that it can be seen.

The log and supplementary form must be kept at every physical location where the employer's operations are performed.[14] The records must be retained at these establishments or they must be made available at the establishments. The log may be prepared at a supplementary location or by means of data processing equipment. In such event, a hard copy of the log current to 45 days must be maintained at the establishment, and sufficient information to complete the log at the supplementary location must be provided so that it is completed within six work days.

The compliance problems associated with employees who do not perform their work at fixed establishments are covered in the BLS Guideline and in the new rule. In general, when employees work out of an establishment or are paid out of an establishment but do not report to it, then that is the establishment where the records are to be kept.

7.6 Access

Currently, an employee, former employer, their personal representative, or authorized employee (union) representative may request and have access to the OSHA Log and Supplemental Form under § 1904.7.

When the new rule goes into effect, when an employee, former employee, or their personal representative, or their authorized employee (union) representative requests a copy of his or her record on the OSHA 300 Log, employers will have until the end of the next business day to provide a free copy. Employees, former employees, or their personal representatives will be entitled to a copy of the OSHA 301 Incident Report by the next business day. Employers will have seven-calendar days to provide copies of OSHA 301 Incident Reports to authorized employee representatives. Upon request from OSHA, employers will have four business hours to provide Form 300 and/or 301 to authorized government representatives. Previously, these records only had to be made "available" to the inspector.

OSHA has finally acknowledged that access to the required injury and illness records raises legitimate privacy concerns, and in some cases constitutes an improper and possibly unlawful invasion of privacy. The new rule addresses this issue by creating a limited class of privacy concern cases for which the employer is prohibited from listing the employee's name on the OSHA 300 and 301, and "may" remove other information that would identify the person. This category is limited to, for example, conditions involving intimate body parts, reproductive organs, HIV, mental illness, or contaminated sharps independently and voluntarily identified by the affected worker as a privacy case. Employers will need to maintain a "privacy

---

[14] *Secretary v. Price Chopper Supermarkets*, 15 OSHC 1518 (No. 900552, 1992).

concern" case log for these cases. OSHA has not specified a form for this log, which appears to provide employers with some flexibility in recording cases with sensitive medical conditions while protecting the privacy of the affected employee.

7.7 Information to Be Recorded

What must be recorded is frequently straightforward, but can also be confusing and subject to debate. What is a recordable event is defined in § 1904.12 of the OSHA regulations in the following terms:

Recordable occupational injuries or illnesses are any occupational injuries or illnesses which result in:

- Fatalities, regardless of the time between the injury and death or the length of the illness;

- lost workday cases, other than fatalities, that result in lost workdays;

- nonfatal cases without lost workdays which result in transfer to another job or termination of employment; require medical treatment (other than first aid); or involve loss of consciousness or restriction of work or motion.

The definition of what is recordable goes on to state that this last mentioned category of recordable events includes any diagnosed occupational illnesses which are reported to the employer but are not classified as fatalities or lost workday cases.

OSHA defines in the recordkeeping regulations the terms "medical treatment," "lost workdays" and "first aid." The term "medical treatment" means treatment by a physician or by registered professional personnel under the "standing orders" of a physician, but does not include "first aid." First aid, however, means any one time treatment and any follow-up visit for the purpose of observation, minor scratches, cuts, burns, splinters, and so forth, "which do not ordinarily require medical care." Lost workdays include all lost work days whether consecutive or not, excluding the day the injury occurred.

The definitions quoted or paraphrased above are also included on OSHA's forms, and the forms can be referred to for the complete definitions. As defined on the forms, an occupational injury is any injury which results from work. That does not make the injury "recordable" unless more than first aid is required. The term "occupational illness" is defined on the forms in the following terms: "any abnormal condition or disorder, other than one resulting from an occupational injury, caused by exposure to environmental factors associated with employment. It includes acute or chronic illnesses or diseases which may be caused by inhalation, absorption, ingestion, or direct contact. Examples of such illnesses and disorders include skin diseases, dust diseases of the lungs, respiratory conditions due to toxic agents, poisoning, disorders due to physical agents, disorders associated with repeated trauma, and all other occupational illnesses."

Under the new rule, the same criteria generally apply in determining whether an injury or illness is recordable. In addition, there is an apparently limited category of significant injuries or illnesses that are recordable if diagnosed by a physician or other licensed health care professional that would not otherwise be recordable (examples include conditions such as cancer, chronic irreversible diseases, broken bones, and punctured eardrums). Moreover, the new rule identifies the following conditions as specifically recordable: an active case of tuberculosis or a positive TB skin test, a needlestick or sharp injury, a (permanent) standard threshold shift, averaging a 10 dB change at 2000, 3000, 4000 hertz, in either or both ears, and a required (not voluntary) medical removal under a medical removal requirement of an OSHA standard.

The foregoing makes it clear that when an employee suffers a traumatic injury on the job which results in a lost work day, recording must be entered on the OSHA Form or its equivalent. Who is an employee and what is work related are not so clear. Indeed, for recordkeeping purposes, what is an "injury" and what is an "illness" also is not clear.

Even under the new rule, OSHA has continued the practice of distinguishing injuries from illnesses largely by relying on an extensive list of examples in the forms that are not part of the rule. In general, under the new rule, first aid will be defined by a list of 14 specified treatments. The significant changes are that use of hot or cold therapy, use of massages, and use of butterfly bandages or Steri-Strips™ are now considered first aid. Conspicuously absent is the single dose of a prescription drug. The revised regulation now regards this, as well as the administration of oxygen, as medical treatment. Medical treatment will not include visits to a physician or other licensed health care professional solely for observation or counseling, the conduct of diagnostic procedures such as x-rays or blood tests, or administration of prescription medications used solely for diagnostic procedures (i.e. eye drops to dilate the pupil).

As stated above, an injury which results only in first aid treatment is not considered recordable. The BLS Guidelines state that "minor scratches, burns, splinters and so forth, which do not ordinarily require medical care," are considered first aid injuries which do not require recording.

While this distinction seems clear, how does OSHA explain why it considers a needle stick recordable as an injury? The natural reaction is that a needlestick injury ordinarily receives first aid treatment. That is, the ordinary reaction would be that a needlestick wound should be cleaned, an antiseptic such as iodine might be applied and a bandage used—all of which are the usual home remedies, and all of which are and have been ordinarily considered first aid treatment. OSHA wants this type of event recorded, however, in case the wounded employee, as a result of the needle-stick or other minor cut, contacts a disease transmitted by a bloodborne pathogen, e.g., the Hepatitis B virus. The disease is, of course, an illness and logically should be recorded as such if it occurs and if it is work related. OSHA, however, focuses on the event which may cause the illness and states that the event is recordable as a work related injury. That makes no logical sense, but the OSHA recordkeeping requirements do not make sense in all respects.

The "employee" question is one of the easier ones to answer. In general, OSHA will consider the person who has the authority to direct and control the activities of a worker as the employer. The person who supervises the "employee" on a day to day basis is usually considered the employer. This means that temporary and part time employees may be considered employees for recordkeeping purposes, but not independent contractors.

When is an illness or injury work-related? Under the new rule, OSHA specifies that work-relationship will be established if work either caused or significantly contributed to the injury or illness, or "significantly aggravated" a pre-existing injury or illness, whereas currently, any contribution or aggravation is enough to establish recordability.

As a general matter, if the injury or illness occurs on the employer's premises, OSHA will presume the event is work-related. What this means is that OSHA will deem the injury or illness is work related and the employer has the burden of demonstrating that it was not work related. If, for example, an injury occurs in an employer-controlled eating facility, a hallway or a rest room, the BLS guidelines indicate it is on the premises and considered work related. On the other hand, parking lots, ball fields, gyms and other such facilities will not ordinarily give rise to the presumption that an injury was or is work related. However, if the employee's duties are such that the relationship to these kinds of facilities appears work related, e.g., if the employee is a parking attendant in the parking lot, or a physical therapist working in the gym, then the injuries will be considered work related. If the employee is involved in a recreational activity that is not employee-sponsored, then the activity is not work related and the injury is not recordable.

There are several new exceptions to the work-relatedness criteria for recordability under the new rule. Among them: injuries or illnesses caused by voluntary participation in wellness programs, eating and drinking food or beverages for personal consumption, intentionally self-inflicted wounds, personal grooming, and the common cold / flu are not considered work-related and are not recordable.

Under the new rule, restricted work will be distinguished from lost workdays and will need to be separately recorded. It occurs when an employer decides, or a physician or other licensed health care professional recommends, on a temporary or interim basis, to keep the employee from performing one or more routine job functions or from working a full shift.[15] A case will no longer involve restricted work if the employer permanently modifies the employee's job in a way that eliminates the restricted job function. Job transfer and restricted work cases will be recorded in the same box.

The focus will be on whether or not the employee is permitted and able to perform his or her routine job functions (defined as the duties performed at least once per week prior to the injury / illness.) Unlike the current rule, if a work restriction or time away from work is limited to the day of the injury or illness and none of the other recording criteria are met, the case will not be recordable.

Days away from work or restricted work / job transfer beyond the day of onset resulting from recordable injuries or illnesses will need to be counted. However, the days are counted using calendar days and employers may stop the count at a cap of 180 calendar days. In addition, the employer will be able to cease counting restricted days if the employer permanently modifies the employee's job in a way that eliminates the restricted job function.

One of the more controversial of the new revisions is the requirement for cases to be recorded using the total number of calendar days a physician stipulates an employee was restricted / unable to work regardless of it being a scheduled holiday, vacation, or days off (including weekends). 29 C.F.R. § 1904.7(b)(3) will require the employer to count the number of calendar days the employee was unable to work because of a work-related illness or injury, regardless of whether or not the employee would have been scheduled to work on those calendar days. Days that a physician recommends that the worker stay at home must be recorded whether or not the employee stays home. Employers still can begin counting days away or restricted on the day after the injury occurred or the illness began. While use of calendar day counting will ease some of the administrative difficulties of log maintenance, the effect of these changes is to artificially inflate the severity rates of individual establishments by skewing the actual number of days away from work or restricted work.

7.7 Musculoskeletal Disorders ("MSDs")

OSHA indicates that injuries which result from the man-to-machine relationship, i.e., ergonomic injuries, are not injuries but are illnesses. OSHA has cited companies for egregious violations for failure to record back sprains and other muscle

---

[15] 29 C.F.R. § 1904.7(b)(4)(i).

related injuries.[16] It should be observed that some of these alleged "illnesses" are of a kind that ordinarily would be considered "first aid" cases, e.g., back sprains.

Under the new illness and injury rule, there is no change in the way a recordable MSD is evaluated. To be recordable, MSDs still need to be work-related and meet one or more of the general recordkeeping criteria (days away from work, restricted work or transfer, or medical treatment beyond first aid). Subjective symptoms (*e.g.* pain or tingling) will still be relevant if work-related and accompanied by one or more general recordkeeping criteria. MSDs are not in OSHA's category of cases that will need to be recorded as privacy concern cases in the future.

OSHA and BLS have taken the position that the Act requires that all occupational illnesses be recorded. That is, OSHA takes the position that there is no first aid exception from recordkeeping requirements for illnesses. This position is not supported by the legislative history of the Act. The committee reports, including the report of the conference committee, all indicated that Congress was fully aware that recordkeeping could be burdensome and they determined to make that burden as light as possible consistent with the goals of the Act. There was no exception for illnesses. The Senate Committee explained "[the committee recognizes the fact that some work-related injuries or ailments may involve only a minimal loss of work time or perhaps none at all, and may not be of sufficient significance to require their being recorded or reported".[17] The following statement appears in the report of the managers for the House following the conference of both Houses over the disagreeing bills:

> Employer Reports. A Senate bill provision without a counterpart in the House amendment permitted the Secretary to require an employer to keep records and make reports on all work related deaths, injuries and illnesses. The House receded with an amendment limiting the reporting requirement to injuries and illnesses other than of a minor nature, with a specific definition of what is not of a minor nature.[18]

The importance of this disagreement between OSHA's interpretation of the Act and what was said by Congress is of greatest importance in the area of ergonomic injuries. First, it should be noted that OSHA has determined that an ergonomic injury is an illness. That is, the agency believes that any sprain, strain, or other cumulative trauma disorder ("CTD") which occurs from repetitive motion and

---

[16] *See Caterpillar*, 1987-1990 CCH OSHD 28,289 (No. 87-0922, 1988), OSHRC No. 87-0922; *Kohler Co.*, 1987-1990 OSHD 28,922 (No. 88-237, 1990), OSHRC No. 88-237; *Pepperidge Farms Co.*, OSHRC No. 89-265).

[17] *Legislative History*, p. 157.

[18] *Legislative History*, p. 1190.

takes a period of time to develop is by virtue of those facts an illness and not an injury. A June 4, 1991, OSHA's Director of Field Programs Memorandum for Regional Administrators contains the following statement:

> However, most cases of...cumulative trauma disorder result from exposures that are not instantaneous, and are considered occupational illnesses. The current guidelines and instructions on the back of the OSHA 200 log define recordable occupational illnesses as any abnormal condition or disorder. These are work related if an exposure in the work environment either caused, aggravated or contributed to the case. Some employers and others are unsure about what constitutes an abnormal condition or disorder for these conditions. As a result, the following criteria have been developed:
>
> - Upper Extremity Cumulative Trauma Disorders (CTDs): (excludes back cases) OSHA will issue citations to employers for failing to record work related CTDs on the OSHA 200 log that are evidenced by:
>
> * at least one physical finding, (i.e., an objective symptom);
>
> - subjective symptom coupled with either medical treatment or lost workdays, (i.e., days away from work and / or days of restricted work activity.

The problems associated with the foregoing instruction are evident. It clearly confuses non-work related activity with work related activity or allows of such confusion. For example, employees who report to work on Mondays with sprains, strains and the like acquired from their weekend activities might well aggravate those conditions on the job. Are they recordable CTDs? The instruction would seem to make them recordable, particularly if a restriction is imposed on the work activity of the employee. It seems clear, however, that OSHA will cite employers who do not record these "illnesses."

There is at least one additional problem presented by the aforenoted instruction, and it is probably best described by the term "catch 22." Assuming employers obey this instruction in the hope that they will thereby avoid citations for recordkeeping violations, they run the risk of amassing evidence of potential ergonomic injuries and potential violations for not avoiding those injuries. There simply is no good way out of the situation. The problem is compounded by the fact that the Review Commission generally defers to OSHA interpretations citing the Supreme Court's *Martin v. OSHRC (CF&I Steel Corp.)* decision. Obviously, the employer will have to make its own decision concerning whether to comply in the way that OSHA is demanding by instructions and interpretations of this kind. The alternative is to contest the citations and be prepared to take the matter up through the Commission and the courts.

8.0 Recordkeeping Requirements for the Hazard Communication Standard

A discussion of the Hazard Communication Standard (HCS) (§1910.1200) illustrates the points in this chapter about the variety and types of records required. The HCS covers virtually all employers in the private sector in the U.S. with exception of the mining industry. The standard requires employers to address all hazardous chemicals to which employees may be exposed. All employers are required to communicate to their employees information about hazardous substances which are known to be present at the worksite. This requirement applies regardless of whether the employer created the exposure. The issue for all employers is whether they "know" that their employees are exposed.

Chemical manufacturers and importers are required to perform hazard determinations on all chemicals they produce or import. If downstream employers, such as wholesalers and distributors, re-label products or in any other way choose not to rely on the manufacturer's determination of hazard, they too must perform the determination. Labels, MSDS, and employee training requirements are specified by the standard.

There are four explicit kinds of records that must be kept under the HCS. First, there is the written program. This must contain sections which address specific subjects listed in the standard. Secondly, every employer must have a list of hazardous chemicals to which employees may be exposed. This list must contain the names of the hazardous products found on the label and MSDS.

The third kind of record that must be kept is the MSDS itself. The ostensible purpose of the MSDS is to provide information to employees about the hazardous chemicals to which they are exposed. However, OSHA attributes a secondary purpose to both the list and the MSDS: to document the exposures of employees over time for use as a tool in conducting epidemiological studies. Thus, these records are considered records of exposure which are subject to the record retention provisions of 29 C.F.R. § 1910.20(d)(1)(ii).

Labels may also be considered records, although they are arguably not exposure records. Companies will want to keep records of what label statements were used and in particular the reasons for making the choices. The standard requires that "appropriate hazard warnings" be used on labels on all containers in the workplace. OSHA recognizes that appropriate labels need not include warnings about every toxic effect of every component in a product. Therefore, where labels evolve as manufacturers learn more about their products, their inherent dangers, and the usage characteristics of customer's operations, the "appropriate" warnings will change. It is important to document this evolution, both for OSHA as well as other legal purposes, and such records will thus become part of the business' ordinary records system.

These are all the records that are explicitly mandated by the HCS, and not all of them are subject to any particular retention policy. In addition, implicit in the requirements of the standard, such as training, hazard communication, and others, is the need to document compliance activities.

In the HCS, the definition of the written program elements is detailed in very simple language: "[e]mployers shall develop, implement, and maintain a written hazard communication program for their workplaces which at least describes how the criteria specified in paragraphs (f), (g), and (h) of this section...will be met..." To meet the standard of describing how these requirements are met requires and extensive written program, however. Simply paraphrasing the language of the standard is not sufficient. OSHA expects to see details such as responsible managers and procedures written out. In OSHA Compliance Instruction (CPL) CPL 2-2.38D, *Inspection Procedures for the Hazard Communication Standard*, OSHA defines what it expects to see in a written Hazard Communication Program (HCP). It must contain the following elements:

- Designation of person(s) responsible for ensuring labeling of in-plant containers.
- Designation of person(s) responsible for ensuring labeling on shipped containers.
- Description of labeling system(s) used.
- Description of written alternatives to labeling of in-plant containers, where applicable.
- Procedures to review and update label information when necessary.
- Designation of person(s) responsible for obtaining / maintaining the MSDS.
- How such sheets are to be maintained (e.g., in notebooks in the work area(s), via a computer terminal), and how employees obtain access to them.
- Procedure to follow when the MSDS is not received at the time of the first shipment.
- For chemical manufacturers or importers, procedures for updating the MSDS when new and significant health information is found.
- Designation of person(s) responsible for conducting training.

- Format of the training program to be used (audiovisuals, classroom instruction, etc.).

- Elements of the training program—compare to the elements required by the HCS (paragraph (h)).

- Procedures to train new employees at the time of their initial assignment and to train employees when a new hazard is introduced into the workplace.

- Procedures to train employees of new hazards they may be exposed to when working on or near another employer's worksite (i.e., hazards introduced by other employees).

- Methods the employer will use to inform employees of the hazards of *nonroutine* tasks and unlabeled chemicals in their work areas.

- Methods the employer will use at multi-employer worksites to inform other employers of any precautionary measures that need to be taken to protect their employees.

- For multi-employer workplaces, methods the employer will use to inform the other employer(s) of its labeling system.

- Whether the written program will be made available to employees and their designated representatives.

The HCS therefore imposes a comprehensive duty on employers to develop written documents both explicitly and implicitly.

Moreover, implementation of the written program necessitates the generation of additional records, such as historical records of program reviews, notes on hazard determinations, drafts of labels and MSDS, and training records. These records are almost always necessary to document that the program in fact is implemented. For example, only in the case of the employer who relies solely on his suppliers for MSDS and labels will there be no hazard determinations. And unless an employer can demonstrate that employees are well trained by other means, records of training will be necessary.

The last kind of record the HCS implicitly requires is the response of the employer to requests for copies of MSDS or other information by medical specialists in an emergency. Under the standard, manufacturers may withhold chemical composition information from labels and MSDS, but must disclose the information when a bona fide request is received from a physician or other health care professional. These requests must be in writing and must contain specific information, and the responses must be prepared accordingly. Of course, such correspondence inevitably creates a record of the transaction which then must be maintained.

It is not always obvious that records are being developed when managers perform their jobs. The process of requesting MSDS generates additional records that OSHA inspectors wish to see. These documents are evidence of good faith in carrying out an employer's responsibilities under the HCS. In addition, they leave an audit trail for in-house verifications that compliance programs are effective.

In developing records under the HCS, it becomes important to consider what information must be kept. With regard to hazard determinations, chemical manufacturers should document the decision-making process which they follow to prepare MSDS and labels. The record should reflect the specific sources of information considered, the issues related to selection of hazard warning statements, and considerations of normal use or foreseeable emergency.

The importance of these documents is that they establish both the employer's good faith efforts to evaluate the hazard as well as the rationale for selecting the particular content and wording of labels and MSDS. In an inspection situation, the compliance officer is at a distinct disadvantage because he does not have reference sources available. Where the procedures and decisions are adequately documented, he is not in a position to question them unless he is willing to spend a significant amount of time researching the issues. Moreover, the judgment of hazard is the employer's, which, of course, can be subject to review by OSHA, but the burden is then on OSHA to demonstrate that the employer's determination is incorrect.

9.0 Enforcement of OSHA Recordkeeping Requirements

Section 17(g) of the Act imposes criminal sanctions for a person or persons who knowingly make false recordings. With respect to criminal law enforcement under the Act, it suffices here to state that misdemeanor penalties are provided for under §17(g), and OSHA has not often enforced the provision.

The more common problem encountered in OSHA enforcement is that the agency usually starts its inspection by requesting and reviewing the records an employer is required to keep. The inspector will also review any particular compliance plans, e.g., the written hazard communication plan, which apply to the workplace. The purpose of this review is to determine where and what will be inspected in the employer's establishment.

Recordkeeping errors, or violations as the agency describes them, play an important role in OSHA's egregious enforcement policy. The policy is set forth in OSHA Instruction CPL 2.80 issued October 1, 1990, entitled, *Handling Cases To Be Proposed For Violation-By-Violation Penalties*. The background statement for the Instruction states that OSHA historically had, in certain cases, proposed separate penalties for each instance of an alleged recordkeeping violation. The egregious policy expands this practice add instance-by-instance penalties for violations of the general duty clause and OSHA's chemical and safety standards. While OSHA and the Review Commission historically have "grouped" instances of violations for penalty purposes, there is nothing in the Act which prevents the imposition of instance-by-instance sanctions or penalties.

This means, for example, that each instance of an alleged failure to record injuries or illnesses on the OSHA Log will be treated as a separate violation and penalized accordingly. Given the fact that penalties of up to $7,000 can be assessed for each "other than serious" violation and up to $70,000 can be assessed for each repeated and willful violation, the egregious policy can and does result in very high penalty proposals. OSHA uses this policy in cases where there is some evidence that the employer is in willful violation of the Act.[19]

10.0 Access to Records

There is a single provision at 29 C.F.R. § 1910.1020 that addresses retention periods and authorized access to medical records. The records access regulation requires employers to inform employees of their right of access at least annually (1910.20(g)), to maintain exposure records for thirty (30) years, and to maintain medical records for the duration of employment and for 30 years thereafter (1910.20(d)). It has a number of paragraphs for employee access as well as access by the Secretary to the retained records. The regulation provides for the transfer of records to the National Institute for Occupational Safety and Health (NIOSH) in the event the employee intends to dispose of them or ceases to do business.

Medical records are defined to exempt health insurance claims records, first aid records of one-time treatment and subsequent observation of cases not involving medical treatment, and to records of employees employed for less than one year who are given the records on termination of employment. The exemption for first aid records applies essentially to those cases which would otherwise not be recordable on the OSHA Log under Part 1904. The employer should be aware that §1904.7 also provides employees and their representatives with a right to access to illness and injury records.

In general, the records access regulations require employers to maintain exposure and employee medical records which they have developed regardless of whether

---

[19] OSHA's use of the egregious policy can be examined in detail in *Caterpillar,* 1987-1990 CCH OSHD 28,289 (No. 87-0922, 1988), OSHRC No. 87-0922, *Kohler Co.,* 1987-1990 OSHD 28,922 (No. 88-237, 1990), OSHRC No. 88-237, and *Pepperidge Farms Co.,* OSHRC No. 89-265).

there is a standard in existence which requires them to generate the records. For example, employers who monitor workplaces for the presence of airborne quartz, i.e., silica, are required by the records access regulation to keep or maintain the records they generate. At present, there is no standard which requires these employers to generate records for silica. Similarly, there is no standard which requires these employers to generate medical records for employees who may be exposed to airborne silica. If, however, these employers have chest X-ray pictures made of their employees or if they have the employees tested for lung volume and capacity, they voluntarily generate medical records which must be maintained under the regulation.

The hazard communication standard requires that material safety data sheets be made available in the affected employee's work area, and they have to be made promptly accessible. It should also be noted at this juncture that material safety data sheets under the hazard communication standard are treated as employee exposure records for purposes of the records access regulation.

OSHA usually addresses employee access in the standards as they are promulgated, and these provisions typically refer back to the records access regulation of §1910.1020. Paragraph (e) of §1910.1020 establishes the right of employees and their designated representatives to obtain access to and copies of medical and exposure records. Employers may not charge for initial copies of records, and must provide access within fifteen working days of a request. Employee medical records are subject to a provision that can limit access by the employee if, in the opinion of an employer's physician, there is information regarding a terminal illness or psychiatric condition that could be detrimental to the employee's health. In such cases, the employer may provide the information to another physician of the employee's choosing after denying the employee access in writing to the detrimental information. Confidential information identifying persons who have provided information about the employee may be excised from the record provided to the employee.

Section 1910.1020 provides for access by employees, whether currently employed, formerly employed, or who may be prospectively exposed, to copies of their exposure and medical records. The word "exposure" in this sense means exposure to substances, agents and other causes of disease or illnesses.

Labor unions, relatives and even attorneys may under certain circumstances have access to employee exposure and medical records. In general, to the extent that unions are the collective bargaining representative for affected employees, they have a right of access both under the regulation and under the National Labor Relations

Act (NLRA).[20] The right of access is virtually absolute insofar as exposure records are concerned. It is qualified with respect to employee medical records.

Under the NLRA decisions, representatives of employees can have access to employee medical records to the extent that markers which would identify individual employees are deleted. Under the regulation, unions and other representatives of employees may have access to medical records if they have a signed authorization from each employee whose records they desire to access.[21]

The various regulations and standards specify differing periods for which employers are to retain the records they generate under the Act. Part 1904 requires that the Log and supplementary information be retained for a period of five years following the year in which the log is generated. As noted, the medical records access regulation states 30 year retention requirements, and this is the standard period specified by most of OSHA's standards. Indeed, the Agency responded to objections filed in the bloodborne pathogen proceeding that 30 years is too long by saying that this is OSHA's standard practice.

OSHA may have access on request to medical records. Regulations found at 29 C.F.R. 1913.10 define agency practice and procedure for gaining access. The rules, issued to satisfy privacy concerns, limit the types of requests that can be made, the agency personnel who may be granted authority to access the records, and limits the uses to which the records are put. OSHA, upon showing an order issued pursuant to the requirements of 29 C.F.R. Part 1913, may have access to employee medical records. There is authority for the additional proposition that the Secretary of Health and Human Services may obtain employee medical records for research purposes.[22] The *Westinghouse* decision recognizes a limited employer right of privacy in such records but that right is in the nature of a guardian's right to temporarily protect the employee's right of privacy, but it is the employee who owns right of privacy to the records. As the court also indicated in *Westinghouse,* the employee's right of privacy may have to give way to the government's interest in protection of the public health.

While the government does have a right of access, it should also be noted that the employer has a right of privacy which it can exert with respect to records it is required to maintain.[23] In general, this means the employer can require the government to present a subpoena or warrant before providing access to records. The

---

[20] See 29 C.F.R. § 1904.7(b); 29 C.F.R. § 102.117.
[21] See, e.g., Secretary v. Wyman-Gordon Co., 15 BNA OSHC 1433 (OSHRC 1991).
[22] See United States v. Westinghouse Electric Corp., 538 F.2d 570 (3rd Cir. 1980).
[23] See McLaughlin v. Kings island, Div. of Taft Broadcasting Co., 849 F.2d 995 (6th Cir. 1988), affirming a decision of the Review Commission, 13 BNA OSHC 1137.

employer may want to assert this right if the records contain information outside the scope of the Act, are privileged, or if necessary to assert privacy interests. However, the Commission has upheld a warrantless intrusion into logs on the basis that the employer has no right of privacy as to such records.[24]

11.0 Conclusion

As OSHA compliance becomes more complex, increasing numbers of standards require the completion and maintenance of implicit as well as explicitly required records. The challenge for employers today is to assure that the records are being kept and that they are accurate, and it is by no means an easy task. Records of activities and programs are always a double-edged sword. On the one hand, they document the reasonable and prudent actions of company officials and employees in conducting the business lawfully. They also often document the failure of company personnel to perform specific tasks, or of the corporate organization to respond adequately to problems. This tension between the utility of records and the risks inherent in keeping them may create a significant dilemma for responsible corporate citizens. Nevertheless, the maintenance of good records is, on balance, positive.

---

[24] *Secretary v. Monfort of Colorado Inc.*, 14 BNA OSHC 2055 (OSHRC 1991).

Chapter 6

Employers' and Employees' Rights

Stanley M. Spracker, Esq.
John B. O'Loughlin, Jr., Esq.
Weil, Gotshal & Manges, LLP
Washington, D.C.

1.0 Overview

The OSH Act and its implementing regulations vest employers and employees with a wide range of rights and protections. The purpose of this chapter is to provide a concise summary of those rights. The chapters throughout this Handbook contain descriptions of employers' and employees' rights in the context of the procedural or substantive issues addressed within each chapter. For example, Chapter 10 contains a discussion of employers' and employees' rights during an inspection, while Chapters 11 and 13 discuss in more detail the due process rights of employers to challenge an enforcement action.

As will be seen, some of what have become known as rights of employees are not expressly set forth in the OSH Act or implementing regulations as "rights" per se, but instead are affirmative regulatory obligations imposed on employers. Consequently, the OSH Act does not allow employees to sue their employers for failing to meet those obligations.

The following discussion does not address certain workplace rights and protections provided by other statutes or common law legal theories, such as protection against unfair labor practices or job discrimination based on race, gender, age, religion, national origin, or other unlawful grounds.

2.0 Employers' Rights

2.1 Inspections and Warrants[1]

OSHA inspections of workplaces are the primary mechanism for enforcing safety standards and the general duty clause and identifying imminent hazards. Section 8(a) of the OSH Act provides OSHA's authority to enter and inspect worksites during reasonable times. Section 8(a) has been interpreted as requiring either a warrant or the employer's consent to the inspection.[2] Warrants can be broad or narrow in scope, and the employer may insist that the inspection of its premises be limited to those areas or work practices specified in the warrant. Similarly, the employer has the right to place conditions on its voluntary consent to an inspection. The employer may insist on a warrant at any time, even after it has consented initially to a search. Although employers do not have a right to participate in warrant proceedings before a federal magistrate, an employer can challenge the validity of a warrant during a subsequent enforcement proceeding. An employer can face contempt proceedings and fines for failing to permit an inspection pursuant to a warrant.

Under limited circumstances, OSHA has the authority to enter and inspect a worksite without obtaining a warrant or the employer's express consent. For example, the worksite may be controlled by a third party who may consent to the inspection. In addition, because the standard for consent to administrative searches is less stringent than that required for consent to a criminal search, courts have held that the mere failure of the employer to object constituted consent.[3] Finally, inspectors may gather information about hazardous conditions that are in plain view even if the purpose of their otherwise lawful presence is unrelated to the alleged violation they observe.[4]

If an inspection is scheduled in response to a complaint rather than as a matter of routine oversight, employers are entitled to receive a copy of the complaint submitted by an employee or employee representative. If an inspection reveals that there is an imminent hazard, the inspector must notify the employer immediately and request that the employer remove its employees or immediately abate the hazard. Inspectors do not, however, have the authority to shut down a facility or

---

[1] *See* Chapters 10 and 14 of this text.
[2] *Marshall v. Barlow's Inc.*, 436 U.S. 307 (1978).
[3] *United States v. Thriftmarts Inc.*, 429 F.2d 1006 (9th Cir. 1970).
[4] *Lake Butler Apparel Co. v. Secretary of Labor*, 519 F.2d 84, 88 (5th Cir. 1975).

order an employer to abate a hazard. If the employer does not abate a hazard voluntarily, the inspector must institute an imminent hazard proceeding and seek a court order to compel the employer to abate the hazard or shut down the facility. The employer has the right to appear and be heard if OSHA seeks a temporary restraining order in a court proceeding.

Even if an employer immediately abates an imminent hazard, the inspector must nevertheless issue a citation and penalty.[5] In addition, if an inspector observes violations that do not qualify as imminent hazards, the inspector usually will issue a citation shortly after the inspection. As discussed below, employers have the right to challenge all citations.

2.2 Challenging Citations and Civil Penalties[6]

Employers may challenge any adverse citation, civil penalty, or abatement order by filing a Notice of Contest (NOC) with the Commission within 15 business days of receipt of the violation. The filing of an NOC by an employer begins the formal process leading to a full evidentiary hearing before an administrative law judge (ALJ). During the trial, employers enjoy many procedural and substantive protections designed to preserve their right to due process. In addition, employers may seek a discretionary review by the Commission of an adverse ALJ decision. Employers also may submit a Petition for Modification of Abatement (PMA) if the required abatement cannot be completed on time due to factors beyond the control of the employer or if abatement would cause significant financial hardship to the employer.[7]

Employers may enter into settlement negotiations with the OSHA area director to reclassify the seriousness of an alleged violation (*e.g.*, from serious to non-serious) or reduce the amount of the penalty. Settlements are contingent upon the employer correcting the violation voluntarily. As a contingency, employers are advised for two reasons to file an NOC even if settlement negotiations are already underway. First, the requirement to file the NOC within 15 days is not tolled during negotiations, so the failure to file can result in the citation becoming a final order not subject to review by the Commission if the matter is not settled. Second, the filing of an NOC delays the start of the abatement period until the Commission issues a final

---

[5] OSHA is not required to initiate an imminent danger proceeding if the employer voluntarily abates the hazard. The employer retains the right to challenge the citation through a Notice of Contest.

[6] *See* Chapter 11 of this text.

[7] 29 C.F.R. § 1903.14a.

order. Employers also may withdraw a challenge any time during the process. Therefore, other than administrative and legal costs, there appears to be no downside to preserving the employer's rights by filing an NOC. Similarly, if the employer files a timely PMA requesting additional time to implement the abatement, the beginning of the abatement period is tolled until the Commission issues a final order.

2.3 Judicial Review[8]

Any employer adversely affected by a final Commission order may file an appeal in the appropriate U.S. circuit court of appeal. The federal circuit courts have exclusive jurisdiction to hear appeals of final orders of the Commission. Review by a circuit court panel is generally restricted to the issues preserved for appeal and is decided on the basis of the written record of the proceedings below, the briefs submitted by the parties, and oral argument. Employers may also petition the U.S. Supreme Court for a discretionary review of an adverse decision by the court of appeals.

2.4 Participation in Rulemakings[9]

Under Section 6 of the OSH Act and the Administrative Procedure Act, employers and their trade associations must have notice of any proposed OSHA rule, an opportunity to provide comments on the proposed rule, and an opportunity to challenge a final rule in court before it goes into effect.[10] The Act also requires OSHA to hold a formal hearing if one is requested during the period prior to the promulgation of a final rule. As a practical matter, OSHA usually includes hearing dates in the *Federal Register* notice of proposed rulemaking because at least one affected party usually requests a hearing. Employers and other parties may preserve their right to participate in rulemaking hearings by filing a Notice of Intent to Appear. Once filed, the Notice gives the party the right to testify, question other witnesses, and submit post-hearing comments and briefs. A party waiving its right to testify at the hearing may still submit comments and briefs after the hearing.

An employer may challenge a final OSHA safety standard or other regulation prior to 60 days after the date the rule is published in the *Federal Register*.[11] Challenges can attack not only the substantive basis for the rule (*e.g.*, by claiming that it

---

[8] *See* Chapter 13 of this text.
[9] *See* Chapter 2 of this text.
[10] 29 U.S.C. § 655. *See also* the Administrative Procedure Act, 5 U.S.C. § 553.
[11] 29 U.S.C. § 655(f). *Associated Indus. of New York State, Inc. v. U.S. Dep't. of Labor*, 487 F.2d 342 (2d Cir. 1973).

is not technically or economically feasible), but may also focus on any procedural deficiencies in the rulemaking process (*e.g.*, by claiming that OSHA failed to provide adequate opportunity for notice and comment, or that the rulemaking record did not contain adequate substantiation for the rule).

Employers have the right to seek temporary relief from compliance with OSHA standards if they make a good faith showing that they need more time to comply. In addition, if an employer can show that alternative practices and control measures provide protection to employees that is equivalent to that provided by the standard in question, OSHA may grant the employer a permanent variance or issue a letter of interpretation that the violation will not trigger enforcement action.[12]

2.5 Protection of Trade Secrets[13]

Section 8(c) of the OSH Act requires employers to "make, keep and preserve, and make available" for inspection records prescribed in the OSHA regulations.[14] Under certain circumstances, employers can request that OSHA protect proprietary and sensitive business information they submit to the Agency. An employer can demand that OSHA obtain a subpoena or warrant before providing access to records.[15] On the other hand, access to records that employers must maintain in accordance with the OSHA regulations, such as access to employee injury logs, may not require a warrant.[16]

If an employer is obligated to submit information to the Agency, it may nevertheless designate the information as confidential, which restricts the extent to which OSHA may disseminate the information to third parties. For example, during a contest challenging a final Commission order, an employer may seek the court's protection of information provided in the discovery process so that it is not made part of the public record or otherwise subject to release upon requests submitted under the Freedom of Information Act.[17] Similarly, employers can designate confidential records submitted during an inspection and can inform the inspector that certain areas of the workplace are competition sensitive to prevent potentially harmful disclosure.[18] To prevent disclosure to third parties, the employer should

---

[12] 29 U.S.C. § 655(b)(6) and (d); 29 C.F.R. Part 1905.
[13] *See* Chapters 5 and 8 of this text.
[14] 29 U.S.C. § 657(c).
[15] *McLaughlin v. Kings Island*, 849 F.2d 991, 995 (6th Cir. 1988).
[16] *Secretary v. Monfort of Colorado, Inc.* 14 BNA OSHC 2055 (OSHRC 1991).
[17] 29 C.F.R. § 2200.52(d).
[18] 29 C.F.R. § 1903.9.

clearly designate on documents and other records that they contain confidential business information.

The Hazard Communication Standard requires employers to provide safety and health information about hazardous chemicals in the workplace.[19] So long as the employer provides the required material safety data sheet (MSDS) for each hazardous chemical and the MSDS adequately describes the known risks and symptoms of exposure to a substance and the proper medical or emergency response methods, an employer does not have the authority to violate the confidentiality of chemical identities that are justifiably withheld by their suppliers. Notably, this provision is protective not of the employer's rights but of its suppliers' trade secrets.

3.0 Employees' Rights

Employees enjoy significant rights under the OSH Act and the implementing regulations. These include the right of employees to complain about health and safety conditions at the worksite, the right to remove themselves from hazardous situations, the right against retaliation for complaining or refusing to work, the right to have an employee representative accompany OSHA personnel during workplace inspections, and the right to contest the time permitted for an employer to abate a violation. As mentioned at the outset, the so-called employee "right" to obtain access to workplace information is more accurately characterized as an affirmative regulatory obligation imposed on employers.

3.1 Complaints[20]

Employees have the right to file complaints about worksite conditions or practices that they believe violate OSHA regulations or pose an imminent danger to their health and safety. Although at one time OSHA distinguished between formal and informal complaints, current practice requires only that the employee or employee representative give reasonably specific notice of the alleged violation and request an inspection by the Area Director or a Compliance Safety and Health Officer.[21] In addition, an employee may, prior to or during a workplace inspection by OSHA personnel, notify the inspector in writing of violations the employee believes to exist in the worksite.[22] All complaints must be signed by the employee. OSHA must

---

[19] 29 C.F.R. § 1900.1200.
[20] *See* Chapter 7 of this text.
[21] 29 U.S.C. § 657(f)(1); 29 C.F.R. § 1903.11(a).
[22] 29 U.S.C. § 657(f)(2); 29 C.F.R. § 1903.11(c).

provide a copy of the complaint to the employer no later than at the time of inspection, but OSHA must withhold the identity of the employee filing the complaint if so requested. Employees also may seek a writ of mandamus if they believe the Secretary of Labor has wrongly declined to seek a temporary restraining order to abate an imminent hazard, although this is a rarely—if ever—used provision.[23]

3.2 Refusal to Work[24]

Employees have the right to refuse to perform a task if they have a good faith basis for believing they would be exposed to an imminent hazard.[25] Although this right is not expressly set forth in the OSH Act, the Supreme Court has upheld the OSHA regulation establishing this important right.[26] The right of self-help is particularly important because employees are powerless to force an employer to abate an imminent hazard and, under the federal program, OSHA must obtain a court order to compel an employer to abate an imminent hazard. Before refusing to perform a task, however, the employee should ask the employer to eliminate the hazard. If a reasonable person would agree that there is a real risk of serious injury or death, if there is insufficient time for the employee to seek an OSHA inspection, and if the employer fails to address the problem, the employee will have a good faith basis for refusing to perform a task.[27] This right does not, however, give employees the unlimited right to walk off the job, and it does not require an employer to pay employees for time not worked.

3.3 Protection from Discrimination[28]

Employers may not discriminate against employees who have complained or testified about workplace conditions or who in good faith refuse to perform a task because of an imminent danger. Section 11(c)(1) of the OSH Act provides:

> No personal shall discharge or in any manner discriminate against any employee because such employee has filed any complaint or instituted or caused to be instituted any proceeding under or related to this Act or has testified or is

---

[23] 29 U.S.C. § 662(d).
[24] *See* Chapter 7 of this text.
[25] 29 C.F.R. § 1977.12(b)(2).
[26] *Whirlpool Corp. v. Marshall*, 445 U.S. 1 (1980).
[27] 29 C.F.R. § 1977.12.
[28] *See* Chapter 7 of this text.

about to testify in any such proceeding or because of the exercise by such employee on behalf of himself or others of any right afforded by this Act.[29]

Conduct short of termination can qualify for discriminatory conduct. For example, suspension, time off without pay, demotion, or reduction in responsibility or pay can constitute discrimination.[30] Section 11 does not create a private right for employees to take legal action against their employers. Instead, employees who believe that they have been unfairly disciplined for asserting their rights under the Act (*e.g.*, by refusing to work, requesting an inspection, participating in inspections or enforcement) must file a complaint with the Secretary of Labor within 30 days of the alleged violation.

3.4 Participation in Inspections and Enforcement[31]

Employees have a right to participate in an inspection.[32] OSHA inspectors routinely include employees or their representatives in the opening and closing conferences with management as well as during the walk-through inspection of the facility. An employee representative may be an employee selected by fellow employees, a union official, industrial hygienist, safety engineer, or other consultant designated by employees to represent their interests. The inspector has the right to limit the number of participants to prevent the inspection party from becoming unwieldy. OSHA encourages its inspectors to interview individual employees as part of the inspection, and employees may initiate contact with inspectors and may volunteer to meet with inspectors in confidence outside of the workplace. Employers are not required to pay employees for time they spend participating in inspections.[33]

In addition to the right of employees to request an inspection leading to possible enforcement actions, employees become aware of their opportunity to participate in enforcement proceedings because employers are required to post citations at a conspicuous place in the worksite. During the 15-day period following the receipt of a citation by the employer, employees or their representatives may request an informal conference with the OSHA area director to discuss the citation, penalty, or abatement, and OSHA has the authority to adjust the citation as a result of the

---

[29] 29 U.S.C. § 660(c)(1); 29 C.F.R. § 1903.11(d).
[30] *Marshall v. Firestone Tire & Rubber Co.*, 8 BNA OSHC 1637 (D.C. Ill. 1980).
[31] *See* Chapter 10 of this text.
[32] 29 U.S.C. § 657(a)(2), (e), and (f)(2).
[33] *Chambers of Commerce v. OSHA*, 636 F.2d 464 (D.C. Cir. 1980).

informal conference. Employees may also testify in court during an imminent hazard proceeding.[34]

Section 11(a) of the OSH Act authorizes employees and employee representatives to challenge formally the time permitted for an employer to abate a violation or unsafe workplace condition.[35] The Act does not, however, grant employees a right to directly challenge a citation or penalty.[36] Instead, employees can elect to intervene in the employer's notice of contest (NOC). Employers must post the NOC in the same location as the citation to provide employees or their representatives the opportunity to become a party to the NOC proceedings. By electing to become a party to the NOC proceeding, the employee or employee representative has the right to introduce evidence and cross-examine witnesses called by the employer or OSHA. Once an employee or employee representative elects party status before the Commission, he or she can also appeal final decisions of the Commission to an appropriate U.S. circuit court of appeals and seek discretionary review by the Supreme Court of an adverse circuit court decision. Even if they have not elected party status, employees must be provided written notice of a proposed settlement so that they can file objections with the ALJ within 10 days after the proposed settlement has been reached. Employees do not have the authority to continue an enforcement action if OSHA withdraws a citation.[37] Finally, if an employer seeks a variance from compliance with an OSHA standard, employees and their representatives must be given advance notice and an opportunity to participate in a hearing.

3.5 Access to Information

Employees have a number of ways to learn of workplace hazards. Section 8(c) of the OSH Act and its implementing regulations require employers to keep and maintain a variety of records, including records of workplace injuries and fatalities.[38] Any current or former employee, or their designated representative must be provided access to these records.[39] In addition, employers are required to post an annual summary of workplace injuries for the previous year in a conspicuous place accessi-

---

[34] 29 C.F.R. § 660(c)(1).
[35] 29 U.S.C. § 659(c); 29 C.F.R. §§ 2200.20-22.
[36] See, Marshall v. Sun Petroleum Prods. Co., 622 F.2d 1176 (3d Cir.), cert. denied, 449 U.S. 1061 (1980).
[37] Cuyahoga Valley Ry. v. United Transp. Union, 474 U.S. 3 (1985).
[38] 29 C.F.R. Part 1904.
[39] 29 C.F.R. § 1904.7(b).

ble to employees.[40] If an employer maintains medical records of employees, the regulations require the employer to provide access to the affected employee or his or her representative upon the employee's instruction.[41]

Section 8(c)(3) of the OSH Act requires employers to allow employees or their representatives to observe the sampling or monitoring of hazards in the workplace. Some OSHA standards include express sampling provisions.[42]

The Hazard Communication Standard requires employers to provide employees with information about the substances to which employees are exposed on the job. This rule requires the employer to prepare a written hazard communication program and maintain material safety data sheets (MSDS) for each chemical in the workplace. These materials must be accessible to employees at the worksite.

4.0 Conclusion

The OSH Act and implementing regulations create a number of rights for employers and employees alike. Both employers and employees may participate in rulemakings and in the inspection and enforcement processes. The rights of employees are somewhat narrower during enforcement, however, because the obligations and restrictions created by the OSH Act fall primarily on employers, and the notion of due process requires OSHA to afford employers an opportunity to defend themselves when charged with a violation. The regulations also create a number of obligations for employers to provide workplace health and safety information to employees and the government. Strictly speaking, these are not "rights" of employees since only OSHA can punish violating employers. Nevertheless, the increasingly widespread notion of the employee's "right to know" has made such obligations an undeniable expectation of workers in most industries.

---

[40] 29 C.F.R. § 1904.5.
[41] 29 C.F.R. § 1900.1020.
[42] *See, e.g.*, 29 C.F.R. § 1910.95(f)(employees permitted to observe taking of noise measurements).

Chapter 7

Refusal to Work and Whistleblower Protection

Frank F. Murtha, Esq.
William J. Rodgers, Esq.
Collier Shannon Scott, PLLC
Washington, D.C.

1.0 Overview

This chapter will discuss the nature and extent of protection afforded by the law to employees who either refuse to perform an assigned task on the grounds that it presents a danger to safety or health,[1] or who register a complaint, i.e., "blow the whistle" concerning a hazardous condition. While there is some overlap, in an effort to minimize confusion, refusal to work and whistleblowing will be treated as separate topics.

Any such protection against retaliation by employers represents an exception to the doctrine of employment at will, which, for over a hundred years, has governed the fundamental relationship between employers and employees in the United States. Under this doctrine, in the absence of a contract of employment for a fixed term, an employee is free to leave his employment at any time, and the employer is free to discharge an employee at any time for any reason, good or bad, or indeed for no reason at all.

While the doctrine of employment at will is still the foundation of the employer-employee relationship, it has undergone continuing erosion not only by state and federal statute but also by court decisions that have found or created exceptions under various common law theories. With respect to union represented employees

---

[1] Unless stated to the contrary, references to "safety" encompass "health."

collective bargaining agreements usually provide that an employee can only be terminated for cause.

2.0 Refusal to Work

2.1 Federal Statutes

There are a number of federal statutes that protect an employee, under certain circumstances, from being discharged or disciplined for refusing to perform an assigned task.[2] Our discussion will be limited to the protection afforded under the Occupational Safety and Health Act ("OSH Act")[3] and the National Labor Relations Act ("NLRA"), as amended.[4]

2.1.1 Occupational Safety and Health Act

While the OSH Act does not expressly confer upon employees a right to refuse to work, Section 11(c) of the Act does prohibit discrimination against an employee because he has filed a complaint, taken part in any legal proceeding brought under the OSH Act or exercised "on behalf of himself or others...any right afforded" by the Act.[5]

The Secretary of Labor ("Secretary") has found in the language of Section 11(c) a right to refuse work if certain criteria are met. These criteria are set forth in a Regulation found at 29 CFR § 1977.12. Under this Regulation, before an employee will be protected from discipline for refusing to perform work it must be found that such refusal was made in good faith and not for some ulterior purpose; the alleged dangerous condition which precipitated the refusal is such that a reasonable person when faced with the same situation would conclude that there is a real danger of death or serious injury; there was not sufficient time to deal with the hazard through the use of the ordinary enforcement mechanisms provided by the Act, and, "where possible," the employee has first tried to get his employer to eliminate or correct the perceived dangerous condition.

---

[2] *See, e.g.*, Surface Transportation Assistance Act, 49 U.S.C. § 31105(b) (commercial driver may refuse to operate vehicle because of reasonable apprehension of serious injury to self or to the public); Federal Railroad Safety Act, 49 U.S.C. §§ 20101, 20109 (protects an employee's good-faith refusal to work based on a reasonable apprehension of death or serious injury if certain criteria are met).

[3] 29 U.S.C. § 651 *et seq.*

[4] 29 U.S.C. § 141, *et seq.*,

[5] 29 U.S.C. § 660(c).

The leading case supporting an employee's right under the OSH Act to refuse to perform hazardous work is *Whirlpool Corp. v. Marshall*, 445 U.S. 1 (1980). In *Whirlpool*, two employees refused to work on wire mesh screening suspended high above the ground. Several employees had previously fallen through the screening, and less than two weeks earlier another employee had been killed when he fell through the screening. The employer issued discipline to the two employees for their refusal to work. While upholding the validity of the Secretary's Regulation, the Supreme Court noted that the self-help remedy of refusing to work was extraordinary and should be available only when the criteria contained in the Regulation were met. Moreover, since the OSH Act was not intended to authorize "strikes with pay," an employer has a right to assign an employee to alternative, non-hazardous work and to discipline the employee if he refuses to perform it.

Recognizing the potential for abuse, courts that have had occasion to confront the issue subsequent to Whirlpool have generally required some objective evidence that the assigned task posed a real danger of death or serious injury before finding that an employee's apprehensions were reasonable.[6]

2.1.1.1 Enforcing Rights under the OSH Act

There is no private right of action for discrimination under the OSH Act. Only the Secretary of Labor can bring an action to enforce an employee's Section 11(c) right to refuse to perform hazardous work.[7]

An employee who believes he has been disciplined in violation of the OSH Act must file a complaint with the Secretary within 30 days of the alleged violation.[8] However, this period may be extended or "tolled" "on recognized equitable principles or because of strongly extenuating circumstances.[9] While the Secretary is supposed to notify the complaining employee of his determination within 90 days,[10]

---

[6] *See, e.g., Marshall v. Natl. Indus. Constructors*, 8 OSHC 1117 (D.Neb. 1980). (There was no objective evidence of danger sufficient to justify refusal to work).

[7] *See, e.g., Ellis v. Chase Communications, Inc.*, 63 F.3d 473 (6th Cir. 1995).

[8] 29 U.S.C. § 660(c)(2).

[9] 29 C.F.R. § 1977.15(d)(3). The Regulation cites as an example of such an extenuating circumstances an employer's misleading of an employee regarding the grounds for discharge. *See, e.g., Donovan v. Hahner, Foreman & Harness, Inc.*, 11 OSHC 1081 (D. Kan. 1982), *aff'd* 736 F.2d 1421 (10 Cir.1984) holding that the 30-day period was tolled because the employer led the employee to believe he was laid off, when in fact he had been fired. Reversing the Secretary's prior interpretation, the Regulation expressly states that pendency of a grievance proceeding under a collective bargaining agreement or a complaint before another agency does *not* justify tolling the 30-day period.

[10] 29 U.S.C. § 660(c)(3).

this deadline is not mandatory and the Secretary often takes longer in deciding whether to institute litigation[11]

2.1.1.2 Secretary's Burden in Litigation

The Secretary has the burden of proving that an employer violated Section 11(c). To be successful, the Secretary must produce evidence that the employee had a reasonable fear of serious injury or death, that there was not sufficient time to eliminate the danger by using the regular statutory enforcement mechanisms and that, circumstances permitting, the employee tried unsuccessfully to have the employer correct the dangerous condition[12]

2.1.1.3 Shifting Burden Analysis

In determining whether a violation has been established, the courts apply a "shifting burden" analysis, adopted from Title VII discrimination cases.[13] Under this analysis, as adapted for OSH Act, Section 11(c) cases, the Secretary must first demonstrate that the employee engaged in protected activity, and that as a result of doing so, the employee suffered an "adverse employment action." Once the Secretary has made this showing, the burden shifts to the employer to articulate a non-discriminatory reason for the adverse employment action, such as, for example, excessive absenteeism or running bad product.

If the employer articulates a non-discriminatory reason, the burden shifts back to the Secretary to prove that this reason is a pretext for discrimination.

2.1.1.4 Remedies

The OSH Act provides that where an employee has been discriminated against in violation of Section 11(c), the court may restrain all such violations and "order all appropriate relief including rehiring or reinstatement...with back pay."[14] At least one court has interpreted this language to authorize punitive damages.[15]

---

[11] *See, e.g., Donovan v. Freeway Construction Co.*, 551 F. Supp. 869 (D. R.I. 1982).

[12] *See, Secretary of Labor v. H.M.S. Direct Mail*, 752 F. Supp. 573 (W.D. N.Y. 1990), affirmed in pertinent part, 936 F.2d 108 (2d Cir. 1991), holding that it is not enough for the Secretary to show that the employee's refusal to operate a dangerous machine was a reason for his discharge, but rather that the employee would not have been discharged "but for" his refusal.

[13] *See, e.g., Reich v. Hoy Shoe Co.*, 32 F2d 361, 16 OSHC 1937 (8th cir. 1994).

[14] 29 U.S.C. § 660(c)(2).

[15] *See Reich v. Cambridgport Air Systems, Inc.*, 26 F.3d 1187 (1st Cir. 1994), in which the First Circuit Court of Appeals sustained the District Court's award of double damages to two

(continued on the following page)

2.1.2 National Labor Relations Act [16]

Although the principal purpose of the NLRA is to encourage the resolution of workplace disputes through collective bargaining, there are two sections that provide protection, under certain circumstances, to employees who refuse to perform unsafe work.

The first protection is found in Section 7 of the Act,[17] which, in addition to authorizing self-organization and collective bargaining, provides that employees have the right to "engage in other concerted activities for the purpose of...other mutual aid or protection...."

The second protection is found in Section 502 of the Labor Management Relations Act of 1947 (LMRA),[18] which amended the NLRA. Section 502 provides among other things that the "quitting of labor by an employee or employees in good faith because of abnormally dangerous conditions for work at the place of employment [shall not] be deemed a strike under this Act."

Since most labor agreements contain a no-strike clause under which employees who cause a work interruption can be disciplined, the principal impact of Section 502 is to protect unionized employees by creating a statutory override of such collective bargaining agreements.

2.1.2.1 Protection under Section 7

While Section 7 is most often applied to protect union organizational efforts, it also prohibits an employer from disciplining employees for engaging in concerted activity to protest unsafe working conditions. The seminal case is *NLRB v. Washington Aluminum*, 370 U.S. 9 (1962), in which seven employees had walked off the job to protest extremely cold conditions in the work area. The employer fired them. The court supported the National Labor Relations Board's (NLRB) ruling that the employer had committed an unfair labor practice under Section 8(a)(1) of the NLRA by interfering with the employees' exercise of their Section 7 right to act in concert for "other mutual aid or protection." In doing so, the court held that it had "long been settled that the reasonableness of workers' decisions to

---

employees who had been discharged for complaining about safety and health problems at the company's Salisbury, Massachusetts plant.

[16] 29 U.S.C. § 151. *et seq.*
[17] 29 U.S.C. § 157.
[18] 29 U.S.C. § 143.

engage in concerted activity is irrelevant to the determination of whether a labor dispute exists or not."[19]

Accordingly, the protection afforded by Section 7 depends not on the reasonableness of the employees' refusal to work, but on whether the refusal can be considered "concerted" activity.

The question of what constitutes concerted action may differ depending on whether or not employees are represented by a union. Where employees are represented by a union, if a single employee refuses to perform work that he believes to be unsafe, even without prior consultation with other employees, this generally constitutes concerted action protected by Section 7, since it represents an attempt to enforce provisions of the labor agreement and hence is an extension of the collective bargaining process, itself a concerted activity.[20]

However, where a single, unrepresented employee refuses to perform allegedly unsafe work, unless he is clearly speaking on behalf of other employees who have authorized or at least consented to his doing so, such action is not concerted and is not protected by Section 7.[21]

In *Meyers*, an employee was discharged for, among other things, refusing to drive a truck that he claimed to be unsafe. The Board ultimately held that an essential element of "concerted activity" is its "collective" nature and that "concerted activity" is distinguishable from "mutual aid and protection." The driver's refusal was found to be for his own individual interest and was not collective in nature, hence not "concerted activity."

Since *Meyers*, a refusal to work by a single employee who is not asserting rights under a collective bargaining agreement has generally been found to be unprotected under Section 7. However, even absent union representation, where there is some evidence that the single employee is raising an issue of concern to other employees, it may satisfy the "concerted activity" requirement and thus be protected under Section 7.[22]

---

[19] 370 U.S. at 16.

[20] *NLRB v. City Disposal Systems*, 465 U.S. 822 (1984). An employee based his refusal to drive a vehicle he believed unsafe on an express provision of the labor agreement.

[21] *Meyers Industries (Meyers I)*, 268 NLRB 493, 115 LRRM 1025 (1984), *rev'd sub nom. Prill v. NLRB*, 755 F 2d 941 (DC Cir.), *cert. denied*, 474 U.S. 971 (1985) decision on remand sub nom. Meyers Industries (Meyers II), 281 NLRB 882, 123 LRRM 1137 (1986), *aff'd* 835 F.2d 1481 (DC Cir. 1987), *cert. denied* 487 U.S. 1205 (1988).

[22] *See, e.g., NLRB v. Jasper Seating Co.*, 857 F.2d 419 (7th Cir. 1988), enforcing 285 NLRB 550, 127 LRRM 1119 (1987). Two employees left their jobs to protest a cold working environment caused by an open door, which the two employees were demanding be closed. The other employees who remained on the job wanted the door kept open. The NLRB found that this

(continued on the following page)

Nor does the lack of specific authorization by the other employees necessarily cast a single employee's protest outside the protection of Section 7.[23]

2.1.2.2 Comparison of Section 7 and Section 502

There are several key differences between Section 7 and Section 502. Under Section 7, a good-faith belief that danger exists is sufficient to protect action, so long as it can be considered "concerted." Under Section 502, in addition to a good-faith belief, there must be "ascertainable objective evidence" of a danger to safety.[24] Moreover, Section 502 only applies to a refusal in the face of an "*abnormally dangerous*" condition. (Emphasis added). An ironworker whose job requires him to walk on steel beams 30 stories above the ground may not refuse to work simply because there is danger of falling. Although, if the appropriate safety equipment was not available, his refusal to work would not be a strike. Finally, the employee's refusal must be based on an "immediate or presently existing threat"[25] not some vague general safety concern.

2.1.2.3 Cooperation between OSHA and the NLRB

In order to avoid unnecessary litigation and possibly conflicting enforcement efforts, the NLRB and OSHA have agreed that when a charge of retaliation that would be covered by Section 11(c) is filed with the NLRB and a corresponding complaint is filed with OSHA, the Board will either dismiss or defer the charge. Where such a charge is filed only with the NLRB, the Board will advise the charging party of his right to file a Section 11(c) complaint with OSHA. When it is determined the charge falls within the NLRB's exclusive jurisdiction but relates to discrimination based on safety and health issues, the Board and OSHA will confer.[26]

---

situation involved a continuing dispute over working conditions and fell within the scope of Section 7.

[23] *Rockwell International Corp. v. NLRB*, 814 F.2d 1530 (11th Cir. 1987), *enforcing* 278 NLRB 55, 122 LRRM 1285 (1986). An employee who raised complaints that the employer knew were shared by the other employees was engaged in activity protected by Section 7.

[24] *Gateway Coal Co. v. U.M.W.*, 414 U.S. 368 (1974).

[25] *Id.* at 386.

[26] Memorandum of Understanding between OSHA and NLRB. 40 *Fed. Reg.* 26,083-26,084.

2.1.3 Arbitration and Collective Bargaining Agreements

2.1.3.1 Collective Bargaining Agreements

Most labor agreements have provisions that require the employer to provide a safe workplace. Some agreements expressly provide that an employee may refuse to perform work that he believes to be unsafe. Usually such refusal can be made only after certain steps have been followed, e.g., requesting the union steward and supervisor to review the situation, and accepting alternative work until the dispute is resolved.

In addition, virtually all labor agreements provide that an employee can be disciplined only for cause and where the union disagrees with the employer's issuance of discipline, the matter will ultimately be resolved through arbitration.

Ordinarily, an employee can not engage in self-help by refusing to perform work. For example, an employee who believes that his seniority entitles him to a different job than the one he is being asked to perform is entitled to file a grievance under the agreement, but must perform the assigned task in the meantime. "Work now - grieve later."

In such a case, the employee who refused to work would be subject to discipline, including discharge "for cause," i.e., insubordination. But arbitrators have carved out an exception when the refusal is based on safety concerns. In deciding such cases, "arbitral reasoning appears to range from the purely subjective test of what the particular employee 'honestly' or 'sincerely' believed as to the presence of a hazard, to the 'cold facts' approach of requiring a showing of actual danger with 'real and imminent' hazard to life and limb, and all the shadings between these two extremes."[27]

It has frequently been said that the arbitrator is a "creature of the contract," and as such cannot exceed the authority that the parties have mutually agreed to give him. Generally, his authority is limited to interpreting the labor agreement. Consequently, the language of the agreement and principles of contract interpretation govern an arbitration decision.[28]

---

[27] Elkouri and Elkouri, *How Arbitration Works*, Fifth Ed. (BNA 1997), p. 979 (cases cited).

[28] *See, e.g., Barrentine v. Arkansaw Freight Sys.*, 101 S. Ct. 1437, 1446-47 (1981), a wage and hour case in which the court held that an arbitrator had no general authority to invoke public laws that conflict with agreement, but must base his decision on the intent of the parties, rather than attempt to enforce the statute.

At the same time, where statutory rights may be in question, arbitrators will sometimes find that they have the authority, granted either explicitly or by implication, to consider statutory requirements in rendering their decision.[29]

2.1.3.2 Arbitration not under a Collective Bargaining Agreement

In recent years, many employers, in an effort to limit litigation, have required employees to sign an agreement that all complaints will be resolved through arbitration as a condition of employment. Other employers have provided for voluntary arbitration as a part of an internal employee complaint procedure, often set forth in an employee handbook or policy manual.

These attempts have received a mixed reception from the courts. In a landmark decision, the Supreme Court upheld enforcement of an agreement requiring arbitration of an Age Discrimination in Employment Act (ADEA) discrimination claim.[30]

There are two issues: 1) can the employee who signed the agreement to arbitrate be foreclosed from initiating litigation to enforce a statutory right, and 2) can an agency like OSHA be precluded from exercising its enforcement powers by an employee's agreement to arbitrate. Court decisions are mixed. While most courts have found that an agreement to arbitrate is enforceable against the parties to such agreement, either on the theory of knowing, voluntary waiver of statutory rights, or promissory or equitable estoppel from asserting such rights,[31] an agreement to arbitrate can not prevent OSHA from choosing to exercise its independent investigation and enforcement authority.[32]

2.1.3.3 Referral

A related issue is the extent of deference that an agency such as OSHA or the NLRB is likely to give to an arbitration award. The NLRB has a clear policy that it will defer to a collectively bargained arbitration process in most instances.[33] The Board will also defer to an arbitrator's decision if certain conditions are met.[34] But

---

[29] *See* Elkouri and Elkouri, *supra*, pp. 969-989 for examples of arbitration decisions involving the impact of the OSH Act and the NLRA in labor arbitration.

[30] *Gilmer Interstate / Johnson Lane Corp.*, 500 U.S. 20 (1991).

[31] *See, e.g., Prudential Ins. Co. v. Lai*, 42 F.3d 1299 (9th Cir. 1999), *cert. denied*, 516 U.S. 812 (1995) and *Rosenberg v. Merrill Lynch*, 170 F.3d 1 (1st Cir. 1999).

[32] *See, e.g., Reich v. Sysco Corp.*, 870 F. Supp. 777 (S.D. Ohio 1994).

[33] *See, Collyer Insulated Wire*, 192 NLRB No. 837 (1971) and cases following.

[34] *See, Spielberg Mfg. Co.*, 112 NLRB 1080 (1955), and cases following.

in all cases the Board reserves the right to assume jurisdiction whenever it believes that the purposes of the NLRA require it.[35]

The Secretary of Labor has promulgated a regulation authorizing OSHA to postpone its own determination and defer to the decision of an arbitrator or other agency where it concludes that the rights conferred by Section 11(c) have been adequately protected.[36] However, there is some judicial authority rejecting OSHA's authority to suspend action on a complaint pending the NLRB's decision on a corresponding charge.[37]

A definitive analysis of the unsettled and evolving state of the law with respect to the affect of an agreement to arbitrate upon the enforcement of statutory rights is beyond the scope of this chapter. But the existence of an agreement to arbitrate should signal the need for informed legal guidance.

2.2 State Statutes

An employee's right not to be disciplined for refusing to perform allegedly hazardous work is protected in many state statutes or constitutions. Under some of the state laws a private right of action is permitted, but most mirror the OSH Act provision that only the authorized state official may initiate litigation.[38]

2.3 Common Law

In addition to statutory protections, many state courts recognize a tort action for wrongful discharge based on a "public policy" exception to the doctrine of employment at will. Under this exception, an employee discharged for refusing to perform illegally hazardous work may bring a civil action to get his job back and to recover compensatory and in some cases punitive damages.[39]

---

[35] Section 10(a) of the NLRA provides that the Board's authority "shall not be affected by any other means of adjustment...that has been...established by agreement, law, or otherwise. 29 U.S.C. § 160(a); *See NLRB v. C & C Plywood Corp.*, 385 U.S. 421 (1967) upholding the Board's jurisdiction in the face of an arbitration under a collective bargaining agreement.

[36] 29 C.F.R. § 1977.23

[37] *See, Newport News Shipbuilding v. Marshall*, 8 OSHC 1393 (E.D. Va. 1980).

[38] *Compare, e.g., Brevik v. Kite Painting, Inc.*, 416 N.W. 2d 714 (Minn. 1987) - (Minnesota State OSH Act creates individual cause of action for discriminatory discharge) with Silkworth v. Ryder Truck Rental, 70 Md. App. 264 (1987) - (Maryland state act tracks federal law and thus creates no private right of action)..

[39] *See, e.g., Cabesuela v. Browning-Ferris Industries of Calif. Inc.*, 68 Cal. App. 4th 101, 80 Cal. Rptr. 2d 60 (6th Dist. 1998), in which the court stated "An employee who fires an employee in retaliation for protesting unsafe working conditions violates fundamental public policy, and
(continued on the following page)

3.0 Whistleblowing [40]

As mentioned previously, there is some overlap between legal protections for employees who refuse to perform allegedly hazardous work and those who complain about unsafe conditions, either to the employer (internally) or to outside authorities (externally). In many instances the protection for refusal to work or whistleblowing emanates from the same statutory language or common law principle, e.g., See 11(c) of the OSH Act, 29 USC § 660(c), which prohibits an employer from discriminating in any way against an employee, either for filing a complaint or refusing to work.

In some instances, a statute prohibits discrimination based on either refusal to work or whistleblowing but not both. For example, § 502 of the Labor Management Relations Act, 29 USC § 143, protects only the right to refuse to work.

3.1 Federal Statutes

There are a number of federal statutes that expressly prohibit retaliation against employees who complain of hazardous conditions or illegal actions attributable to their employers.[41]

3.1.1 Occupational Safety and Health Act

The anti-retaliation provisions of the OSH Act are found in Section 11(c) discussed above in connection with refusal to work.

Although on its face, Section 11(c) appears to be limited to protecting those who file complaints with OSHA or testify in OSHA proceedings, it has been given a broad interpretation. Hence, even employees who complain only to their employ-

---

the discharged employee may bring a tort action for wrongful discharge in addition to his or her statutory remedies.

[40] For a comprehensive discussion of decisions in which courts have discussed or determined whether or under what circumstances an employer may be held liable for discriminating against employees for raising public complaints concerning health or safety, *see* 75 ALR 4th 13, Gregory G. Sarno, J.D., and *Annotation: Liability for Retaliation Against At-Will Employee for Public Complaints or Efforts Relating to Health or Safety.*

[41] *E.g.,* Surface Transportation Act, 49 U.S.C. § 31105(b) (commencing, participating or testifying in proceedings relating to violations of commercial motor vehicle safety provisions); Federal Railroad Safety Act, 49 U.S.C. § 20101 (refusal to work or reporting safety violations); Energy Reorganization Act, 32 U.S.C. § 5851(a) (commencing, participating, or testifying in proceedings under the ERA or Atomic Energy Act).

ers are protected from retaliation under Section 11(c), so long as such complaints are made in good faith.[42]

Employees are also protected under Section 11(c) from retaliation based on their having informed journalists of safety and health concerns at the workplace.[43]

3.1.1.1 Enforcement Procedure

Enforcement of Section 11(c) has already been discussed above with respect to refusal to work cases.

3.1.1.2 Preemption

In some cases, courts have ruled that a given state statute or legal action is preempted by one or more federal statutes.[44] An extended discussion of the complex question of preemption is beyond the scope of this chapter, but anyone involved in a potential state-whistleblowing case should be aware of the issue.[45]

3.2 State Statutes

Most states have statutes that protect employees who complain of hazards that threaten employees or the public in general. Some of these protect both private and

---

[42] *See, e.g., Marshall v. Klug & Smith Co.*, 7 OSHC 1162 (D.N.D. 1979); *Marshall v. Power City Elec.*, (1979) OSHD § 23,947 (E.D. Wash. 1979); *Marshall v. P & Z Co.*, 6 OSHC 1587 (D.D.C. 1978), *aff'd* 600 F.2d 280, 7 OSHC 1633 (D.C. Cir. 1979); *Marshall v. Wallace Bros., Mfg. Co.*, 7 OSHC 1022 (M.D. Pa. 1978); *Marshall v. Springville Poultry*, 445 F. Supp. 2, 5 OSHC 1761 (M.D. Pa. 1977).

[43] *See, e.g., Donovan v. R.D. Andersen Constr. Co.*, 10 OSHC 2025 (D. Kan. 1982).

[44] For example, a state wrongful termination action by a former employee was held preempted by the Energy Reorganization Act (ERA), 42 USC § 5851(a) which expressly prohibits retaliation for surfacing complaints concerning safety in a nuclear power plant. *Snow v. Bechtel Constr. Inc.*, 647 F. Supp. 1514 (CD Cal. 1986). The court held that the federal remedy created by the anti-retaliation provision in the ERA preempted the state's wrongful discharge claim.

But see, *English v. General Electric Co.*, 496 U.S. 72(1990), holding that an employee's state law action for intentional infliction of emotional distress in retaliation for whistleblowing did not fall within that area of nuclear safety that was held preempted in earlier cases and did not present a conflict with the anti-retaliation provision of the ERA.

[45] For a fuller discussion, see Gregory G. Sarno, J.D., *Annotation: Federal Pre-Emption of Whistleblower's State-Law Action for Wrongful Retaliation,* 99 ALR Fed. 775.

public employees who blow the whistle on unsafe conditions,[46] while others protect only public employees.[47]

3.3 Common Law

In addition to state statutes that protect the right of employees to complain about hazardous conditions, the courts of most states have created an exception to employment at will in the form of an action in tort[48] for violation of public policy.

The essentials necessary to establish a successful case under the public policy exception vary from state to state but are similar in nature. Most require that the employee demonstrate that his complaint relates to something that is in violation of a clearly defined public policy, often as contained in a specific statute. Thus, e.g., under Missouri law, the at-will employee must prove (1) that he raised with his employer or public authorities serious misconduct, (2) that there was a violation of the law or of well established public policy, (3) the employer discharged him and (4) "but for" his reporting of the violations, he would not have been discharged.[49] Similarly, e.g., under Kentucky law, when a discharge violates public policy "clearly defined" by statute, an employee may prevail under the public policy exception.[50]

A New Jersey court has held that an individual employee has a private right of action under the public policy exception for retaliatory discharge for reporting a violation of OSHA because, in addition to being a violation of OSHA, the discharge violated New Jersey's policy favoring safety in the workplace.[51]

In some states, a tort brought under the public policy exception is only permitted where there is no statutory remedy available.[52]

---

[46] *See, e.g.*, Mass. Ann. Laws Ch. 149, § 185; N.H. Rev. Stat. Ann. § 275-E1 to 7; N.J. Stat. Ann. § 34:19-1 - 34:19-8; Ga. Code Ann. § 34(d); Haw. Rev. Stat. Ann. § 378-61-69; Ct. Gen. Stat. Ann. § 31-51m; Ohio Rev. Code Ann. § 4113.52; Or. Rev. Stat. Ann. § 659.550.

[47] *See, e.g.*, Alaska Stat. § 39.90.100 to 39.90.150; Del. Code Ann. Tit. 29 § 5115.

[48] A tort is "a violation of a duty imposed by general law or otherwise upon a person under given circumstances." See *Black's Law Dictionary*, 6th ed., West Publishing at p. 1489."

[49] *Bell v. Dynamite Foods*, 969 SW 2d 847 (Mo. Ct. App. E.D. 1998); *Porter v. Reardon Mach. & Co.*, 962 SW 2d 932 (Mo. Ct. App.. W.D. 1998).

[50] *Hines v. Elf Atochem N.Am., Inc.*, 813 F. Supp. 550 (1993, WD Ky.).

[51] *See, e.g., Cerracchio V. Alden Leeds, Inc.*, 223 N.J. Super. 435 (1988).

[52] *See, e.g., Miles v. Martin Marietta Corp.*, 861 F. Supp 73, (DC Colo. 1994), holding that, under Colorado law, a separate public policy exception tort is not permissible where the statute that sets forth the public policy contains a remedy for retaliation.

In other states, a separate tort action based on the public policy exception is available in addition to the statutory remedy.[53]

Finally, some states do not recognize the public policy exception and only an express statutory provision will suffice to support an action for retaliatory discharge for reporting hazardous conditions.[54]

4.0 Conclusion

In light of all the possible sources of potential liability, an employer should refrain from discriminating against an employee in any manner, including imposing discipline because that employee has, in good faith, raised an issue about safety or health, unless there are unrelated circumstances that clearly justify such action against the employee.

---

[53] *See, e.g., Gutierrez v. Sundance Indian Jewelry*, 117 NM 41 (App. 1993), *cert. den.* 117 N.M. 121, holding that the state legislature did not intend the New Mexico OSH Act to provide the exclusive remedy for retaliatory discharge.

[54] *See, e.g., Dray v. New Market Poultry Products, Inc.*, 518 S.E. 2d 312 (Va. 1999), pointing out that in Virginia there is no common law tort for retaliatory discharge for whistleblowing. Consequently, an employee who was discharged for reporting unsanitary conditions at a poultry farm had no remedy other than that provided by statute.

Chapter 8

Hazard Communication and Employee Right-to-Know

Chris S. Leason, Esq.
Scott E. Kauff, Esq.
McKenna & Cuneo, L.L.P.[1]
Washington, D.C.

1.0 Overview

Congress, through the Occupational Safety and Health Act ("the Act"),[2] gave the Occupational Safety and Health Administration ("OSHA") authority "to assure so far as possible every working man and woman in the Nation safe and healthful working conditions."[3] The Act imposes very few restrictions or conditions on American workplaces; rather, it is OSHA's implementation of the statute, through its regulations, that impose requirements and restrictions designed to protect the health and safety of American workers. These standards are codified, in large part, at 29 C.F.R. Part 1900, the most notable of which is the Hazard Communication Standard (hereinafter referred to as "HAZCOM Standard," "HAZCOM," or "Standard").

---

[1] The authors appreciate the assistance of Leah Warnick, an associate at McKenna & Cuneo, in preparing this chapter.
[2] 29 U.S.C. §§ 651-678.
[3] 29 U.S.C. § 651(b).

Section 6(b)(7) of the Act requires OSHA to issue a mandatory standard, requiring employers to disclose to their employees risks inherent in handling hazardous chemicals and additional information regarding how to mitigate those risks.[4] Under this statutory authority, OSHA promulgated the HAZCOM Standard, which went into effect on November 25, 1983.[5] Often referred to as an employee right-to-know regulation, the HAZCOM Standard is designed to "inform...employees properly, and to design and implement employee protection programs...[so] that [employees] can meaningfully participate in, and support, the protective measures instituted in their workplaces...to reduce the incidence of chemical source illnesses and injuries...."[6]

This section requires chemical manufacturers or importers to assess the hazards of chemicals which they produce or import, and all employers to provide information to their employees about the hazardous chemicals to which they are exposed, by means of a hazard communication program, labels and other forms of warning, material safety data sheets, and information and training.[7]

OSHA has estimated that more than 32 million workers are exposed to 650,000 hazardous chemical products in more than 3 million American workplaces.[8]

HAZCOM regulates manufacturers and importers of hazardous chemicals; it also regulates employers whose employees come into contact with hazardous chemicals at the workplace. Manufacturers and importers are required to perform hazard determinations, described in detail below, a decision making process designed to identify which chemicals manufactured or imported impose health (e.g., irritation, sensitization, carcinogenicity) or physical (e.g., flammability, corrosion, reactivity) hazards to exposed employees.[9] The decision that a chemical is hazardous is the trigger for the bulk of the HAZCOM Standard's requirements: once a chemical is deemed hazardous under HAZCOM, a manufacturer or importer must adhere to labeling, training, and informational requirements regarding that chemical in order to equip workers with the information they need to protect themselves against the hazards posed by that chemical. The hazard determination also triggers the imposition of requirements upon employers. While

---

[4] 29 U.S.C. § 655(b)(7).
[5] 48 *Fed. Reg.* 53,280 (1983)(codified at 29 C.F.R. § 1910.1200).
[6] 48 *Fed. Reg.* 53,280 (1983)(codified at 29 C.F.R. § 1910.1200).
[7] 29 C.F.R. § 1910.1200(b)(1).
[8] <Http://www.osha-slc.gov/OshDoc/Fact_data/FSNO93-26.html>; <http://www.osha-slc.gov/SLTC/hazardcommunications/index.html>.
[9] *See* 29 C.F.R. § 1910.1200(c).

employers are not required to make independent hazard determinations and may rely on those done by the manufacturers or importers of the chemicals the employer uses, the employer must ensure that the labeling, training, and informational requirements are satisfied with regard to the hazardous chemicals to which its employees are exposed.

The HAZCOM Standard itself is not particularly detailed or comprehensive in its explanation of the requirements it imposes on American manufacturers, importers, and employers. Much of the detail regarding the requirements of the Standard is found in the Standard's Appendices, as well as OSHA directives, information sheets, and opinion letters. These documents set forth the details of HAZCOM's major requirements, including assessing the hazard, establishing a written hazard communications program, labeling, training, providing information to employees, creating and disseminating material safety data sheets ("MSDSs"), and protecting trade secrets. Another layer of complexity is brought on by state employee right to know laws and regulations. OSHA has approved programs from approximately half of the states to operate in lieu of the federal HAZCOM Standard. In addition, the possibility exists that American employers will face the requirements of international HAZCOM provisions in the near future.

The HAZCOM Standard is organized according to the steps a manufacturer, importer, distributor, or employer would take to comply with the Standard's requirements. The first sections of the regulation describe the hazard determination. The remaining sections of the Standard set forth the requirements imposed where employees will be exposed to a chemical that a manufacturer or importer has deemed hazardous. At its heart, HAZCOM regulates the dissemination of information. After a manufacturer determines that a chemical in its workplace is hazardous and not within any of the exemptions provided by the Standard, it must develop labels to be attached to containers of the chemical, create a MSDS containing more detailed information regarding the hazards presented by the chemical, organize an employee training program, and provide employees with information regarding the chemical's hazards. After an employer learns that a chemical to which its employees are exposed is hazardous, it must fulfill similar requirements. The difference is that employers may rely on the hazard determination of manufacturers and importers and use the MSDSs that the manufacturers and importers provide with the chemical. However, employers must independently provide an adequate written hazard communication program, information, and training for their employees.

In addition to the previously described regulatory requirements, the HAZCOM Standard provides protection for some trade secrets; manufacturers, importers, and employers are not necessarily required to divulge all of the information the regulation would otherwise require.

In addition to requirements imposed by the federal HAZCOM Standard, manufacturers, importers, and employers may find themselves subject to additional requirements imposed by states and perhaps, in the future, by international agreement. There have been a number of cases regarding federal preemption of state level regulations that also seek to control workplace hazards, including a recent controversy in California regarding the relationship between the HAZCOM Standard and that state's Proposition 65. Finally, in the last decade there has been an international movement to standardize information, labeling, and MSDSs regarding hazardous chemicals in workplaces worldwide. The Globally Harmonized System ("GHS") is still being negotiated among participating countries, including the United States, in the hope that internationally uniform labels and MSDSs will promote worker safety worldwide and stimulate trade.

2.0 Assessing the Hazard

The HAZCOM Standard is designed to ensure that workers who are exposed to hazardous chemicals in their workplaces have sufficient information to participate meaningfully in their employers' safety programs and to protect themselves.[10] To accomplish this goal, the HAZCOM Standard requires manufacturers, importers, and employers to provide information regarding the hazards imposed by certain chemicals that are deemed hazardous. Thus, the first step toward compliance with the HAZCOM Standard is to determine whether a chemical known to be present in the workplace is hazardous and whether employees will be exposed to it. This "hazard determination," the decision whether a chemical to which workers may be exposed imposes health or physical hazards, is the trigger for the requirements set forth in the HAZCOM Standard.

2.1 The Hazard Determination

Manufacturers and importers are required to make a hazard determination for each chemical they produce or import.[11] Initially, this requires manufacturers and importers to determine whether a chemical is hazardous.[12] Employers may initiate an independent hazard determination or may rely on that done by manufacturers or importers.[13] The HAZCOM Standard is not detailed in its description of this hazard determination; it simply mandates that chemical manufacturers, importers,

---

[10] <Http://www.osha-slc.gov/SLTC/hazardcommunications/index.html>.

[11] 29 C.F.R. § 1910.1200(d)(1).

[12] 29 C.F.R. § 1910.1200(c).

[13] 29 C.F.R. § 1910.1200(d)(1).

or employers "identify and consider the available scientific evidence concerning such hazards."[14] There are two kinds of hazards that manufacturers and importers should consider: health and physical.

OSHA defines a health hazard as "a chemical for which there is statistically significant evidence based on at least one study conducted in accordance with established scientific principles that acute or chronic health effects may occur in exposed employees."[15] Also by definition, the term health hazard includes carcinogens, toxic agents, "nephrotoxins, neurotoxins, agents which act on the hematopoietic system, and agents which damage the lungs, skin, eyes, or mucous membranes."[16] The HAZCOM Standard is not as specific in its treatment of physical hazards; the Standard merely defines such a hazard as "a chemical for which there is scientifically valid evidence that it is a combustible liquid, a compressed gas, explosive, flammable, an organic peroxide, an oxidizer, pyrophoric, unstable (reactive) or water-reactive."[17]

OSHA relies on Appendices A and B of the regulation to provide further clarification of the scope and adequacy of health hazard determinations. Appendix A provides:

> Although safety hazards related to the physical characteristics of a chemical can be objectively defined in terms of testing requirements..., health hazard definitions are less precise and more subjective. Health hazards may cause measurable changes in the body—such as decreased pulmonary function...Employees exposed to such hazards must be apprised of both the change in body function and the signs and symptoms that may occur to signal that change.[18]

In order to encourage the objective classification of health hazards, OSHA requires manufacturers and importers to classify health hazards as either chronic or acute. Rapid effects resulting from short-term exposure are "acute" (e.g. irritation, corrosivity, sensitization, lethal dose); effects of longer duration resulting from long-term exposure are "chronic" (e.g. carcinogenicity, teratogenicity, mutagenicity).[19]

---

[14] 29 C.F.R. § 1910.1200(d)(2). HAZCOM does not require testing of chemicals to be performed as part of the hazard determination. OSHA, Directive Number CPL 2-2.38D, Inspection Procedures for the Hazard Communication Standard, 29 C.F.R. §§ 1910.1200, 1915.99, 1917.28, 1918.90, 1926.59, and 1928.21, Appendix A, § (b)(3) at 24-25 (1998) (hereinafter OSHA Directive CPL 2-2.38D).

[15] 29 C.F.R. § 1910.1200(c).

[16] 29 C.F.R. § 1910.1200(c).

[17] 29 C.F.R. § 1910.1200(c).

[18] 29 C.F.R. § 1910.1200, Appendix A, "Health Hazard Definitions."

[19] 29 C.F.R. § 1910.1200, Appendix A, "Health Hazard Definitions."

OSHA lists categories of health effects in Appendix A. If a chemical falls into any of the following categories, it is automatically treated as a health hazard under the HAZCOM Standard—carcinogen, corrosive, toxic or highly toxic, irritant, sensitizer, or if it causes "target organ effects."[20] Chemicals that cause target organ effects, according to the regulation, are those that cause damage to blood, skin, eyes, major organs, or systems of the body.[21]

Appendix B outlines the principles and procedures of the hazard assessment. OSHA provides no specific methodologies or procedures for assessing hazards. OSHA simply requires that the manufacturers and importers be able to demonstrate that their hazard determinations have "adequately ascertained the hazards of the chemicals produced or imported."[22] This "performance-oriented" process relies heavily upon the professional judgment of those making the determination. The evaluator must conduct a "scientifically defensible evaluation," considering four criteria: (1) carcinogenicity (if certain organizations determine that a chemical is a carcinogen, it is conclusive evidence that the chemical is hazardous); (2) human data (epidemiological studies); (3) animal data (when human data is unavailable); and (4) adequacy and reporting of data (established scientific principles and statistically significant conclusions).[23]

Often, the HAZCOM Standard takes the hazard assessment authority out of the hands of the manufacturers and importers: the HAZCOM Standard contains several provisions that establish a baseline list of chemicals that automatically are deemed hazardous. Sections (d)(3) and (d)(4) of the regulation require chemical manufacturers and importers preparing to assess the hazards of a chemical to look first at the determination of various organizations.[24] Although the chemical manufacturer or importer making the hazard determination is still responsible for identifying the hazards imposed by chemicals on these lists, OSHA prescribes that the manufacturers and importers automatically treat chemicals as hazardous if listed in 29 C.F.R. § 1910, Subpart Z, and on the Threshold Limit Values ("TLVs") for Chemical Substances and Physical Agents in the Work Environment list published by the American Conference of Governmental Industrial Hygienists ("ACGIH").[25] In addition, OSHA also requires manufacturers and importers to treat the following sources regarding the carcinogenicity of chemicals in making their hazard determi-

---

[20] 29 C.F.R. § 1910.1200, Appendix A, §§ (1)(a) through (7)(h), "Health Hazard Definitions."
[21] 29 C.F.R. § 1910.1200, Appendix A, §§ (7)(a) through (7)(h), "Health Hazard Definitions."
[22] 29 C.F.R. § 1910.1200, Appendix B, "Hazard Determination."
[23] 29 C.F.R. § 1910.1200, Appendix B, "Hazard Determination."
[24] 29 C.F.R. §§ 1910.1200(d)(3) and (d)(4).
[25] 29 C.F.R. §§ 1910.1200(d)(3)(i)-(ii).

nations as conclusive: the National Toxicology Program's ("NTP's") latest edition of its Annual Report on Carcinogens; the International Agency for Research on Cancer's ("IARC") latest edition monographs; and OSHA's regulations at 29 C.F.R. Part 1910, Subpart Z, Toxic and Hazardous Substances.[26]

2.2 Employee Exposure

The second factor to consider when determining whether the HAZCOM Standard applies to a certain chemical is whether employees will be exposed to the chemical. Manufacturers, importers, and employers "must assess the hazards associated with the chemicals including hazards related to any anticipated or known use which may result in worker exposure."[27] This assessment of exposure must also consider employee exposure to intermediates, by-products, and decomposition products" of the chemical.[28] If a chemical is deemed hazardous, any employee exposure is sufficient to trigger the requirements of HAZCOM, which applies to "any chemical which is known to be present in the workplace in such a manner that employees may be exposed under normal conditions of use or in a foreseeable emergency."[29] OSHA's examples of foreseeable emergencies include "equipment failure, rupture of containers, [and the] failure of control equipment which could result in an uncontrolled release of a hazardous chemical into the workplace."[30] Therefore, any employee exposure, with the exception of unforeseeable emergencies, is sufficient to trigger the HAZCOM Standard. If an employee is subjected, through any route of entry, to the potential for exposure, accidental or otherwise, such potential exposure is sufficient to trigger the regulation.[31] If there is no potential for exposure, for example, if the chemical is inextricably bound and cannot be released, the HAZCOM Standard is not triggered, even if the chemical is hazardous.[32]

2.3 Mixtures

Manufacturers, importers, and employers often use hazardous chemicals in mixtures. In the HAZCOM Standard, OSHA defines mixtures as combinations of two

---

[26] *See* 29 C.F.R. § 1910.1200(d)(4).
[27] OSHA Directive CPL 2-2.38D, Appendix A, § (d)(1), at 29.
[28] OSHA Directive CPL 2-2.38D, Appendix A, § (d)(1), at 29.
[29] 29 C.F.R. § 1910.1200(b)(2).
[30] 29 C.F.R. § 1910.1200(c).
[31] 29 C.F.R. § 1910.1200(c); OSHA Directive CPL 2-2.38D, Appendix A, § (c), at 28.
[32] 29 C.F.R. § 1910.1200, Appendix E, "Guidelines for Employer Compliance"; OSHA, Directive CPL 2-2.38D, Appendix A, § (d)(2), at 30.

or more chemicals, not the result of a chemical reaction.[33] "A hazardous chemical is considered a mixture if the components retain their chemical identity after being combined."[34] Mixtures pose special problems for manufacturers and importers conducting a hazard determination. For example, manufacturers often inquire as to whether a mixture is automatically deemed hazardous if one of its components is a hazardous chemical. OSHA provided special provisions in Subsection (d)(5) of the HAZCOM Standard to guide manufacturers and importers, as well as employers, in making hazard determinations for mixtures.

There are two ways to test a mixture to determine whether it presents a physical or health hazard to exposed employees. First, the mixture could be tested as a whole; if tested as a whole, the results of this testing should be used to assess whether the chemical is hazardous.[35] If the mixture has not been tested as a whole but, rather, only its components have been tested, the hazard determination will rely on the test results of the components.

If a mixture has not been tested as a whole to determine whether the mixture is a health hazard, the mixture shall be assumed to present the same health hazards as do the components which comprise one percent (by weight or volume) or greater of the mixture...[36]

The one exception to the above rule occurs when any component comprising 0.1 percent concentration or greater of the mixture is considered a carcinogen under the Standard; in that case, OSHA will treat the mixture as presenting a carcinogenic, health hazard.[37] Thus, any mixture containing 0.1 percent or greater of a carcinogenic component, will be treated as a hazardous chemical under the Standard.

Similarly, if a mixture has not been tested as a whole, its hazard determination for physical hazards will be determined by the physical hazards imposed by its components. In such a case, "the chemical manufacturer, importer, or employer may use whatever scientifically valid data is available to evaluate the physical hazard potential of the mixture."[38]

---

[33] 29 C.F.R. § 1910.1200(c).

[34] OSHA, Standards Interpretation Letter, Definition of mixture under the Hazard Communication standard (Oct. 13, 1998).

[35] 29 C.F.R. § 1910.1200(d)(5)(i).

[36] 29 C.F.R. § 1910.1200(d)(5)(ii).

[37] 29 C.F.R. § 1910.1200(d)(5)(ii).

[38] 29 C.F.R. § 1910.1200(d)(5)(iii).

The exposure factor of the hazard determination with regard to mixtures is found in Subsection (d)(5)(iv) of the HAZCOM Standard. If a component in a mixture in concentrations of less than one percent (or 0.1 percent for carcinogenic components) could be released in concentrations exceeding OSHA permissible exposure limits or the ACGIH TLVs, OSHA will treat the mixture as posing the same hazards as the component.[39]

3.0 Written Hazard Communication Program

Employers whose employees will be exposed to hazardous chemicals in the workplace and who do not fall within one of HAZCOM's exemptions, must create a written hazard communication program.[40] A written hazard communication program is the employer's blueprint for compliance, describing the employer's plans for complying with the labeling, MSDS, and employee information and training requirements imposed by the HAZCOM Standard.[41] Each employer must take care to tailor its hazard communication program to reflect the chemicals, hazards, and processes that occur at the employer's workplaces.[42] The written program should include the employer's plan to achieve compliance, a list of the hazardous chemicals known to be present in the workplace, and how the employer will inform employees about the hazards associated with non-routine tasks and with "chemicals contained in unlabeled pipes in their work areas."[43] When compiling the list of hazardous chemicals at their workplaces, employers must include hazardous chemicals in all of their forms, including liquids, solids, gases, vapors, fumes, and mists.[44] Employers are required to make their written hazard communication programs available, upon request, to employees and the employees' designated representatives.[45]

Employers of employees who travel between worksites need only make available the written hazard communication program at the primary worksite.[46]

---

[39] 29 C.F.R. § 1910.1200(d)(5)(iv).

[40] Only laboratories and workplaces where employees handle hazardous chemicals in sealed containers are exempt from the written hazard communication program. *See* 29 C.F.R. §§ 1910.1200(b)(3) and (b)(4), respectively.

[41] 29 C.F.R. § 1910.1200(e)(1).

[42] 29 C.F.R. § 1910.1200, Appendix E, "Guidelines for Employer Compliance," § 3, "Identify Hazardous Chemicals in the Workplace."

[43] 29 C.F.R. §§ 1910.1200(e)(1)(i) and (ii).

[44] 29 C.F.R. § 1910.1200, Appendix E, "Guidelines for Employer Compliance," § 3, "Identify Hazardous Chemicals in the Workplace."

[45] 29 C.F.R. § 1910.1200(e)(4).

[46] 29 U.S.C. § 1910.1200(e)(5).

4.0 Labeling and Other Forms of Warning

The written hazard communication program serves as a blueprint for employers to explain the methods they will use to implement the substantive requirements of the HAZCOM Standard. The first of these substantive requirements is labeling, the means by which an immediate warning regarding hazards is provided to employees. Subsection (f) of the HAZCOM Standard directs manufacturers, importers, and distributors to label, tag, or mark each container of hazardous chemicals leaving their workplaces with specific information.[47] Accordingly, "[e]mployers purchasing chemicals can rely on the labels provided by their suppliers."[48] The labels must include the identity of the hazardous chemicals enclosed in the labeled container,[49] the appropriate hazard warnings, and the name and address of the chemical manufacturer, importer, or other responsible party.[50] This information can also be conveyed through the use of signs, placards, process sheets, batch tickets, operating procedures, or other written materials in lieu of labels.[51] Regardless of the format for the information, the HAZCOM Standard is satisfied as long as this information is legible, in English, and prominently displayed or "readily accessible to the employees in their work area throughout each work shift."[52]

OSHA defines a hazard warning as "any words, pictures, symbols, or combination thereof appearing on a label or other appropriate form of warning which convey the specific physical and health hazard(s), including target organ effects, of

---

[47] 29 C.F.R. § 1910.1200(f)(1). The labels must accompany each shipment: "A label may not be shipped separately, even prior to shipment of the hazardous chemical, since to do so defeats the purpose of providing an immediate hazard warning." OSHA Directive CPL 2-2.38D, Appendix A, § (c), at 33.

[48] 29 C.F.R. § 1910.1200, Appendix E, "Guidelines for Employer Compliance," § 4(a), "Preparing and Implementing a Hazard Communication Program: Labels and Other Forms of Warning."

[49] According to OSHA's definition of "identity," the name appearing on the label must conform to that on the MSDS and on the list of chemicals in the written hazard communication program so that employees can link these three sources of information. 29 C.F.R. § 1910.1200(c); 29 C.F.R. § 1910.1200, Appendix E, "Guidelines for Employer Compliance," § 4(a), "Preparing and Implementing a Hazard Communication Program: Labels and Other Forms of Warning."

[50] 29 C.F.R. §§ 1910.1200(f)(1)(i)-(iii).

[51] 29 C.F.R. § 1910.1200(f)(6). In a recent OSHA Standards Interpretation Letter, OSHA said that "stick-on" labels, labels that were permanently affixed to the piece of equipment in which the subject hazardous chemical solvents were used, satisfied HAZCOM's labeling requirements. OSHA, "Standards Interpretation Letter – Using 'stick-on' labels to meet the requirements of 1910.1200" (Oct. 20, 1999).

[52] 29 C.F.R. §§ 1910.1200(f)(6) and (f)(9).

the chemical(s) in container(s)."[53] Manufacturers, importers, and distributors creating labels for hazardous chemicals should use whatever combinations of words, symbols, and pictures necessary to convey general information regarding the hazards. This information, combined with the more specific information conveyed through the remainder of the hazard communication program (i.e. the written hazard communication program, MSDSs, and employee information and training), should give employees specific information regarding the health and physical hazards they face when using hazardous chemicals known to be present at their worksite.[54] Employers must include information regarding hazards to which employees would be exposed during normal conditions of use or in foreseeable emergencies.[55] Once a manufacturer, importer, or distributor becomes aware of additional significant information regarding a hazardous chemical, it is required to review its labels for that chemical within three months[56] of becoming aware of the information.[57] Information regarding precautionary measures that may be taken when using the hazardous chemical can be provided voluntarily but is not required by the HAZCOM Standard.[58]

There are a few exceptions to HAZCOM's requirement that all containers be labeled. First, there is an exception for solid materials, such as metal, wood, plastic items (not articles, see *infra*), and whole grain. Only the initial shipment of these

---

[53] 29 C.F.R. § 1910.1200(c). The 1998 OSHA Instruction clarifies OSHA's requirement that target organ effects be described in the hazard warning on labels.

The hazard warning must convey the particular hazards of the chemical, including target organ effects. Statements such as "Hazardous if Inhaled," "Caution," "Danger," are precautionary statements and are not to be considered appropriate hazard warnings. If, when inhaled, a chemical causes lung damage, then the appropriate hazard warning is "lung damage," not inhalation.

OSHA Directive CPL 2-2.38D, Appendix A, § (f)(1), at 32. OSHA requires target organ effects to be included in the label in order to provide employees with additional information regarding the health and physical hazards presented by the chemicals known to be present at their worksites. *Id.* Employers should refer to OSHA's definitions of health and physical hazards in the HAZCOM Standard to obtain a more complete understanding of what constitutes an adequate target organ effect warning. 29 C.F.R. § 1910.1200(c).

[54] 29 C.F.R. § 1910.1200(f)(5)(ii).

[55] 29 C.F.R. § 1910.1200, Appendix E, "Guidelines for Employer Compliance," § 4(a), and Implementing a Hazard Communication Program: Labels and Other Forms of Warning."

[56] "A stay of enforcement has been placed on the requirement for revision of container labels within three months of becoming aware of significant hazard information. OSHA will alert the regulated community at the time that the stay is lifted." OSHA Directive CPL 2-2.38D, Appendix A, § (f)(11), at 36.

[57] 29 C.F.R. § 1910.1200(f)(11).

[58] 29 C.F.R. § 1910.1200, Appendix E, "Guidelines for Employer Compliance," § 4(a), "Preparing and Implementing a Hazard Communication Program: Labels and Other Forms of Warning."

items must be labeled; subsequent shipments to the same employer need not contain the label, unless the information on the label changes.[59]

A second significant exception to HAZCOM's labeling requirements is the portable container exception. Generally, if an employer transfers a hazardous chemical from one container to another, he will have to place an appropriate label on the new container. An employer, however, "is not required to label portable containers into which hazardous chemicals are transferred from labeled containers, and which are intended only for the immediate use of the employee who performs the transfer."[60]

The labeling requirements of OSHA's comprehensive substance specific standards, found in Subpart Z of Section 1910, may supersede HAZCOM's labeling provisions. Any hazardous chemicals labeled in accordance with the provisions of subpart Z will be deemed to comply with the health effects labeling requirements of HAZCOM.[61]

5.0 MSDS

HAZCOM's main vehicle for manufacturers and importers to communicate detailed information regarding the hazards of hazardous chemicals to employers and employees is the Material Safety Data Sheet ("MSDS"). Unlike a label, a MSDS is a very detailed, written statement describing the health and physical hazards posed by a chemical and includes a host of additional information regarding the chemical. Manufacturers and importers are required to obtain or develop an MSDS for every hazardous chemical they produce or import.[62] Employers and distributors may choose to rely on the MSDSs provided to them by manufacturers and importers: "Distributors and employers who in good faith choose to rely upon the sheets provided to them by the chemical manufacturer or importer assume no responsibility for the content and accuracy of the MSDSs."[63] They are responsible, however, for ensuring that they have received MSDSs from their suppliers for each hazardous chemical that is used at their workplaces.

---

[59] 29 C.F.R. § 1910.1200(f)(2)(i).
[60] 29 C.F.R. § 1910.1200(f)(7).
[61] OSHA Directive CPL 2-2.38D, Appendix A, Subpart Z, at 34.
[62] 29 C.F.R. § 1910.1200(g)(1).
[63] OSHA Directive CPL 2-2.38D, Appendix A, § (g)(1), at 36.

Although OSHA does not require a certain format for MSDSs,[64] it does impose minimum requirements regarding the information manufacturers and importers must include in MSDSs. At a minimum, manufacturers and importers must provide, in English,[65] the following information:

- Identity of the hazardous chemical as it appears on the chemical's label;[66]
- Physical and chemical characteristics of the hazardous chemical;[67]
- Physical hazards of the hazardous chemical;[68]
- Health hazards of the hazardous chemical;[69]
- Primary route(s) of entry;[70]
- Exposure limits, including the OSHA permissible exposure limit, ACGIH Threshold Limit Value, and any other exposure limit used or recommended by the chemical manufacturer, importer, or employer who is preparing the MSDS;[71]
- The NTP, IARC, or OSHA's finding that the chemical is a potential carcinogen, if applicable;[72]

---

[64] OSHA has developed a non-mandatory format, OSHA Form 174, which may be used by chemical manufacturers and importers to comply with the rule. 29 C.F.R. § 1910.1200, Appendix E, § 4(b), "Material Safety Data Sheets." OSHA also recommends use of the American National Standards Institute ("ANSI") Z400.1-1993 standard in developing MSDSs. The ANSI standard is becoming internationally accepted and provides a uniform form for providing the information required on an MSDS. OSHA Directive CPL 2-2.38D, Appendix A, § (g)(2), at 37-38.

[65] 29 C.F.R. § 1910.1200(b)(2). While manufacturers and importers must provide MSDSs in English, this requirement does not prohibit a manufacturer, importer, or employer from also providing translated versions of the MSDSs to assist non-English speaking employees in understanding the health and physical hazards to which they are exposed in the workplace. OSHA Directive CPL 2-2.38D, Appendix A, § (g)(2), at 38.

[66] 29 C.F.R. § 1910.1200(g)(2)(i)(A); *see* 29 C.F.R. §§ 1910.1200(g)(2)(i)(B) and (C) for specific requirements for identifying mixtures and their ingredients on MSDSs.

[67] 29 C.F.R. § 1910.1200(g)(2)(ii). Examples of such characteristics include vapor pressure and flash point. *Id.*

[68] 29 C.F.R. § 1910.1200(g)(2)(iii). This should include the potential for fire, explosion, and reactivity. *Id.*

[69] 29 C.F.R. § 1910.1200(g)(2)(iv). This should include "signs and symptoms of exposure, and any medical conditions which are generally recognized as being aggravated by exposure to the chemical." *Id.*

[70] 29 C.F.R. § 1910.1200(g)(2)(v).

[71] 29 C.F.R. § 1910.1200(g)(2)(vi).

[72] 29 C.F.R. § 1910.1200(g)(2)(vii).

- Precautions regarding safe handling and use;[73]
- Control measures;[74]
- Emergency and first aid procedures;[75]
- Date of the MSDSs preparation or its last change;[76] and
- Name, address, and telephone number of the manufacturer, importer, employer, or other responsible party who prepared or distributed the MSDS.[77]

Because the goal of the HAZCOM Standard is for hazard information to be available to employees, employers are required to keep the MSDSs in the workplace, readily accessible[78] to employees during their work shifts, in their work areas.[79] If an employee has to ask for access to the MSDS, the MSDS is not readily accessible.[80]

Manufacturers and importers are required to provide distributors and employers the MSDS with or prior to the initial shipment of the hazardous chemical.[81] Distributors who act as suppliers of hazardous chemicals, must ensure that the MSDS and any updated information are provided to other distributors and employers with or prior to the initial shipment.[82] After an MSDS has been updated with new information, the manufacturer, distributor, or importer is obligated to send the updated MSDS with the next shipment of the hazardous chemical.[83]

---

[73] 29 C.F.R. § 1910.1200(g)(2)(viii). This should include "appropriate hygienic practices, protective measures during repair and maintenance of contaminated equipment, and procedures for clean-up of spills and leaks." *Id.*

[74] 29 C.F.R. § 1910.1200(g)(2)(ix). This should include "appropriate engineering controls, work practices, or personal protective equipment." *Id.*

[75] 29 C.F.R. § 1910.1200(g)(2)(x).

[76] 29 C.F.R. § 1910.1200(g)(2)(xi).

[77] OSHA requires this information to be included so employers and employees may contact these sources for additional information regarding the hazardous chemical and appropriate emergency procedures for specific scenarios. 29 C.F.R. § 1910.1200(g)(2)(xii).

[78] OSHA interprets "readily accessible" to mean immediately accessible. OSHA Directive CPL 2-2.38D, Appendix A, § (g)(8), at 39.

[79] 29 C.F.R. §§ 1910.1200(g)(1) and (g)(8). Employers must also make the MSDSs available to employees' designated representatives and to OSHA. 29 C.F.R. § 1910.1200(g)(11).

[80] OSHA, "Standards Interpretation Letter – Employee access to MSDSs required by 1910.1200 vs. 1910.1020" (Dec. 7, 1999).

[81] 29 C.F.R. §§ 1910.1200(g)(6)(i) and (ii).

[82] 29 C.F.R. §§ 1910.1200(g)(7)(i) and (ii). Sections (7)(iii) through (7)(vii) describe the MSDS requirements OSHA imposes upon retail and wholesale distributors.

[83] 29 C.F.R. § 1910.1200(g)(6)(i).

Employers and distributors are obligated to make sure that they actually receive the MSDS. If they do not receive one with or prior to the initial shipment, they must request one from their supplier as soon as possible;[84] the Standard requires manufacturers and importers to provide MSDSs upon request.[85]

Consistent with HAZCOM's goal of ensuring worker safety through the dissemination of information, OSHA requires manufacturers and importers to update MSDSs within three months of becoming aware of significant, new information regarding a chemical's hazards or precautionary measures. Manufacturers and importers are always under the obligation to ensure that their MSDSs accurately reflect "the scientific evidence used in making the hazard determination."[86]

A recent trend in American workplaces is the attempt to make MSDSs available to employees electronically. OSHA permits "electronic access, microfiche, and other alternatives to maintaining paper copies of the material safety data sheets," as long as such alternatives do not create barriers to employees' immediate access to the MSDSs in their workplaces.[87] In a more recent Instruction, OSHA included computers with printers, the Internet, CD-ROMS, and fax machines as acceptable alternatives.[88] In an appendix to the HAZCOM Standard regulations, OSHA said that "[a]s long as employees can get the information when they need it, any approach may be used" to give employees access to MSDSs.[89] In other words, no matter how employers provide access, the MSDSs must always be readily accessible to employees, at their work stations, during their work shifts. To ensure access, employers will have to train their employees on how to use these alternative devices, including specific software, and ensure that there is an adequate backup system that also allows for ready access to the MSDSs in the event of an emergency such as a power outage, equipment failure, and on-line access delays.[90]

OSHA recently approved electronic provision of MSDSs by manufacturers, importers, and distributors to downstream users. The downstream user, however, must not have to take the initiative and ask for the MSDS; it is the responsibility of the manufacturer or importer to provide the MSDS to the downstream user. Thus, the manufacturer, distributor, or importer can provide the MSDS electronically, but he must take "some positive action to ensure that the downstream users are

---

[84] 29 C.F.R. § 1910.1200(g)(6)(iii).
[85] 29 C.F.R. § 1910.1200(g)(6)(iv).
[86] 29 C.F.R. § 1910.1200(g)(5).
[87] 29 C.F.R. § 1910.1200(g)(8).
[88] OSHA Directive CPL 2-2.38D, Appendix A, § (g)(8), at 39.
[89] 29 C.F.R. § 1910.1200, Appendix E, § 4(b) "Material Safety Data Sheets."
[90] OSHA Directive CPL 2-2.38D, Appendix A, § (g)(8), at 39.

willing and able to receive the information electronically."[91] Such "positive action" could include a specific letter to the customer telling him how he can obtain the MSDS electronically and asking him if he would like to do so.[92] If the downstream user does not wish to receive the MSDS electronically, a hard copy must be provided.[93]

Manufacturers and importers often seek to avoid the burden of preparing MSDSs for each hazardous chemical they produce or import through the use of generic MSDSs. OSHA permits the use of generic MSDSs for complex mixtures where each batch of the mixture may not contain exactly the same ingredients but the chemical ingredients and the health hazards associated with the mixtures are essentially the same.[94] In such cases, the chemical manufacturer or importer may use one MSDS for all of these similar mixtures.[95] Similarly, OSHA allows manufacturers and importers to streamline compliance with HAZCOM's MSDS requirements by designating an MSDS to cover the hazards of a process, rather than to cover each individual chemical involved in that process. When groups of hazardous chemicals are used in a work area in a single process, often it is more appropriate to create an MSDS for the process. In order for employees to have the information needed to protect themselves, such MSDSs must still include, however, all of the information that would be contained in individual MSDSs for all of the hazardous chemicals involved.[96]

Employees who must travel between workplaces during their shifts are still entitled to access the MSDSs of hazardous chemicals with which they come into contact. Employers are not required to maintain copies of the MSDSs at each worksite, only at the primary workplace, but they must ensure that traveling employees can obtain the information contained in the MSDSs in an emergency.[97]

---

[91] OSHA, "Standards Interpretation Letter – Chemical suppliers must ensure downstream flow of hazard information (MSDSs)" (April 21, 1999).
[92] OSHA, "Standards Interpretation Letter – Chemical suppliers must ensure downstream flow of hazard information (MSDSs)" (April 21, 1999).
[93] OSHA, "Standards Interpretation Letter – Chemical suppliers must ensure downstream flow of hazard information (MSDSs)" (April 21, 1999).
[94] OSHA Directive CPL 2-2.38D, Appendix A, § (g)(4), at 39.
[95] 29 C.F.R. § 1910.1200(g)(4).
[96] 29 C.F.R. § 1910.1200(g)(10).
[97] 29 C.F.R. § 1910.1200(g)(9).

6.0 Employee Information and Training

The goal of HAZCOM's information and training requirements is to change employee behavior to reduce the incidence of chemical source illness and injuries.[98] To that end, OSHA requires employers to "provide employees with effective information and training on hazardous chemicals."[99] Employers must provide information to employees, at the time of their initial assignment, about the requirements of the HAZCOM Standard, "operations in their work area where hazardous chemicals are present," and the location and availability of the written hazard communication program and MSDSs.[100] Employers must train an employee regarding the hazards presented by hazardous chemicals in his workplace at the time the employee is assigned to work with a hazardous chemical.[101] Employers must provide employees information regarding the hazardous chemicals they are exposed to or potentially exposed to in their workplaces *before* they encounter them.[102]

In addition to providing employees with information and written warnings about hazardous chemicals through the written hazard communication program, labels, and MSDSs, employers must also teach and train their employees how to use the information contained in these warnings.

> [T]hrough effective information and training, workers will learn to read and understand such information, determine how it can be obtained and used in their own workplaces, and understand the risks of exposure to the chemicals in their workplaces as well as the ways to protect themselves.[103]

As mentioned above, training is required prior to an employee's potential exposure. Employers must provide additional training for employees whenever a new physical or health hazard is introduced into the employees' work area.[104] Notice that the requirement for additional training is triggered by the introduction of a new hazard, not a new hazardous chemical. If the addition of a new hazardous chemical

---

[98] 29 C.F.R. § 1910.1200, Appendix E, "Employee Information and Training."

[99] 29 C.F.R. § 1910.1200(h)(1).

[100] 29 C.F.R. §§ 1910.1200(h)(2)(i)-(iii).

[101] 29 C.F.R. § 1910.1200(h)(1). OSHA Directive CPL 2-2.38D, Appendix A, explains employers' information and training obligations toward new hires, employees at multi-employer worksites, and temporary employees.

[102] OSHA Directive CPL 2-2.38D, Appendix A, § (h), at 40. "Each employee who may be 'exposed' to hazardous chemicals when working must be provided information and trained prior to initial assignment to work with a hazardous chemical, and whenever the hazard changes." 29 C.F.R. § 1910.1200, Appendix E, § C, "Employee Information and Training."

[103] 29 C.F.R. § 1910.1200, Appendix E, § C, "Employee Information and Training."

[104] 29 C.F.R. § 1910.1200(h)(1).

into the work area introduces no new hazards into the workplace, employers are not obligated to provide additional training.

Employers must train employees about the following:

- The methods through which employees can detect the presence or release of a hazardous chemical;[105]
- The physical and health hazards of the hazardous chemicals known to be present in the workplace;[106]
- The precautionary measures the employees can take in order to protect themselves from these hazards;[107]
- The details of the employer's hazard communication program, including the labeling system, MSDSs, and how employee's can obtain and use the information contained in these warnings.[108]

OSHA expects HAZCOM training to be proportional to the hazards to which employees are exposed in the workplace.[109] If a chemical to which employees will be exposed is particularly hazardous, the training should reflect that increased risk by including more extensive training about emergency procedures.[110]

The HAZCOM Standard permits employers to streamline the training process by training employees by categories of hazards, such as flammability, carcinogenicity, or corrosivity, rather than by specific chemical.[111] Allowing training according to hazard category does not hinder OSHA's goal of informing employees about the hazards they will face in the workplace because employees will have access to specific information about each chemical through labeling and MSDSs.[112] Many employers who use numerous hazardous chemicals in the workplace or who change chemicals frequently find it more efficient to train employees according to hazard category instead of individual chemical.[113] Training by hazard category further streamlines

---

[105] 29 C.F.R. § 1910.1200(h)(3)(i).

[106] 29 C.F.R. § 1910.1200(h)(3)(ii).

[107] 29 C.F.R. § 1910.1200(h)(3)(iii).

[108] 29 C.F.R. § 1910.1200(h)(3)(iv).

[109] OSHA Directive CPL 2-2.38D, Appendix A, § (h), at 42.

[110] OSHA Directive CPL 2-2.38D, Appendix A, § (h), at 42. This OSHA Instruction also addresses the interaction between HAZCOM's training requirements and those of OSHA regulation 29 C.F.R. § 1910.120.

[111] 29 C.F.R. § 1910.1200(h)(1) and Appendix E, "Employee Information and Training."

[112] 29 C.F.R. § 1910.1200, Appendix E, "Employee Information and Training."

[113] 29 C.F.R. §1910.1200, Appendix E, "Employee Information and Training."

the training process when new hazardous chemicals are introduced into the workplace. If a new hazardous chemical is introduced that does not introduce a new hazard into the workplace, OSHA does not require employers to provide additional training to employees regarding that new chemical.[114] Employers are simply required to inform employees as to which hazard category the new hazardous chemical belongs and provide the appropriate chemical-specific warnings through labels and MSDSs.[115]

Notwithstanding the explicit language allowing training by hazard category, OSHA has issued citations for failing to provide chemical-specific training. Many OSHA compliance officers have read language in Subsection (h) of the HAZCOM Standard to require chemical-specific training, despite a provision in the same section specifically stating that training can be done by hazard category. The language to which these compliance officers are looking is the portion of Subsection (h) that sets forth the requisite components of a HAZCOM training program. Because these provisions require employers to teach employees about "[t]he physical and health hazards of the chemicals in the work area," the "measures employees can take to protect themselves from these hazards," and "[m]ethods and observations that may be used to detect the presence or release of a hazardous chemical," compliance officers believe that employers must train for specific chemicals and often issue citations accordingly.

The Occupational Safety and Health Review Commission ("OSHRC") issued decisions in 1993 and 1995 that OSHA believes interpret the HAZCOM Standard to require chemical-specific training.[116] However, the regulated community does not agree with OSHA's conclusion. In 1999, the OSHRC agreed to review Cagle's Inc.,[117] a case that promises to settle this controversy. An OSHA compliance officer found Cagle in violation of HAZCOM because its employees were not knowledgeable regarding the hazards of one hazardous chemical in the workplace, carbon dioxide, ignoring any hazard-category training provided by Cagle. A coalition of seven trade associations filed an amicus brief in the case, challenging the legality of OSHA's most recent Instruction and Compliance Directive because it is contrary to the language of the HAZCOM Standard and its regulatory history.

---

[114] "For example, if a new solvent is brought into the workplace, and it has hazards similar to existing chemicals for which training has already been conducted, then no new training is required." OSHA Directive CPL 2-2.38D, Appendix A, § (h), at 41.

[115] OSHA Directive CPL 2-2.38D, Appendix A, § (h), at 41.

[116] *In re Safeway Store No. 914*, 16 BNA OSHC 1504, 1513-14 (OSHRC 1993) and *Well Solutions Inc.*, 17 BNA OSHC 1212, 1215 (OSHRC 1995).

[117] OSHRC Docket No. 98-485.

7.0 Exemptions

The HAZCOM Standard provides various exemptions for certain workplaces, conditions, and materials. In most cases, these exemptions are limited and, often, only relieve manufacturers, importers, or employers from a portion of the Standard's regulatory requirements. Generally, only laboratories and work places where hazardous chemicals are only handled in sealed containers receive the benefit of having to comply with reduced HAZCOM requirements.[118] OSHA lists several categories of products and materials that are exempt from its labeling requirements, as well.[119] Finally, there are categories of materials that receive complete exemption from the HAZCOM Standard.[120]

7.1 Partially Exempted Work Places / Conditions

7.1.1 Laboratories

Although the requirements of the HAZCOM Standard are less burdensome for laboratories, most of the Standard's requirements still apply to quality control laboratories, laboratories whose function is to produce commercial quantities of materials, and all laboratories connected with production processes."[121] The main benefit to laboratories from the Laboratories provision in Subsection (b)(3) is that laboratories are not required to produce, maintain, or make available a written hazard communication program. However, under the HAZCOM Standard, OSHA still requires laboratories to ensure that: the labels on incoming hazardous chemicals not be removed or defaced;[122] employers maintain MSDSs that accompany incoming chemicals and make them readily available to employees;[123] and employers comply with HAZCOM's information and training requirements.[124] Laboratory employers who ship hazardous chemicals are considered manufacturers or distributors under the Standard and must meet all of its requirements for labels under Subsection (f)(1) and MSDSs under Subsections (g)(6) and (g)(7).[125] When

---

[118] 29 C.F.R. §§ 1910.1200(b)(3) and (b)(4).
[119] 29 C.F.R. §§ 1910.1200(b)(5) and (b)(6).
[120] 29 C.F.R. § 1910.1200(b)(6).
[121] OSHA Directive CPL 2-2.38D, Appendix A, § (b)(3), at 24.
[122] 29 C.F.R. § 1910.1200(b)(3)(i).
[123] 29 C.F.R. § 1910.1200(b)(3)(ii).
[124] 29 C.F.R. § 1910.1200(b)(3)(iii).
[125] 29 C.F.R. § 1910.1200(b)(3)(iv).

employees are trained regarding the hazards of chemicals they use in the laboratories, they must also be informed of the absence of a written program.[126]

Note that manufacture of chemical specialty products, as well as pilot plant operations, dental, photo finishing, and optical laboratories, are not considered laboratory operations for the purposes of HAZCOM.[127] Because employers engaged in each of these activities are producing a finished product, they are subject to all requirements of the HAZCOM Standard. Laboratories where relatively small quantities of hazardous chemicals are used on a non-production basis are subject to the "Occupational Exposure to Hazardous Chemicals in Laboratories" Standard ("Laboratory Standard"), 29 C.F.R. § 1910.1450, in lieu of the HAZCOM Standard.[128] Many of the requirements of the Laboratory Standard are more comprehensive than the HAZCOM requirements for laboratories.[129] Accordingly, it is very important for an employer to evaluate whether its laboratories are subject to the reduced HAZCOM requirements, the entire HAZCOM Standard, or the Laboratory Standard to ensure compliance with OSHA's regulations.

7.1.2 Sealed containers

The HAZCOM Standard also provides reduced requirements for employers whose employees handle hazardous chemicals in sealed containers.[130] OSHA defines a container as "any bag, barrel, bottle, box, can, cylinder, drum, reaction vessel, storage tank, or the like that contains a hazardous chemical."[131] Pipes, piping systems, engines, fuel tanks, and other operating systems inside a vehicle are not considered containers by OSHA.[132] To be sealed, a container must not be opened under normal conditions of use. Activities such as "marine cargo handling, warehousing, or retail sales" are activities during which sealed containers are handled and are subject to reduced requirements.[133]

Because any container can be broken or can leak, employees handling sealed containers still face potential exposure to any hazardous chemicals within a sealed container. Thus, employers whose employees handle sealed containers are still

---

[126] OSHA Directive CPL 2-2.38D, Appendix A, § (b)(3), at 24.
[127] OSHA Directive CPL 2-2.38D, Appendix A, § (b)(3), at 24.
[128] OSHA Directive CPL 2-2.38D, § XIII (B), at 22; 29 C.F.R. §§ 1910.1450(a)(1) and (b).
[129] OSHA Directive CPL 2-2.38D, § XIII(B), at 22.
[130] 29 C.F.R. § 1910.1200(b)(4).
[131] 29 C.F.R. § 1910.1200(c).
[132] 29 C.F.R. § 1910.1200(c).
[133] 29 C.F.R. § 1910.1200(b)(4).

subject to some of HAZCOM's requirements, specifically those that require disseminating information to employees regarding the hazards they may face. Employers whose employee's handle hazardous chemicals in sealed containers must adhere to the HAZCOM Standard in three ways, similar to the requirements imposed on laboratories. First, they must ensure that labels affixed to containers of hazardous chemicals are not removed or defaced.[134] Second, employers shall maintain MSDSs for the sealed containers of hazardous chemicals.[135] Finally, employers must also provide information and training to employees to the extent necessary to ensure that the employees can protect themselves in the event of a spill or leak of a hazardous chemical from a sealed container.[136]

7.2 Partially Exempted Materials (Chemicals Subject to the Labeling Requirements of Other Statutes or Regulations)

Generally, the HAZCOM Standard does not require any hazardous chemicals to be labeled if they are subject to the labeling requirements of other statutes or regulations. Specifically, pesticides do not need to be labeled under HAZCOM if they are subject to the labeling requirements of the Federal Insecticide, Fungicide, and Rodenticide Act (7 U.S.C. § 136 et seq.).[137] Similarly, toxic substances do not need to be labeled under HAZCOM if they are subject to the labeling requirements of the Toxic Substances Control Act (15 U.S.C. § 2601 et seq.).[138] In addition, OSHA does not require HAZCOM labeling for food, food additives, color additives, drugs, cosmetics, medical or veterinary devices, and any materials intended for use as ingredients in any of these products if they are subject to the labeling requirements of the Federal Food, Drug, and Cosmetic Act (21 U.S.C. § 301 et seq.) or the Virus-Serum-Toxin Act of 1913 (21 U.S.C. § 151 et seq.).[139] Beverage alcohol not intended for industrial use receives the same exemption if it is subject to the labeling requirements of the Federal Alcohol Administration Act (27 U.S.C. § 201 et seq.).[140] The same exemptions are also available to consumer products with regard to the Consumer Product Safety Act (15 U.S.C. § 2051 et seq.) and agricultural or

---

[134] 29 C.F.R. § 1910.1200(b)(4)(i).
[135] 29 C.F.R. § 1910.1200(b)(4)(ii).
[136] 29 C.F.R. § 1910.1200(b)(4)(iii).
[137] 29 C.F.R. § 1910.1200(b)(5)(i).
[138] 29 C.F.R. § 1910.1200(b)(5)(ii).
[139] 29 C.F.R. § 1910.1200(b)(5)(iii).
[140] 29 C.F.R. § 1910.1200(b)(5)(iv).

vegetable seeds treated with pesticides with regard to the Federal Seed Act (7 U.S.C. § 1551 et seq.).[141]

7.3 Fully Exempted Materials

Subsection (b)(6) of the HAZCOM Standard provides that several categories of materials are entirely exempt from the Standard's requirements:

- Hazardous wastes as defined by the Solid Waste Disposal Act and the Resource Conservation and Recovery Act ("RCRA," 42 U.S.C. § 6901 et seq.);[142]
- Hazardous substances as defined by the Comprehensive Environmental Response, Compensation, and Liability Act ("CERCLA," 42 U.S.C. § 9601 et seq.) "when the hazardous substance is the focus of remedial or removal action,"[143]
- Cosmetics, food, and drugs brought into the workplace by employees for their personal consumption or that are packaged by the manufacturer for retail sale,[144]
- Food, food products, or alcoholic beverages sold, used, or prepared by retail establishments,[145]
- Tobacco or tobacco products,[146]
- Wood and wood products that will not be processed and which only pose a threat of combustibility to employees,[147]
- Nuisance particulates that do not pose a health or physical hazard, ionizing and nonionizing radiation, and biological hazards;[148] and

---

[141] 29 C.F.R. §§ 1910.1200(b)(5)(v) and (vi).
[142] 29 C.F.R. § 1910.1200(b)(6)(i).
[143] 29 C.F.R. § 1910.1200(b)(6)(ii); *see also* OSHA Directive CPL 2-2.38D, Appendix A, § (b)(6), at 25.
[144] 29 C.F.R. §§ 1910.1200(b)(6)(vi), (vii), and (viii).
[145] 29 C.F.R. § 1910.1200(b)(6)(vi).
[146] 29 C.F.R. § 1910.1200(b)(6)(iii).
[147] 29 C.F.R. § 1910.1200(b)(6)(iv).
[148] 29 C.F.R. §§ 1910.1200(b)(6)(x), (xi), and (xii).

- Consumer and household products, when those products are used in the workplace for similar duration and at similar frequency as a consumer would use them (same duration and frequency of use).[149]

The HAZCOM Standard also exempts "articles" from the Standard's requirements. Many of OSHA's opinion letters are consumed by manufacturers asking OSHA whether a product they produce is an article.

Article means a manufactured item other than a fluid or particle: (i) which is formed to a specific shape or design during manufacture; (ii) which has end use function(s) dependent in whole or in part upon its shape or design during end use; and (iii) which under normal conditions of use does not release more than very small quantities, e.g., minute or trace amounts of a hazardous chemical (as determined under paragraph (d) of this section), and does not pose a physical or health risk to employees.[150]

Imbedded in this definition are the two factors considered during the hazard determination—whether a chemical imposes a health or physical hazard and whether employees are exposed to the chemical. Articles are products that are manufactured to take on a specific shape or design, but, more importantly, articles are products whose chemical components pose no health or physical risks and to which employees are effectively unexposed.[151]

Also important in the definition of "article," is the phrase "normal conditions of use." OSHA includes processes such as cutting, burning, and heating as potential normal conditions of use. If a product produces a hazardous by-product or in any other way creates the potential for employee exposure to a hazardous chemical through its normal conditions of use, such as through inhalation, that product is not an article and is subject to the HAZCOM Standard.[152] Examples of articles include a stainless steel table, vinyl upholstery, tires, and adhesive tape.[153]

---

[149] 29 C.F.R. § 1910.1200(b)(6)(ix). This exemption relieves a large regulatory burden from employers who use consumer and household products, such as glass cleaner, white out, etc., in their workplaces.

[150] 29 C.F.R. § 1910.1200(c).

[151] OSHA intentionally uses the word "risk" instead of "hazard" in its definition of "article." In fact, OSHA changed the language from "hazard" to "risk" in 1994. A hazard is "an inherent property of the chemical and exists no matter what quantity of the chemical is present"; a risk is a function of both the hazard of and the exposure to a chemical. OSHA Directive CPL 2-2.38D, Appendix A, § (d)(5), at 31. "To be exempted as an article, exposure must not pose a risk to employee health." OSHA Directive CPL 2-2.38D, Appendix A, § (c), at 27.

[152] OSHA Directive CPL 2-2.38D, Appendix A, § (b)(6), at 26.

[153] OSHA Directive CPL 2-2.38D, Appendix A, § (b)(6), at 26.

8.0 Trade Secrets

Manufacturers are often reluctant to reveal the detailed information required on an MSDS by the HAZCOM Standard. OSHA has tried to balance an employee's right to know about hazards he will be exposed to in the workplace and an employer's right to maintain trade secrets. OSHA allows manufacturers, importers, and employers to withhold the specific chemical identity of a hazardous chemical when that company can make a valid claim that the information should be protected as a trade secret.[154] OSHA defines a trade secret as "any confidential formula, pattern, process, device, information or compilation of information that is used in an employer's business, and that gives the employer an opportunity to obtain an advantage over competitors who do not know or use it."[155] In addition to providing protection for the chemical identity, OSHA also protects information regarding mixtures: "Nothing in [the HAZCOM Standard] shall be construed as requiring the disclosure under any circumstances of process or percentage of mixture information which is a trade secret."[156] The manufacturer, importer, or employer must still create an MSDS for the chemical or mixture and include in it information regarding its "properties and effects," including exposure limits,[157] and indicate that the chemical's identity is being withheld as a trade secret.[158]

Even when OSHA protects disclosure of certain information as trade secrets, it may require release of that information to particular groups for health and safety reasons. Subsection (i)(2) of the HAZCOM Standard mandates that manufacturers, importers, or employers must disclose a chemical's identity in emergency health situations;[159] Subsection (i)(3) discusses the circumstances under which a manufacturer, importer, or employer must disclose the chemical's identity in non-emergency health situations. Subsection (i)(3) provides that the manufacturer, importer, or employer must provide the chemical identity to a health care professional treating an exposed employee upon the request of that treating health care professional. The request must follow certain procedures outlined in the rule, including requiring the

---

[154] 29 C.F.R. § 1910.1200(i)(1)(i).

[155] 29 C.F.R. § 1910.1200(c).

[156] 29 C.F.R. § 1910.1200(i)(13).

[157] "Despite the claim that a hazardous chemical or a constituent thereof, is a trade secret, the PEL, TLV, or other designated exposure limit must be included in the MSDS." OSHA Directive CPL 2-2.38D, Appendix A, § (i)(1), at 42.

[158] 29 C.F.R. §§ 1910.1200(i)(1)(ii) and (iii).

[159] "For medical emergencies, failure to disclose the information shall result in the issuance of a willful citation, if the elements of a willful citation can be established." OSHA Directive CPL 2-2.38D, § XI(G)(2)(b), at 21.

health care professional to sign a confidentiality agreement.¹⁶⁰ If the manufacturer, importer, or employer denies the request for disclosure, Subsection (i)(7) prescribes the manner in which it must do so. The health care professional can then refer its request to OSHA, which, after considering the list of factors in Subsection (i)(9), can, if it agrees with the health care professional, issue a citation to the manufacturer, importer, or employer who made the denial.¹⁶¹

OSHA sets forth its criteria for evaluating whether a piece of information is a trade secret in Appendix D of the HAZCOM Standard; it reprints section 757, comment b, of the Restatement of Torts' (1939) definition of "trade secret," most of which is also reprinted in HAZCOM Subsection (c)'s definition. Appendix D, however, includes comment b of Restatement section 757, which elaborates that a trade secret is secret information regarding, for example, a formula for a chemical, a manufacturing or treatment process, a pattern, or a list of customers.¹⁶² OSHA incorporates the Restatement's list of factors to consider in determining whether a piece of information is a trade secret:

- The extent to which the information is known outside of the business;
- The extent to which it is known by employees and other involved in the business;
- The extent of measure taken by the business to guard the secrecy of the information;
- The value of the information to the business and its competitors;
- The amount of effort and money expended by the business in developing the information; and
- The ease or difficulty with which the information could be properly acquired or duplicated by others.

9.0 Multiple Employer Worksites

Special problems are posed by multi-employer workplaces (sometimes referred to as "MEWs") at which another employer's employees may be exposed to hazardous

---

[160] 29 C.F.R. § 1910.1200(i)(3)(v).

[161] 29 C.F.R. § 1910.1200(i)(10). "In response to non-emergencies, where OSHA believes that the chemical manufacturer, importer or employer will not be able to support the trade secret claim, the withholding of a specific chemical identity shall be cited as a violation of paragraph (g)(2)." OSHA Directive CPL 2-2.38D, § XI(G)(2)(a), at 21.

[162] 29 C.F.R. § 1910.1200, Appendix D, "Definition of 'Trade Secret'."

chemicals. MEWs occur where there are employees of two or more employers working in the same workplace. An employer in control of such a workplace must take additional steps to ensure that outside employees are also informed of the hazards that exist as a result of potential exposure to hazardous chemicals known to be present at the workplace. The requirements that HAZCOM imposes on employers controlling MEWs focus on communicating hazards to the outside employers, but not to the outside employees.[163] OSHA requires the employer controlling the worksite to make provisions in its written hazard communication program for the methods it will use to make certain information available to outside employers.[164] Controlling employers must describe the methods they will use to inform outside employers of precautionary measures needed to protect their employees during normal operating conditions and in foreseeable emergencies and to inform the other employers about access to MSDSs and the labeling system in use at the workplace.[165] An August 2000 decision by the OSHRC confirms that employers controlling MEWs have no duty to train outside employees; the Standard "requires only that [the outside employer] be supplied with any information it might need to train its own employees."[166] Thus, it is the outside employer's responsibility to inform and train its employees regarding the hazards that they will be exposed to at the shared worksite.

10.0 Preemption and Approval of State Plans

Section 18(b) of the OSHA Act encourages states to adopt their own occupational safety and health plans. Before OSHA promulgated the federal HAZCOM Standard, several states had worker right to know laws. These state worker right to know laws varied widely and often were much broader in scope than the federal HAZCOM Standard. Since the inception of HAZCOM, many manufacturers, importers, and employers have challenged state laws that are more protective than the HAZCOM Standard. Industry groups favored the federal HAZCOM Standard because it offered one uniform rule, rather than many, diverse rules set forth by individual states. Such groups argued that the federal HAZCOM Standard preempted these state laws. In 1992, the United States Supreme Court held that the HAZCOM Standard preempted state worker right to know laws that had the same

---

[163] *Heed the Safety Implications of Recent Court Rulings*, Safety Director's Report, May 2000, at 710-11.

[164] 29 C.F.R. § 1910.1200(e)(2).

[165] 29 C.F.R. §§ 1910.1200(e)(2)(i) – (iii).

[166] *In re Key Energy Services, Inc.*, 2000 OSAHRC LEXIS 95, 2000 OSHD (CCH) P32,212 (2000).

dual purpose as the HAZCOM Standard, employee and public health.[167] Since that time, OSHA has established a system whereby it will assess state worker right to know programs; if OSHA approves of a state's program, the federal HAZCOM Standard does not preempt it.

The HCS preempts all state (in states without OSHA-approved job safety and health programs) or local laws which related to an issue covered by [HAZCOM] without regard to whether the state law would conflict with, compliment, or supplement the federal standard, and without regard to whether the state law appears to be "at least as effective as" the federal standard.[168]

A state can avoid preemption of its worker right to know plan "only if it submits, and obtains Federal approval of, a plan for the development of such standards and their enforcement."[169] Such a plan must be at least as effective in providing safe and healthful workplaces for employees as is the federal HAZCOM Standard.[170] Twenty-five states and territories have their own OSHA-approved worker right to know programs.[171] Manufacturers, importers, distributors, and employers whose operations are in a state with an OSHA-approved state plan must comply with that state's requirements.

The most recent controversy over federal preemption of state worker right to know laws involved California's incorporation of Proposition 65, the Safe Drinking Water and Toxic Enforcement Act,[172] into its state HAZCOM Standard. OSHA approved the amendment of California's state plan in June 1997. Proposition 65 requires the state to maintain a list of chemicals that cause cancer, birth defects, and reproductive harm; persons doing business in California are prohibited from exposing individuals to these chemicals without providing warnings.[173] Before

---

[167] *Gade v. National Solid Wastes Management Association*, 505 U.S. 88 (1992).

[168] OSHA, Fact Sheet No. OSHA 93-26.

[169] 59 *Fed. Reg.* 6,126 (Feb. 9, 1994), Hazard Communication (HAZCOM), VI. Federalism and State Plan Applicability.

[170] 59 *Fed. Reg.* 6,126 (Feb. 9, 1994), OSHA Preambles, Hazard Communication (HAZCOM), VI. Federalism and State Plan Applicability.

[171] Alaska, Arizona, California, Connecticut, Hawaii, Indiana, Iowa, Kentucky, Maryland, Michigan, Minnesota, Nevada, New Mexico, New York, North Carolina, Oregon, Puerto Rico, South Carolina, Tennessee, Utah, Vermont, Virginia, Virgin Islands, Washington, and Wyoming. 59 *Fed. Reg.* 6,126 (Feb. 9, 1994), Hazard Communication (HAZCOM), VI. Federalism and State Plan Applicability.

[172] Cal. Health & Safety Code §§ 25249.6 and 25249.8.

[173] Frederick J. Ufkes and Mark E. Gustafson, *Can Plaintiffs Use Proposition 65 to Require Chemical Manufacturers to Warn Downstream Users?*, Mealey's Emerging Toxic Torts, August 18, 2000.

California incorporated Proposition 65 into its state HAZCOM Standard, the federal HAZCOM Standard preempted almost all of Proposition 65.[174] Because only those portions of Proposition 65 which require employers to warn their employees were incorporated into the state HAZCOM Standard, some argue that those doing business in California workplaces are not required to warn anyone other than their employees, including downstream users.[175]

11.0 The Future of HAZCOM – The Globally Harmonized System ("GHS")

While several nations around the world impose requirements on their manufacturers, importers, and / or employers similar to those imposed by the HAZCOM Standard, there is still significant variation among these nations' requirements and those of the United States. OSHA[176] believes that a "harmonized" system, one in which manufacturers, importers, and employers worldwide adhere to uniform standards, would improve protection to American workers and facilitate international trade.[177] Currently, American workers face hazards from chemicals imported from foreign countries:

> To ensure the protection of workers, American employers need to receive appropriate label and MSDS information from foreign suppliers. With differing requirements worldwide, including many countries with no requirements for classifying and labeling chemicals, this is often difficult and the information may be inadequate to protect workers.[178]

The Globally Harmonized System ("GHS") is "an effort by the U.S. and other countries to promote common, consistent criteria for classifying chemicals accord-

---

[174] *Industrial Truck Ass'n, Inc. v. Henry*, 125 F.3d 1305 (9th Cir. 1997) (applying the Supreme Court's *Gade* decision and holding that the portions of Proposition 65 not incorporated into the state's HAZCOM plan were preempted).

[175] Raphael Metzger, *Proposition 65 in the Workplace: Are Chemical Manufacturers Obligated to Warn Downstream Users of Carcinogenic Hazards?*, September 22, 1999; *see also*, Frederick J. Ufkes and Mark E. Gustafson, *Can Plaintiffs Use Proposition 65 to Require Chemical Manufacturers to Warn Downstream Users?*, Mealey's Emerging Toxic Torts, August 18, 2000.

[176] "OSHA has been an active participant in the interagency and international discussions related to harmonization. In fact, OSHA is the only U.S. federal agency with a public commitment to adopt such a system when it is developed." Jennifer Silk, "A Globally Harmonized System for Hazard Communication," *Job Safety and Health Quarterly*, Fall 1999, at 43.

[177] <Http://www.osha-slc.gov/SLTC/hazardcommunications/global.html>.

[178] Jennifer Silk, "A Globally Harmonized System for Hazard Communication," *Job Safety and Health Quarterly*, Fall 1999, at 40.

ing to their health, physical and environmental hazards, and to develop compatible labeling, material safety data sheets for workers, and other information based on the resulting classifications."[179] The United States' participation in the GHS would be voluntary, enacted through a non-binding instrument, and would not supercede the requirements of, although it may precipitate change to, the current HAZCOM Standard.[180]

In 1992, the United Nations Conference on Environment and Development ("UNCED"), adopted an international mandate to develop a globally uniform system for workplace hazard classification and chemical labeling, including MSDSs.[181] Currently, the international effort to develop a GHS has been delegated among three international organizations: the Organization for Economic Cooperation and Development (dealing with health and environmental hazards),[182] the United Nations' Committee of Experts on the Transport of Dangerous Goods (dealing with physical hazards), and the International Labor Organization (dealing with communicating hazards through labeling and MSDSs).[183] The United States' work on a GHS has been through an interagency committee coordinated by the Department of State, including the work of the Consumer Product Safety Commission, the Department of Commerce, the Department of Transportation, the Food and Drug Administration, the Environmental Protection Agency, OSHA, the Office of the U.S. Trade Representative, the Department of Agriculture, and the National Institute of Environmental Health Sciences.[184]

Beyond requiring a high degree of international, inter-governmental, and interagency cooperation, the biggest obstacle to an internationally uniform standard is developing a system of symbols to be used on labels that could be easily understood in multiple cultures.[185] While the governmental bodies working on developing the GHS do not intend for it to weaken the current hazard communication requirements in any country, it is likely that the GHS will result in changes to "each

---

[179] <Http://www.osha-slc.gov/SLTC/hazardcommunications/global_questions_answers.html>.

[180] <Http://www.osha-slc.gov/SLTC/hazardcommunications/global_questions_answers.html>.

[181] <Http://www.osha-slc.gov/SLTC/hazardcommunications/global.html>.

[182] "The OECD has completed eight health and environmental criteria, including definitions for acute toxicity, irritation, corrosion, sensitization, carcinogenicity, germ cell mutagenicity, reproductive toxicity, and aquatic toxicity." Jennifer Silk, "A Globally Harmonized System for Hazard Communication," *Job Safety and Health Quarterly*, Fall 1999, at 43.

[183] <Http://www.osha-slc.gov/SLTC/hazardcommunications/global.html>.

[184] <Http://www.osha-slc.gov/SLTC/hazardcommunications/global_questions_answers.html>.

[185] *Next Step for Global Hazard System Is Narrowing List of Communication Options*, BNA Daily Labor Report, September 29, 2000; Jennifer Silk, "A Globally Harmonized System for Hazard Communication," *Job Safety and Health Quarterly*, Fall 1999, at 43-44.

existing system of hazard classification and labeling."[186] Participants anticipate that the GHS will be in place sometime in 2001.[187]

---

[186] <Http://www.osha-slc.gov/SLTC/hazardcommunications/global_questions_answers.html>.
[187] <Http://www.osha-slc.gov/SLTC/hazardcommunications/global_questions_answers.html>.

Chapter 9

Self-Audits

Michael T. Heenan, Esq.
Margaret S. Lopez, Esq.
Heenan, Althen & Roles, LLP
Washington, D.C.

1.0 Overview

With the complex array of federal and state regulations that reach nearly every aspect of every company's operations, it is not surprising that most companies find self-auditing of their regulatory compliance to be an essential business practice.[1] This is certainly true with regard to workplace safety and health.

In a survey conducted by OSHA in 1999, the agency found that of a representative sample of 492 employers 85.9 percent had conducted a self-audit in the previous twelve months.[2] Overall, the number of companies conducting self-audits appears to have grown considerably. A comparable survey conducted by the National Institute for Occupational Safety and Health (NIOSH) in the early 1980's found that only 53.1 percent of companies surveyed conducted self-audits.[3]

Quite clearly, safety and health audits have become an integral procedure for a majority of enterprises. The fact is, well planned and executed periodic audits can provide company management with an excellent means of precisely identifying

---

[1] Of course, companies also conduct self-audits for reasons not directly related to monitoring regulatory compliance (e.g., to monitor compliance with company policies or to look for potential problems not related to regulatory concerns). Since this book concerns OSHA law, this chapter will concentrate on audits in the context of OSHA regulatory compliance.

[2] OSHA Survey Finds 85 Percent of Employers Do Self-Audits of Safety and Health Conditions in Workplaces, <http://www.osha.gov/media/oshnews/nov99/trade-19991112.html>, Feb. 1, 2001. The study found that 94.9 percent of companies surveyed in the construction industry had conducted audits while 89.4 percent in general industry with twenty or more employees had done so. In general industry companies with fewer than twenty employees, the survey found that 80.2 percent of companies had self-audited. Most of the forty-one companies responding that they never conducted audits were those with fewer than twenty employees.

[3] Id.

safety and health hazards which will allow management to correct such conditions promptly. Audits may serve as the central monitoring device for a company's safety and health program. When the information from an audit results in workplace improvement, this serves to demonstrate a company's commitment to a safe workplace. Audits also allow companies, over time, to identify trends in their long term efforts to improve safety and compliance with regulations.

With the increasing employer utilization of self-auditing procedures has come concern about the ways in which audit information may be used against the employer in regulatory agency enforcement activity as well as in criminal and civil litigation. This chapter will briefly discuss the broad significance of voluntary self-auditing as a component of a company's safety and health program. The chapter will then consider the legitimate concerns about the potential for self-audit information to be used as adverse evidence against the company. The chapter will describe OSHA's voluntary self-audit policy, which the agency developed to respond to those concerns. Finally, other potential legal protections for audit information will be discussed.

2.0 The Significance of Voluntary Safety and Health Auditing

Safety and health audits are a comprehensive approach to monitoring and assessing safety protections and ensuring compliance with government regulations. There are ample reasons why almost any employer might embark on such a procedure. Reasonable screening for violations and prompt elimination of problem conditions not only will serve to protect safety and health, but will also put the employer in a position of being able to demonstrate diligence in its regulatory compliance efforts. Such a showing of diligence could help counter allegations of negligence in legal proceedings against the employer. In this connection, it is noteworthy that some insurance providers strongly encourage the use of auditing by reducing premiums for companies with a strong safety and health programs that include auditing.

OSHA's 1999 survey found that the three top reasons employers gave for conducting safety and health audits were: (1) to reduce injury and illness rates (83 %); (2) because it was the right thing to do (79.3 %); and (3) to monitor compliance with OSHA regulations.[4] Those who were not auditing reported that they would

---

[4] "OSHA Survey Finds 85 Percent of Employers Do Self-Audits of Safety and Health Conditions in Workplaces," <http://www.osha.gov/media/oshnews/nov99/trade-19991112.html>, Feb. 1, 2001.

consider doing so if it would reduce their insurance premiums, decrease injury and illness rates, or entitle them to a penalty reduction in the event they were cited by OSHA for a violation.[5]

As safety and health regulations have come to cover more and more aspects of the workplace and as workplaces themselves, in many instances, have become more complex, auditing has become a significant tool for monitoring regulatory compliance and identifying potential safety and health hazards in the workplace. Auditing gives the company the ability literally to inspect itself and thereby to find and resolve problems before they become legal liabilities.

2.1 Overview of Audits

2.1.1 The Audit Team

Safety and health audits can take many forms. Often they are conducted by teams composed of company personnel who specialize in safety and health. Many companies also use outside consultants to audit company operations. Outside consultants offer two advantages over exclusive use of in-house personnel. First, consultants can provide technical expertise needed to identify hazards company personnel may miss. Second, they can provide an independent review of the company's operation.

Legal counsel are also often involved in conducting safety and health audits. The company's lawyers can assist in evaluating whether a condition identified by technical members of the audit team actually would constitute a violation under legal precedent. They can also assist, from a legal perspective, in assessing the severity of the conditions observed and can provide guidance, again from a legal perspective, for evaluating potential remedies for the hazards identified. As will be discussed in more detail later in this chapter, the involvement of legal counsel in the audit also may provide the basis for application of evidentiary privileges for protecting certain audit information from mandated disclosure to third parties.

2.1.2 Scope of the Audit

Audits vary considerably in their scope. They may extend from a limited audit (focusing on a part of an operation or compliance with a single regulation) to a

---

[5] This latter incentive can now be found in OSHA's self-audit policy, which was formalized after the survey was conducted.

comprehensive audit covering the entire operation and the full panoply of OSHA regulations. They may be conducted infrequently, or they may be conducted on a regular, scheduled basis. The frequency of audits may be a function of the particular company's size and ability to dedicate resources. Larger companies may be more dependent than smaller companies on audit procedures to ensure that safety requirements are being met. No matter how frequently an employer conducts audits, the important thing is to ensure that a procedure is put in place to promptly respond to any deficiencies, violations or hazards that are discovered. Generally speaking, the scope of an audit should never be greater than the employer's capability for immediate response.

2.1.3 Audit Information

Depending upon their scope and the methods used to conduct them, audits may produce a considerable amount of information. A variety of tools may be used to gather this information. Some of the more common are interviews, first-hand observations, technical tests and measurements, and questionnaires. This information is usually preserved in the form of notes of the audit team members, test reports and other technical data, diagrams, photographs, and videos. Before the audit begins, there are often memoranda and perhaps an audit planning document which detail the scope and outline the steps for conduct of the audit.

Once the information gathering phase is over and the audit team has had a chance to evaluate the data, most audit teams will create a formal report. The report will detail the audit team's findings and recommendations for correction of hazards. The report, therefore, typically contains not only descriptions of what the team found, it also contains their analysis as to the meaning of those findings and their opinion as to what corrective action should be taken.

Other documents may be created after the audit is finished, which reflect the company's response to the audit findings and recommendations. These documents may show management's consideration of alternative actions, either proposed in the audit itself, or beyond those the audit team recommended. These documents may also describe the manner and timing of correction of the problems identified in the audit.

Certainly, much of the information generated in the course of an audit is important to the audit's serving its intended function of improving safety and health at the operation. In many instances, a company's audit information demonstrates that the company has a solid safety and health program in place. It shows that the company regularly conducts thorough audits and in a timely manner addresses the deficiencies found in those audits.

2.2 Auditing Tips

An exhaustive list of "dos" and "don'ts" for auditing are beyond the scope of this chapter. However, a few tips to avoid legal pitfalls are offered here for general guidance.

2.2.1 Take Steps to Protect Confidentiality of Audit Information

In some circumstances, it may be in the company's best interest to resist efforts of outside parties seeking to force disclosure of the company's audit information—not because the company has done anything wrong, but because the audit may be used in a way that may turn the company's good intentions and efforts against itself.[6] Thus, from a legal perspective, part of the audit planning should include consideration of what documents will be created and how their confidentiality will be maintained. The reasons and methods for this will be discussed in further detail later in this chapter. We mention it here, however, because disclosure issues should always be in the forefront of counsel's mind when advising a company on conducting a safety and health audit.

2.2.2 Be Prepared to Timely Respond to Every Hazard Identified in the Audit

It is essential that the company commit, in advance of the audit, to timely remedy of every hazard that will be identified in the audit. For obvious reasons, conducting a voluntary audit and then ignoring the problems found is worse than not auditing at all. Care must also be taken if upper management disagrees with an audit team's conclusion that something is unsafe. Prudence would dictate that management accept and address the conclusions of the audit team without reservation. Once an issue is raised in an audit it cannot be ignored without peril. More than one company has found itself facing charges after a serious accident because audit warnings were ignored or rejected.

2.2.3 Document Every Significant Step Taken to Respond to Hazards

Whenever a company has documented problems found, it is important to also document promptly that the problems were resolved in a timely manner. If the

---

[6] For example, the company may not have had sufficient time to correct an identified condition before it would be required to disclose its audit documents in a government inspection, investigation or legal proceeding.

resolution will take some time, the company should create a contemporaneous record of each step taken to resolve the problem. In this way, the company will be able to show its prudence and good faith in responding to hazards that have come to its attention.

2.2.4 Do Not Censor the Auditors

From the preliminary planning phase through preparation of the last document concerning the audit findings and recommendations or remediation, management must not censor audit documents. It is one thing to decide in advance that, for valid reasons, the audit will only address a limited list of compliance issues, but it is quite another thing to instruct an auditor to ignore a hazard that was found in the audit (or to edit a valid finding out of a draft report).

Again, once the company has committed to conducting an audit, the company must be willing to face up to the findings, whatever they are. Covering up findings will only escalate legal problems for the company should the censorship later come to light in an enforcement or litigation action.

It is also imperative that audit documents not be destroyed as a means of covering up evidence of a violative condition. To avoid even the appearance of possible destruction of evidence, audit documents should only be destroyed in the ordinary course of document destruction in accordance with the company's record retention and destruction program.

2.2.5 Attribute Appropriate Gravity to Audit Findings and Recommendations

The audit report and all other audit-related documents should neither exaggerate nor understate problems found in the audit. Audit documents should treat the findings and recommendations in a manner commensurate with their actual importance. This will provide the proper foundation for later measurement of the appropriateness of the company's response, should that be necessary.

3.0 OSHA's Voluntary Self-Audit Policy

3.1 Purpose

In recognition of the important role that voluntary audits play in maintaining effective workplace safety and health programs (and in response to considerable political pressure),[7] OSHA issued a formal policy on agency use of company audit information.[8] The policy statement, titled *Final Policy Concerning the Occupational Safety and Health Administration's Treatment of Voluntary Employer Safety and Health Self-Audits*, was published in the *Federal Register* in July, 2000. There is insufficient information at this time to say whether the policy is serving its objective of encouraging more employers to conduct audits or to what extent agency enforcement actions are still causing concern among employers about adverse use of company audit information. Still, OSHA's policy directive is a positive official step in support of the agency's stated goal to encourage voluntary auditing.

3.2 Scope

OSHA's audit policy only applies to audits that are "systematic, documented, and objective reviews conducted by, or for, employers to review their operations and practices to ascertain compliance with the Act." Further, the policy only applies to voluntary audits. It does not apply to audits that are required to be conducted by the Act, by agency regulations, or by a settlement agreement.[9]

The policy extends to information gathered in the course of a voluntary self-audit, as well as to "analyses, conclusions, and recommendations" resulting from the audit. Thus, the term "voluntary self-audit report" as used in the policy is given a

---

[7] Numerous bills have been introduced in Congress in recent years which would have created an audit privilege in one form or another and would have required OSHA to employ other incentives to encourage auditing. *See, e.g.*, Safety and Health Audit Promotion Act of 1999, H.R. 1438, 106th Cong. (1999); Safety and Health Audit Promotion and Whistleblower Improvement Act of 1999, H.R. 1439, 106th Cong. (1999); Safety and Health Improvement and Regulatory Reform Act of 1995, H.R. 1834, 104th Cong. (1995).

[8] OSHA, Final Policy Concerning the Occupational Safety and Health Administration's Treatment of Voluntary Employer Safety and Health Self-Audits, 65 *Fed. Reg.* 46,498-46,503, July 28, 2000.

[9] Many standards require companies to self-inspect to monitor compliance and to retain and make available to the agency related records of monitoring. *See, e.g.*, 29 C.F.R. §§ 1910.119, 1910.120, 1910.1025(d),(e), 1926.20(b).

broad definition, including not only the audit report itself but, presumably, other documents containing audit-related information.

3.3 Provisions

3.3.1 Use of Self-Audits in Agency Inspections

There are four major declarations in the policy. The first states that "OSHA will not routinely request voluntary self-audit reports at the initiation of an inspection." In other words, "OSHA will not use such reports as a means of identifying hazards upon which to focus inspection activity." This declaration is OSHA's attempt to respond to a major concern of employers that they were essentially creating a roadmap in company audit records of areas of the operation for OSHA inspectors to focus on in an enforcement inspection. Even so, the word "routinely" has still left some employers nervous that OSHA has left itself a large loophole.

The policy goes on to state that "if the Agency has an independent basis to believe that a specific safety or health hazard warranting investigation exists, OSHA may exercise its authority to obtain the relevant portions of voluntary self-audit reports relating to the hazard." Here lies a potentially even larger loophole.[10]

3.3.2 No Citation for Corrected Conditions

In the second declaration, OSHA states that it "will not issue a citation for a violative condition that an employer has discovered as a result of a voluntary self-audit, if the employer has corrected the violative condition prior to the initiation of an inspection (and prior to a related accident, illness, or injury that triggers the inspection) and has taken appropriate steps to prevent a recurrence of the violative condition that was discovered during the voluntary self-audit." This statement illustrates the importance of documenting actions taken to respond to hazardous conditions identified in an audit, particularly actions that are aimed at preventing recurrence of the hazard.

---

[10] The authority that OSHA is purporting to rely on must be carefully examined in each case. If there is no OSHA regulation requiring a document, such as an audit report, OSHA compliance officers may not be able to compel its production without a search warrant or subpoena. At the same time, it must be remembered that even if the agency does not seek to obtain audit information at the time of inspection, the audit report may be demanded later in discovery during litigation concerning any citations that may be issued based on other evidence.

3.3.3 Protection from Use of Self-Audits to Show Willfulness

The third declaration, the so-called "safe harbor" provision, provides that "if an employer is responding in good faith to a violative condition discovered through a voluntary self-audit and OSHA detects the condition during an inspection, OSHA will not use the voluntary self-audit report as evidence that the violation is willful." (A willful violation is one in which the employer intentionally violated a requirement of the Act and has either shown reckless disregard for the possibility of a violation or has shown plain indifference to safety and health of employees. Penalties for such violations may range from a minimum of $5,000 to $70,000.)[11]

OSHA explains that "this policy is intended to apply when, through a voluntary self-audit, the employer learns that a violative condition exists and promptly takes diligent steps to correct the violative condition and bring itself into compliance, while providing effective interim employee protection as necessary." Again, note the importance of good documentation with respect to intermediate responsive steps.

3.3.4 Penalty Reduction for Good Faith

The final declaration provides that the agency will "treat a voluntary self-audit that results in prompt action to correct violations found [as described in the safe harbor provision], and appropriate steps to prevent similar violations, as strong evidence of an employer's good faith with respect to the matters covered by the voluntary self-audit." Accordingly, the agency will reduce by up to 25 % the amount of the penalty that would otherwise be assessed for the violation.[12] The policy defines "good faith" as "an objectively reasonable, timely and diligent effort to comply with the requirements of the Act and OSHA standards."

3.4 Limitations

In issuing the policy, the agency was careful to state in the policy itself that it is internal guidance only and is not legally binding: "the policy statement is not a final Agency action. It is intended only as a general, internal OSHA guidance, and is to be applied flexibly, in light of all appropriate circumstances. It does not create any

---

[11] 29 U.S.C. § 666(a).

[12] The Act provides that an employer's good faith is to be considered by the Occupational Safety and Health Review Commission in assessing civil penalties. 29 U.S.C. § 666(j). OSHA's *Field Inspection Reference Manual* authorizes the agency to reduce a civil penalty by up to 25 percent when the agency finds the employer acted in good faith.

legal rights, duties, obligations, or defenses, implied or otherwise, for any party, or bind the Agency." Therefore, an employer would find it difficult to successfully hold the agency to the provisions of this policy in any form of legal action.

3.5 Critique

3.5.1 "Routine" Use

The regulated community is concerned about OSHA's commitment to refrain only from "routinely" requesting audit information at the outset of an inspection.[13] The policy is completely silent as to what the agency would consider a "non-routine" situation in which OSHA would ask to see a company's audit report.[14] Indeed, the agency asserts in the supplementary information to the policy that compliance officers must be given discretion in this regard. This leaves employers without guidance as to when the agency will or will not request to see company audit information.

3.5.2 Use of Audit Information to Supplement Other Evidence Already Found

The policy states that OSHA may seek company audit information in circumstances in which the agency already has other evidence of a violation. Thus, there remain many situations in which the agency will still use company audits as further evidence to prove a violation or to establish the gravity of a violation -- all the more reason companies must be diligent about timely correcting hazards identified in audits.

---

[13] See "OSHA Employer Audit Policy Applauded, But Some Questions Remain, ASSE Says", *Occupational Safety and Health* (BNA) Vol. 30, No. 33 Aug. 17, 2000 at 757.

[14] In the draft of the self-audit policy statement that OSHA published for comment in October, 1999, the agency provided examples of situations in which it might seek audit information. Proposed Policy Statement Concerning the Occupational Safety and Health Administration's Use of Voluntary Employer Safety and Health Self-Audits, 64 *Fed. Reg.* 54,358-54,361 (October 6, 1999). The examples provided were where the agency was investigating a fatal or catastrophic accident or where the agency has reason to believe a hazardous violation exists and it is investigating the extent of the hazard.

3.5.3 Penalty Reduction

The 25 % penalty reduction offered for good faith demonstrated by self-auditing is no more than the agency already offered prior to implementation of the policy. The policy simply publicizes the fact that voluntary audits may be used to find good faith. It would have been more effective in terms of encouraging auditing to have offered a greater penalty reduction than that already in place.

4.0 Privileges and Protections from Disclosure of Audit Information

4.1 Introduction

There are certainly many instances in which it is in the company's interest to disclose safety and health audit information to OSHA and other third parties. As already discussed, audit information may be used to establish good faith and thereby avoid a finding of a willful violation. Under OSHA's self-audit policy, the fact the company has performed an audit and taken prompt and effective action to correct the hazards found may result in a 25 percent penalty reduction. Audit information may also be helpful to the company in the civil and criminal litigation context in demonstrating that the company has a strong commitment to safety and health in the workplace.

Unfortunately, there are also many situations in which it is not in the company's best interest to be forced to disclose its audit information. For this reason, company counsel must be knowledgeable about the protections available, how to secure and preserve them, and what their limitations are. This section will discuss three of these protections: the self-audit privilege, the attorney / client privilege, and the attorney work product doctrine.

4.2 The Self-Audit Privilege

The self-audit privilege, at least in its common law version, is seldom found by courts to be available to protect safety and health audits. As is explained in greater detail later in this section, this is partly because courts tend to regard the privilege as unnecessary to encouraging employers to conduct audits and because courts also tend to find that the public good is better served by requiring disclosure of audit information. The rare statutory version of this privilege (where such a statute exists) would presumably be a far more reliable protector of audit information. Whether a company has available to it a statutory audit privilege or must rely on the less dependable common law version, it is important for legal counsel to know what this privilege is and to be thoroughly familiar with its limitations.

4.2.1 The Common Law Audit Privilege

4.2.1.1 General Description

By "common law" audit privilege we mean the privilege that is created by the courts, as opposed to the privilege that is created by the legislature. This is the form of the privilege that we look exclusively to the courts to define.[15]

Courts are very reluctant to create new privileges. This is because generally they do not want to stand in the way of full discovery in litigation.

In the federal courts, privileges are recognized pursuant to Rule 26 of the Federal Rules of Civil Procedure and Rule 501 of the Federal Rules of Evidence.[16] Rule 26 of the Federal Rules of Civil Procedure provides that privileged information is protected from discovery.[17] In other words, a court cannot require a party to reveal to another party privileged information. Rule 501 of the Federal Rules of Evidence provides that evidentiary privileges shall be created and interpreted in the common law, subject to exceptions such as where Congress has created a privilege through statute.

The purpose of the common law audit privilege is to encourage companies to audit their operations concerning subject areas in which there is a strong public interest that accurate and candid self-audits be conducted. In situations where the privilege is applicable, some or all of a company's audit information will be protected from disclosure. In other words, the company will not be required to reveal privileged audit information to an opponent in discovery.

---

[15] Since many of the significant cases concerning evidentiary privileges have arisen outside the safety and health context, but are nevertheless applicable to safety and health audit privilege issues, this section will cite cases that do not necessarily concern safety and health audits per se.

[16] The Occupational Safety and Health Review Commission's Rules of Procedure provide that the Commission will apply the Federal Rules of Civil Procedure, unless the Commission Rules have a specific provision covering a subject, and they provide that the Federal Rules of Evidence apply to Commission proceedings. 29 C.F.R. §§ 2200.2(b), 2200.71.

[17] The analogous Review Commission Rule provides "The information or response sought through discovery may concern any matter that is not privileged...." 29 C.F.R. § 2200.52(b).

4.2.1.2 Factors Used in Determining Whether to Apply the Privilege

Unfortunately, the audit privilege has been inconsistently defined by the courts that do apply it and many courts refuse to even recognize that such a privilege exists. Those courts that do recognize the privilege (often in contexts other than safety and health) will usually consider the following factors in determining whether the privilege will apply: (1) whether the information at issue was generated in the course of a self-audit conducted by the company; (2) whether the company intentionally preserved the confidentiality of the information; (3) whether there is a strong public interest in encouraging audits of this type to be conducted; and (4) whether there is a strong likelihood that not applying the privilege in this context will discourage companies from conducting these types of audits.[18]

This first factor, whether the information at issue was generated in the course of a self-audit conducted by the company, may appear to be an easy one to satisfy, but often it is not. Some courts have found that audits of an investigatory nature (as opposed to a pure compliance review audit) do not qualify for the privilege because the courts find that the company would have conducted these audits regardless of whether a privilege was available.[19] Some courts have refused to apply the privilege to protect self-critical information that was generated in the normal course of business of the company.[20] Again, the courts reason that since the company finds it to be a good business practice to regularly audit its operation, the company does not need a privilege to encourage auditing. Auditing is simply a necessity to running the business.

Confidentiality is a critical factor for securing the privilege. Like other evidentiary privileges, this privilege is considered waived if the information at issue has been disclosed to third parties.[21] Therefore, it is imperative that the company restrict distribution of audit information to those who need that information to assist the company in achieving its objectives in conducting the audit. This may include outside consultants, but even within the company, distribution must be limited. A simple aid to preserving confidentiality and (should the need arise) for demonstrat-

---

[18] *See* Peter A. Gish, "The Self-Critical Analysis Privilege and Environmental Audit Reports," 25 *Envtl. L.* 73, 80-82 (Winter 1995); Donald P. Vandegrift, Jr., "Legal Development: The Privilege of Self-Critical Analysis: A Survey of the Law," 60 *Alb. L. Ref.* 171, 187 (1996).

[19] *See Davidson v. Light*, 79 F.R.D. 137, 139-40 (D. Colo. 1978). *But see Dowling v. American Hawaii Cruises, Inc.*, 971 F.2d 423, 427 (9th Cir. 1992).

[20] *See Dowling*, 971 F.2d at 426. *But see Hickman v. Whirlpool Corp.*, 186 F.R.D. 362 (N.D. Ohio 1999).

[21] *See Peterson v. Chesapeake & Ohio Ry. Co.*, 112 F.R.D. 360, 363 (W.D. Mich. 1986).

ing the company's intent to keep the information confidential is to mark all audit documents "confidential."

The third and fourth factors are related to the purpose of the privilege, which is to protect the public interest in having companies conduct such audits. As discussed above, under these factors, courts will consider the nature of the audit and whether employers will refrain from conducting such audits in the absence of the privilege.[22]

4.2.1.3 Other Limitations in Application of the Audit Privilege

Even where this privilege is found to apply, it is generally limited to protection of opinions and subjective analysis, not facts.[23] Therefore, the common law privilege would most likely not protect information showing that a hazard existed or that management personnel knew the hazard existed, but it may protect opinions of auditors about the scope of the hazard and recommendations for corrective action.

This is a qualified privilege. The privilege can be overcome by the opposing party showing that it has a compelling need for the information, for example, by showing that the information is critical evidence and that there is no other reasonably available means of independently obtaining such information.

The audit privilege generally does not apply to protect against discovery by the government.[24] This is obviously a severe limitation in the safety and health context. This limitation is based on the principal purpose of the common law privilege, which is to further a public interest. Unlike other privileges, this privilege does not serve to protect the owner of the information. Courts, therefore, generally conclude that it is not in the public interest to preclude access to such information by a government agency.

4.2.2 Statutory Audit Privilege

There is no federal statute that provides an audit privilege in the safety and health context. However, there have been multiple bills introduced in Congress in recent years that, if enacted, would have created such a statutory privilege.[25]

---

[22] See ASARCO v. NLRB, 805 F.2d 194 (6th Cir. 1986).
[23] See Price v. County of San Diego, 165 F.R.D. 614, 619 (S.D. Cal. 1996).
[24] See Reich v. Hercules, Inc., 857 F. Supp. 367, 371 (D.N.J. 1994).
[25] See, e.g., Safety and Health Audit Promotion Act of 1999, H.R. 1438, 106th Cong. (1999); Safety and Health Audit Promotion and Whistleblower Improvement Act of 1999, H.R. 1439, 106th Cong. (1999); Safety and Health Improvement and Regulatory Reform Act of 1995, H.R. 1834, 104th Cong. (1995).

At the state level, at the present time, at least one state has enacted a statute creating an audit privilege for safety and health audits.[26] Of course, a state statute would not preclude federal OSHA from obtaining audit information. Neither would it prevent disclosure in federal litigation.

4.3 The Attorney / Client Privilege

Because the self-audit privilege is generally unavailable to protect company safety audits, companies must look to other protections for circumstances in which they seek to avoid having to provide audit information to third parties. The only two other protections likely to be available, however, require attorney involvement in the audit for the protections to apply. Therefore, if the company may want to rely on these other protections, it is necessary before the audit commences to secure the involvement of company counsel in the audit.

The purpose of the attorney / client privilege is to facilitate candid communication between attorneys and their clients, unfettered by concerns that such communications can be subject to discovery. This goal is accomplished by the availability of the attorney / client evidentiary privilege which protects from required disclosure confidential communications between client and attorney the purpose of which were to provide legal services.[27]

This privilege will apply to protect all such communications between the client and attorney, whether those communications consist of fact or opinion. It is important to note, however, that merely sending a document to counsel does not cloak that document with the privilege. The document must have been created for the purpose of assisting the attorney in providing legal advice. Thus, merely copying company counsel on the final audit report, without further involvement from counsel in the planning and conduct of the audit and in making recommendations based on that report, will not be sufficient to protect the report from disclosure.[28]

Counsel must be actively involved in the audit process for audit information to be protected by the attorney / client privilege.[29] It is best if company counsel actually leads the audit, at least by directing that the audit be performed and reviewing the audit findings and recommendations, and adding legal conclusions

---

[26] See Environmental, Health, and Safety Audit Privilege Act, *Tex. Rev. Civ. Stat. Art.* 4447cc (2000).

[27] See *Upjohn v. United States*, 449 U.S. 383, 389-90 (1981).

[28] See *Hercules, Inc.*, 857 F. Supp. at 372.

[29] *Id.*

and advice to the audit report. Since the privilege does not apply to lawyer's providing business advice, it is critical that the lawyer's role in the audit process be to provide legal advice to the company.

This does not mean that counsel must participate in all aspects of the audit. The privilege will extend to communications with and between members of the audit team (even if outside consultants are used) as long as the ultimate purpose of those communications is to enable company counsel to provide legal advice to the company (presumably on compliance and legal liability issues).[30] In this regard, it is best if company counsel is actively involved in the selection and direction of audit team members.

As with the self-audit privilege, the attorney / client privilege will be considered to have been waived if audit information is not kept strictly confidential. The same methods to preserve the confidentiality of audit information for purposes of the audit privilege apply to preserve the attorney / client privilege. Limiting distribution within the company may be as important as preventing disclosure outside the company.

Unlike the audit privilege, the attorney / client privilege is absolute. It cannot be overcome by a claim that the party seeking the information has a compelling need for that information.

4.4 Attorney Work Product Doctrine

The work product doctrine is more limited than the attorney / client privilege. It will protect confidential opinions and analyses of the attorney that appear in audit information only if such material was created in anticipation of litigation.[31] This is a substantial limitation for safety and health audits because they often are not conducted in anticipation of litigation. The fact that they may be conducted with an eye to avoiding the remote possibility of litigation in the future has been found by some courts to not be enough to justify application of the work product doctrine.[32]

---

[30] *See In re Grand Jury Matter*, 147 F.R.D. 82, 84-86 (E.D. Pa. 1992).
[31] *See Hickman v. Taylor*, 329 U.S. 495, 510-11 (1947); Fed. R. Civ. P. 26(b)(3).
[32] *See Martin v. Bally's Park Place Hotel & Casino*, 983 F.2d 1252, 1260-61 (3d Cir. 1993).

The work product doctrine generally does not protect factual information, only opinions and analysis of counsel.[33] Therefore, the work product doctrine will not protect all of a company's audit information.

Like the audit privilege, the protection from disclosure that work product doctrine otherwise may provide can be overcome by a showing of substantial need on the part of the party seeking access to the information.[34] If that party can show that it can obtain the information only through extraordinary means or that it cannot obtain the information from any other source and that the information is critical to its case, a court is likely to find that the information must be handed over.

5.0 Conclusion

Given their increasing reliance on workplace audits to improve safety and health and to monitor regulatory compliance, employers are considering what protections may be available in the event they perceive a need to protect audit information from being used as adverse evidence in an agency enforcement action or in litigation. In most circumstances, safety and health audits will not be protected by a self-audit privilege. Other protections are also limited and usually necessitate substantial involvement of counsel. These include the attorney / client privilege and the attorney work product doctrine.

In view of the foregoing, it is in the interest of every employer to treat all audit activities and reports with a view to their possibly becoming public or at least available to adverse parties in litigation. Therefore, it is critical that companies conducting audits follow through appropriately and ensure prompt correction of every deficiency found. In this way, companies will fully benefit from the auditing process and the safety and health of their workplaces will be enhanced.

---

[33] *See Hickman*, 329 U.S. at 511; *Hercules*, 857 F. Supp. at 373. *But see Bally's Park*, 983 F.2d at 1261-62.
[34] *See Bally's Park*, 983 F.2d at 1255.

Chapter 10

Inspections

Lawrence P. Halprin, Esq.
Keller and Heckman, LLP
Washington, D.C.

1.0 Overview

To achieve the purpose of the OSH Act, Congress authorized the Secretary to conduct workplace safety and health inspections, and to enter workplaces for that purpose. OSHA's inspectors are called compliance safety and health officers (CSHOS) or "compliance" officers. Typically, an inspector focuses on either safety hazards (and is known as a "safety officer" or "safety inspector") or health hazards (and is known as an "industrial hygiene officer" or "health inspector"). Traditionally, the role of the OSHA inspector has been to inspect workplaces, gather evidence of non-compliance, write up citations to be issued by their area directors, participate in informal settlement conferences,[1] and testify for the Secretary in cases where contests are filed.[2] More recently, there has been an effort to have compliance officers advise employers on how they might abate or correct a hazard, or improve their workplace safety and health practices, even in circumstances where no apparent violation is found, or no citation is issued.

OSHA's inspections can be broadly categorized into two types: programmed (scheduled in advance based on a neutral administrative scheme) and unprogrammed (scheduled in response to an accident, complaint, referral, etc.). Prior to 1997, OSHA's inspection-selection plans for programmed inspections were

---

[1] Unfortunately, there are some state plan states that do not require the participation of the compliance officer in the informal settlement conference. Where the inspector is absent but supervisory enforcement personnel view the inspector as a necessary participant and are not willing to engage in good faith discussions in the inspector's absence, the informal settlement conference generally becomes a frustrating waste of time.

[2] Compliance officers also make special inspections to determine the position the Secretary will take with regard to employer petitions for variances, to amend abatement dates and with regard to various other applications by employees.

based primarily on industry hazard rankings,[3] establishment lists,[4] and an inspection register.[5] Since 1997, OSHA has relied primarily on site-specific data (lost workday injury and illness rates) derived from OSHA Form 200 data submitted to OSHA as part of the agency's annual data collection survey. The survey is sent to approximately 80,000 to 120,000 participating employers.

2.0 OSHA's Authority to Inspect Places of Employment

OSHA's authority to inspect workplaces is set forth in Section 8(a) of the Act,[6] as follows:

1. to enter without delay and at reasonable times any factory, plant, establishment, construction site, or other area, workplace or environment where work is performed by an employee of an employer; and

2. to inspect and investigate during regular working hours and at other reasonable times, and within reasonable limits and in a reasonable manner, any such place of employment and all pertinent conditions, structures, machines, apparatus, devices, equipment, and materials therein, and to question privately any such employer, owner, operator, agent or employee.

3.0 The 4th Amendment's Prohibition against Unreasonable Search and Seizure, Warrants, and Probable Cause

On its face, the language of Section 8(a) quoted above seems to authorize inspections without a warrant or its equivalent. However, in *Marshall v. Barlow's Inc.*,[7] the

---

[3] The industry hazard ranking reports were supplied to each local area office by OSHA's National Office in Washington, D.C. It ranked *industries* (by Standard Industrial Classification or SIC) for safety hazards according to their LWDI (lost workday injury) rates, and for health hazards according to the average number of serious citations issued per OSHA inspection.

[4] The *establishment list* for a particular area office contains the names of those establishments in the high hazard SICs located within the local OSHA office's territorial jurisdiction that are a part of those industries. The list is also supplied by the OSHA National Office.

[5] The *"inspection register"* is made up by the local area office. It lists the name of each establishment actually scheduled for inspection in the current cycle, and the order in which the establishments will be inspected. However, strict adherence to that order is not required. Firms may be inspected in any order that makes efficient use of resources. Similar inspection targeting procedures are employed by the states that operate their own OSHA programs.

[6] OSH Act Section 8(a), 29 U.S.C. § 657(a).

Supreme Court pointed out that the basic purpose of the Fourth Amendment to the Constitution of the United States "is to safeguard the privacy and security against arbitrary invasions by governmental officials," and held that:

> The businessman, like the occupant of a residence, has a constitutional right to go about his business free from unreasonable official entries upon his private commercial property. The businessman, too, has that right placed in jeopardy if the decision to enter and inspect for violation of regulatory laws can be made and enforced by the inspector in the field without official authority evidenced by a warrant.[8]

Consequently, OSHA cannot enter any private premises for inspection purposes unless it first obtains either the employer's consent or a warrant issued by a court authorizing the inspection. An application for a warrant will state the reasons why the business was selected for inspection and will be signed under oath. The warrant is usually obtained ex parte,[9] and is issued by a U.S. District Court (usually over the signature of a magistrate), or the corresponding state court in a state plan state, upon a finding of "probable cause" as defined below. A warrant can be requested prior to or at any time during the course of an inspection.

Warrants generally contain limitations on the time, scope and manner of inspection. Some restrict the scope of the inspection to, for example, the subject of an employee complaint; others are as broad as the authority granted in Section 8(a). On the other hand, there are no specific legal restrictions on the time, scope, or manner in which warrantless inspections are conducted unless the employer who consents to the inspection specifies such conditions in his consent and OSHA agrees to the restrictions. In other words, an employer may condition its consent to an inspection on virtually any restriction the inspector will agree to and may withdraw its consent to a warrantless inspection at any time. An employer who permits a warrantless inspection is generally deemed to have consented to it and waived its right to object to the manner in which it was conducted.

If OSHA requests a court to issue a warrant, it must establish probable cause to justify the search. Except in extraordinary cases where OSHA has initiated a criminal investigation, OSHA can gain access to a workplace under the authority of an administrative inspection warrant which is available under a reduced showing of probable cause. There are at least two probable cause tests under which OSHA can

---

[7] *Marshall v. Barlow's Inc.*, 436 U.S. 307 (1978).
[8] *Marshall v. Barlow's Inc.*, 436 U.S. 307, 312 (1978).
[9] The term "ex parte" means without notice to the employer.

obtain an administrative inspection warrant. First, OSHA can establish probable cause for an un-programmed inspection by offering *specific evidence* which indicates that it is likely one or more violations exist in the place to be inspected. These inspections are generally limited to the specific matter which gives rise to them. Second, OSHA can establish probable cause for a programmed inspection by showing that the planned inspection is part of a *neutral administrative selection plan* that satisfies reasonable legislative or administrative standards for conducting an inspection.[10]

There are two initial requirements that must be established by OSHA to satisfy the probable cause test applicable to a neutral administrative plan. The agency's warrant application must: (1) identify the plan that was used to select the establishment targeted for inspection; and (2) demonstrate that the administrative plan was neutral (unbiased), both as written and in its implementation.[11]

Evidence of probable cause under the *specific evidence* test can arise from several sources. The most common sources are written or oral employee complaints, the need for a follow-up inspection, a referral, or a newspaper article. OSHA may also derive evidence of possible violations from the observations of hazardous conditions by a compliance officer, admissions by the employer's representatives, and accident reports.

The information presented in the warrant application must be sufficient to satisfy an impartial judge that a violation of law probably (not possibly) exists on the premises sought to be inspected. The judge must conduct an independent review and evaluation of the asserted basis for the inspection, and may not simply accept OSHA's assertion that a violation exists.[12]

A magistrate reviewing the evidence offered by OSHA to determine whether probable cause exists is supposed to consider the following four factors:

1. *specificity of the information*: at a minimum, the nature of the hazardous condition must be disclosed so the magistrate can determine whether the conditions alleged constitute a violation of the Act; apparently, even a conclusory description will suffice.

---

[10] *Marshall v. Barlow's Inc.*, at 320.

[11] *Establishment inspection of Northwest Airlines*, 587 F.2d 12, 14-15 (7th Cir. 1978).

[12] "We decline to strip the magistrate of his probable cause function, as the very purpose of a warrant is to 'have the probable cause determination made by a detached judicial officer rather than by a perhaps over-zealous law enforcement agency." *Weyerhauser Co. v. Marshall*, 592 F.2d 373, 378 (7th Cir. 1979).

2. *likelihood of a violation*: there must be some plausible basis for believing that a violation is likely to exist; the facts must be sufficient to warrant further investigation or testing.
3. corroboration of the information by other sources.
4. *staleness*: are the conditions alleged of a type that will likely disappear through mere passage of time.

The standard for specificity of the alleged violative conditions is very low. A conclusory description by the compliance officer made in an affidavit to support the warrant is usually sufficient. OSHA usually does not know beforehand whether violations actually exist. The agency is not required to cite specific regulations it believes have been violated in order to obtain a warrant.[13]

4.0 Consensual Searches v. Warrant Requirement

4.1 Background

An important question faced by employers is whether to grant permission for OSHA to conduct a warrantless inspection or to deny entry and force OSHA to obtain a warrant. Few companies have adopted policies that prohibit all warrantless searches of their premises. Of course, once a policy to this effect is established and made known to OSHA, the agency will simply obtain an ex parte warrant before it arrives at the employer's premises to conduct the inspection.

Most companies usually allow warrantless searches and believe that warrants should only be requested in limited circumstances. They feel that requiring a warrant may antagonize the inspectors and affords little protection to the company. Furthermore, the typical boilerplate warrant generally gives OSHA far more investigative power than an employer would permit during a warrantless (consensual) inspection. Companies usually can protect themselves against unreasonable inspections by handling the actual investigation in a prudent manner.

The circumstances will usually dictate whether an employer should consent to a warrantless inspection. For instance, the employer may be more inclined to demand a warrant if the employer perceives this to be a case of harassment or overreaching, or an unjustified effort to expand the scope of a complaint or referral inspection, or

---

[13] *In re Establishment Inspection of at Jeep Corp.*, 836 F.2d 1026 (6th Cir. 1988).

if the employer needs time to assemble its inspection team and OSHA refuses to cooperate.[14]

In the event that an OSHA violation results in a fatality, (or in a serious non-fatal injury in California) the employer must recognize that there is a potential for criminal liability, not only against the corporation, but against individual employees, usually managerial officials, as well. The risk of criminal liability is especially high for employers who have accepted a citation for the same alleged violation in the past or have a history of civil violations involving the same kind of circumstance as those which led to the fatality. The employer must also be aware of the general rule that evidence uncovered by OSHA during an administrative inspection, conducted under the authority of an administrative inspection warrant, may be used in a criminal prosecution. At the point where OSHA decides to treat the matter as a criminal case, it must then proceed under a criminal search warrant. By that time, the agency generally has all the information it needs to pursue a criminal prosecution.

In the past, there had been questions about whether the OSH Act preempts state criminal laws. Now, however, states are generally free to enforce their criminal laws to further valid state interests.[15] Beginning in the mid-1980's, there has been a substantial increase in the number of state criminal prosecutions brought as a result of fatalities or injuries in the workplace.[16] The charges have included reckless endangerment, manslaughter, and murder.

For example, in Texas, a construction company was charged and found guilty of criminally negligent homicide in the death of an employee killed in a trench cave-in and was assessed a fine of $20,000.[17] California has prosecuted many cases including the prosecution of a corporation for manslaughter following an industrial accident.[18]

---

[14] See the discussion in Chapter 1 regarding instance-by-instance or egregious violations.

[15] *See, e.g., Note, Getting Away with Murder: Federal OSHA Preemption of State Criminal Prosecutions for Industrial Accidents,* 101 Harv.L.Rev. 535 (1988); *Michigan v. Hegedus,* 443 N.W.2d 127 (1989); *Illinois v. Chicago Magnet Wire Corp.,* 1989 O.S.H. Dec. (CCH) § 28,421 (S. Ct. Ill. 1989).

[16] *Id.*

[17] *Peabody Southwest, Inc., v. Texas,* 1991 O.S.H. Dec. (CCH) § 29,502 (Tex. Ct. App. 1991); *Sabine Consolidated, Inc. v. State,* 1991 O.S.H. Dec. (CCH) § 29,491 (Tex. Ct. App. 1991).

[18] *Granite Construction Co. v. Superior Court,* 149 Cal. App. 3d 465, 197 Cal. Rptr. 3 (1984). See, OSHA Compliance, Human Resources Management-New Developments (CCH) § 13,569 – Cal / OSHA Reports on Criminal Prosecutions in 1989 (1990).

Business owners and managers should be guided by the knowledge that they may be exposed to criminal liability for workplace conditions. An employer may, therefore, insist that the compliance officer obtain a warrant, and he or she may refuse to answer certain questions or even refuse to speak at all concerning the accident. This would be a valid exercise of an individual's Fifth Amendment right against self-incrimination. A corporation does not have such a Fifth Amendment right; but an individual, whose responses may later be used in a criminal prosecution, does have such a right.

4.2 Challenging a Warrant

An employer must be aware of the potential consequences of refusing to permit inspections pursuant to warrants because the Secretary has brought a number of contempt proceedings (both civil and criminal)[19] against employers for refusing such warrant inspections. If found in contempt of court, the employer may be subject to sanctions, including the assessment of fines and the Secretary's costs in the contempt action.[20]

For example, in *Donovan v. Southwest Electric Co.*,[21] the court found the warrant to have been validly issued and based on a general administrative plan derived from neutral sources. Thus, the employer's refusal to honor the warrant for inspection of its premises was sufficient to hold the employer in civil contempt and to assess costs and expenses as compensation for the Secretary's efforts to inspect the employer's worksite and to enforce the inspection warrant in subsequent court proceedings.[22]

Employers generally have been unsuccessful in contempt hearings. One reason is because courts have held that good faith is no defense for refusing to comply with

---

[19] Civil contempt is a sanction to enforce compliance with an order of the court or to compensate for losses or damages sustained by reason of noncompliance, while the purpose of criminal contempt is to punish past defiance of a court's judicial authority, thereby vindicating the court. *Shillitani v. United States*, 384 U..S. 364, 369-370 (1966); *McComb v. Jacksonville Paper Co.*, 336 U.S. 187, 191 (1949).

[20] *Donovan v. Hackney, Inc.*, 769 F.2d 650 (10th Cir. 1985); *Donovan v. Wollaston Alloys, Inc.*, 11-O.S.H. Cas. (BNA) 1587 (Mass. Dist. Ct.); *Donovan v. Southwest Electric Co., 10* O.S.H. Cas. (BNA) 1849 (Rev. Comm'n. 1982); *Marshall v. Grand River Dam Authority*, 10 O.S.H. Cas. (BNA) 1542 (Rev. Comm'n. 1981) (Ct. has discretion to award those expenses necessary to make the Secretary whole in a contempt action).

[21] *Donovan v. Southwest Electric Co., 10* O.S.H. Cas. (BNA) 1849 (Rev. Comm'n. 1982).

[22] *Southwest Electric Co., 10* O.S.H. Cas. (BNA) at 1849.

a warrant.[23] Another reason employers have lost contempt hearings is due to the fact that the probable cause determination made by the judge who issued the warrant is entitled to great deference and is conclusive in the absence of arbitrariness or clear legal error.[24] Moreover, the Secretary does not need to prove willfulness in order to be awarded costs and fees.[25] If an employer determines that a warrant was improperly issued, it must determine whether to file a motion to suppress the warrant, and whether to refuse to honor the warrant until the motion to suppress is decided, or whether to allow the inspection to proceed under protest and challenge the warrant in any subsequent enforcement proceeding.

4.3 Warrant Requirement Exceptions

There are circumstances under which OSHA does not need a warrant to make an inspection. For example, a third party may control the worksite and give consent.[26] The standard of consent for administrative inspections is less stringent than that required for criminal searches. The consent to an administrative inspection need not be express, and the failure to object to a known search constitutes consent.[27] In addition, OSHA does not need a warrant when unsafe conditions are in plain view of inspectors while lawfully on or off the premises.[28] In theory, OSHA may also be able to gain access under emergency conditions.[29]

---

[23] *Donovan v. Enterprise Foundry Inc.*, 751 F.2d 30 (1st Cir. 1984)); *NLRB v. Maine Caterers, Inc.*, 732 F.2d 689, 690 (1st Cir. 1984).

[24] See *West Point-Pepperell, Inc., v. Donovan*, 689 F.2d 950, 959 (11th Cir. 1982)(citing *Spinelli v. United States*, 393 U.S. 410, 419 (1969)).

[25] *Donovan v. Burlington Northern, Inc.*, 781 F.2d 680 (9th Cir. 1986)(citing *Perry v. O'Donnell*, 759 F.2d 702, 705 (9th Cir. 1985)).

[26] *J.L. Foti Const. Co. v. Donovan*, 786 F.2d 714 (6th Cir. 1986) (consent by worksite's general contractor); *Donovan v. A.A. Beiro Constr. Co.*, 746 F.2d 894 (D.C. Cir. 1984) (prime contractors working at construction site owned by the District of Columbia consented to an inspection); *Stephenson Enterprises, Inc. v. Marshall*, 578 F.2d 1021, 1024 (5th Cir. 1978) (consent by plant manager upheld). Furthermore, voluntary consent by the employer to a search obviates any need for OSHA to obtain a warrant, and the consent prevents the employer from later asserting that the search somehow violated his expectation of privacy.

[27] *United States v. Thriftmarts Inc.*, 429 F.2d 1006 (9th Cir. 1970)).

[28] *Stephenson Enters. v. Marshall*, 578 F.2d 1021, 1024 n.2 (5th Cir. 1978); *Lake Butler Apparel Co. v. Secretary of Labor*, 519 F.2d 84, 88 (5th Cir. 1975).

[29] See, *Camara v. Municipal Court*, 387 U.S. 523, 539 (1967).

5.0 Types of Inspections

OSHA and its state counterparts conduct between 60,000 and 90,000 inspections per year. That means that, on average, 6 million workplaces covered by the Act can be inspected once every 67 to 100 years. Therefore, OSHA has established a system of inspection priorities.

Imminent danger situations are given top priority. Imminent danger means any condition which could reasonably be expected to cause death or serious physical harm immediately or before the danger can be eliminated through normal enforcement procedures.[30] Second priority is given to investigation of fatalities and catastrophes resulting in-patient hospitalization of three or more employees. Such situations must be reported to OSHA by the employer within 8 hours. Third priority is given to employee complaints of alleged violation of standards or of unsafe or unhealthful working conditions.[31]

Next in priority are programmed, or planned, inspections aimed at specific high-hazard workplaces, industries, occupations, activities or toxic chemicals. For example, OSHA conducts special emphasis inspections on a national, regional or area office basis in one or more industries in which special or particular hazards (*e.g.*, PSM in the chemical industry, bloodborne pathogens in the health care industry, crystalline silica exposure control in the sand, gravel, forging and sandblasting industries). In general, the lowest priority is given to follow-up inspections, which determine whether previously-cited violations have been corrected. If an employer has failed to abate a violation, the compliance officer informs the employer that he or she is subject to further citations ("Notification of Failure to Abate") and proposed daily penalties for each day such failure or violation continues.

6.0 Inspection Procedures

6.1 General Requirements

As noted above, the OSH Act mandates that each OSHA inspection must be conducted in a "reasonable manner," at reasonable times and within reasonable limits; that it must be preceded by presentation to the employer of the inspector's

---

[30] 29 C.F.R. § 1903.13.
[31] *Id.* § 1903.11.

credentials; and that representatives of both the employer and the employees must be given the opportunity to accompany the inspector on his rounds.[32] The on-site procedures followed by OSHA during an inspection are set forth in detail in Chapter 3 of the OSHA *Field Inspection Reference Manual which generally replaced the OSHA Field Operations Manual.*

Generally, the employer will not have advance notice of an inspection. Section 17(f) of the OSH Act and the regulation at 29 C.F.R. § 1903.6 generally prohibit the giving of advance notice of inspections, except as authorized by the Secretary.[33] Exceptions to this rule occur when advance notice is necessary to conduct an effective investigation or to quickly identify and eliminate an imminent danger. Cases requiring advance notice to conduct an effective investigation generally involve situations where: (1) it is necessary to conduct the investigation after regular business hours; (2) it is necessary to ensure the presence of employer and employee representatives; or (3) it is necessary to ensure the employer's cooperation. The last situation might involve the sampling of airborne exposure levels or noise levels where the employer insists on conducting side-by-side samplings.

6.2 Opening Conference

An inspection is an evidence gathering expedition. The purpose from OSHA's point of view is to gather evidence to support a prosecution (civil or criminal or both). *It is not a consultation.*[34]

The initial contact between the inspector and the employer is known as the "opening conference," and is mandated by OSHA regulations.[35] The "opening conference" begins with the presentation of the compliance officer's credentials to

---

[32] OSH Act, §§ 8(a) and (e); 29 U.S.C. §§ 657(a) and (e).

[33] Occupational Health and Safety Administration *Field Inspection Reference Manual*, U.S. Department of Labor, OSHA Instruction CPL 2.103, at I(E)(3) - Advance Notice of Inspections, p. 1-11 (Sept. 26, 1994).

[34] All too often employers report that OSHA inspectors gave the company's representative the impression that no penalties or citations will issue if the employer corrects conditions the inspector states are hazardous. The impression is not correct. Citations and proposed penalties always issue if the inspector believes the conditions constitute violations of a standard or the General Duty Clause.

[35] 29 C.F.R. § 1903.7(a). That does not mean, however, that the opening conference must be held, or that an employer is automatically entitled to have any citation vacated because no opening conference was held. A citation will not be vacated unless the failure to hold an opening conference resulted in substantial prejudice to the employer.

the employer or its agent.[36] At the opening conference, the compliance officer explains the purpose of the investigation (*i.e.*, to investigate an alleged imminent danger situation, a complaint, a fatality, or to conduct a programmed health or safety inspection). If the inspection is the result of an employee complaint, the employer is entitled to a copy of the complaint, and should ask for a copy if the inspector does not promptly provide one. The employee's name will be withheld if the employee has requested anonymity.

During the opening conference, the employer is told of the planned scope of the inspection. Generally, it will include a physical inspection of the workplace and records, may include private employee interviews, and remains subject to change depending on what occurs during the course of the inspection. The employer may also be given copies of applicable laws, standards and regulations. At this time, the compliance officer also is supposed to determine whether or not the employer is covered by any of the exemptions or limitations that might apply – *e.g.*, participation in the OSHA state consultation program, VPP or another partnership program, or pre-emption by virtue of the exercise of safety and health authority by another agency.

At the opening conference, the compliance officer will inquire whether the employees are represented by a union. If they are, the inspector will request that a union representative be summoned to act as the "employee representative" during the inspection.[37] If the employees are not represented by a union the inspector may ask to speak with members of a site safety committee (if one exists) or simply seek to interview a representative number of employees during the course of the inspection.

During the opening conference, the compliance officer will ask questions about the employer's business to confirm that the employer is covered by the OSH Act and to get other information (*e.g.*, number of employees at the site and within the corporate family) that OSHA keeps in its database and may bear on penalty adjustments and referral inspections.[38] More recently, the opening conference, particularly for a programmed inspection, has included an inquiry into the content

---

[36] *Id.*

[37] *See* 29 C.F.R. § 1903.8.

[38] OSHA maintains a data base of inspections it conducts. The database may be called upon by any area office to determine an employer's past history. The data base should reveal,, whether the employer has previously been cited and it should indicate the disposition made of the citations.

and effectiveness of the employer's safety and health management program at the site.[39]

Any safety and health hazards that may be encountered by the inspector during the inspection should be reviewed at this time so that appropriate safety precautions are followed and appropriate personal protective equipment is identified and available for use.[40] Also, during the opening conference, the employer representative should identify areas in the establishment that might reveal trade secrets and discuss any objections to or special procedures for the taking of photographs or videos with the compliance officer.[41] In carrying out their enforcement responsibilities, the compliance officer and all other OSHA personnel are required to preserve the confidentiality of all information and investigations that might reveal an employer's trade secrets.[42] The U.S. Criminal Code and OSHA regulations specify the penalties for unauthorized disclosure of trade secret information.[43]

The employer should also mention whether areas the compliance officer has selected to inspect are subject to government security regulations. The compliance officer has to have security clearance to inspect such areas. If the compliance officer is not cleared, arrangements must be made with the area office to obtain an inspector who is properly cleared to make the inspection.[44]

The employer has a right to participate in the inspection process and should diligently exercise that right. The employer should feel free to ask questions at this time as well as express its preferences and concerns as well as any restrictions regarding the manner in which the inspection is to be conducted. In many cases, OSHA will be receptive to reasonable suggestions. An inspection procedure can be worked out that truly is mutually beneficial to both OSHA and the company. For example, the inspector may agree to a timetable suggested by the employer.

---

[39] This inquiry generally addresses the topics covered by OSHA's Program Evaluation Guidelines (PEG), also known as the Program Evaluation Profile (PEP).

[40] 29 C.F.R. § 1903.7(c).

[41] 29 C.F.R. § 1903.7(b).

[42] 29 C.F.R. §1903.9(c). Trade secrets are any confidential business device or process which gives an employer an advantage over competitors. Occupational Safety and Health *Field Inspection Reference Manual, supra* note 33, at II(A)(3)(i) and II(A)(4)(g) - Trade Secrets (1994).

[43] 18 U.S.C. 1905; 29 C.F.R. § 1903.9(b).

[44] Occupational Safety and Health *Field Inspection Reference Manual, supra* note 33, at I-12- Personal Security Clearance. It is the Regional Administrator's duty to ensure that an adequate-number of compliance officers with appropriate security clearances are available within the Region and that their security clearances are current. *Id.*

6.3 Records Inspection / Subpoenas

Section 8(c) of the Act requires employers to "make, keep and preserve, and make available" such records as the Secretary "may prescribe by regulation as necessary or appropriate for the enforcement of OSHA."[45] This section also requires employers to maintain accurate records of and to make periodic reports on "work-related deaths, injuries and illnesses."[46] Relying on that statutory authority, the Secretary has issued regulations which specify records that must be made and maintained.[47] In addition, OSHA has adopted regulations which purport to authorize a compliance officer, during the inspection, to review records required by the Act or OSHA regulations, and other records which are directly related to the purpose of the inspection.[48]

A review of the employer's injury and illness records is conducted whenever an establishment is scheduled for an inspection. The purpose of this review is to identify activities, programs, conditions and / or individuals for follow up investigation and calculate a Lost Workday Injury and Illness ("LWDII") case rate for the establishment. The results of the records review, including the calculated LWDII rate, are given to the employee representative.

Accordingly, during the opening conference, the inspector will ask to review the OSHA Form 200 and the corresponding OSHA Form 101s (or Worker Compensation First Report of Injury Forms) for the current year, and quite likely the Form 200 for at least two prior years. The inspector will also ask for the total "hours worked" figures for those years.[49] The information provided on the OSHA 200 (or 101) may provoke questions in the inspector's mind or indicate certain machines, processes or personnel he / she will want to observe or investigate.

During the records inspection, the OSHA compliance officer will also start a review of the employer's hazard communication program, lockout / tagout program, personal protective equipment program, and in some cases, the employer's

---

[45] OSH Act, § 8(c)(1); 29 U.S.C. § 657(c)(1).
[46] 29 U.S.C. § 657 (c)(2).
[47] *See* 29 C.F.R. §§ 1904.2, 1904.4, 1904.6.
[48] 29 C.F.R. §§ 1903.3(a), 1904.7.
[49] OSHA Form No. 200 is required by 29 C.F.R. § 1904.2(a). On the last *Federal Register* publication date of the Clinton Administration, OSHA finalized a substantially revised version of 29 C.F.R. § 1904. Occupational Injury and Illness Recording and Reporting Requirement, 66 *Fed. Reg.* 5416 (January 19, 2001). Assuming the rule is not revoked or delayed, it is scheduled to go into effect on January 1, 2002. The new rule would replace OSHA Forms 200 and 101 with 300, 300-A, and 301.

confined spaces and bloodborne pathogens programs.[50] Usually, the inspector will ask for a copy of the written plans which the employer is required to keep. In addition to the records required to be maintained by OSHA rules, compliance officers routinely ask to examine other employer records during a worksite inspection because the records may contain information showing knowledge of an OSHA standard and possible violations of the standard. Such documents would then provide damaging evidence to be used against the employer.

This is a split of authority as to whether a compliance officer is required to obtain a warrant or a subpoena in order to obtain records that an employer is required to keep by regulation. Some courts have held that OSHA cannot gain access to the records on workplace injuries and illnesses without a warrant or subpoena because the employer maintains a privacy interest in the records.[51] For instance, the court in *Brock v. Emerson Electric Co.* quoted *See v. City of Seattle*, which stated that "the agency has the right to conduct all reasonable inspections of such documents which are contemplated by statute, but it must delimit the confines of a search by designating the needed documents in a formal subpoena."[52] Other courts, however, have concluded that the employer has no reasonable expectation of privacy in providing OSHA with documents it is required by regulation to maintain and therefore that OSHA need not procure a warrant or administrative subpoena in order to obtain access to these records.[53]

Compliance officers are authorized to issue subpoenas, but they have no legal effect unless OSHA persuades a court to enforce them.[54] The employer is protected because he or she is allowed to question the reasonableness of the subpoena and raise any available objections to the petition filed by OSHA to enforce the subpoena before suffering any penalties for refusing to comply with it.[55] Unlike inspection

---

[50] Occupational Safety and Health *Field Inspection Reference Manual, supra* note 33, II(A)(4)(a), P. II-14 - Hazard Communication.

[51] *McLaughlin v. Taft Broadcasting Co., King's Island Div.*, 849 F.2d 990 (6th Cir. 1988); *Brock v. Emerson Elec. Co.*, 834 F.2d 994 (11th Cir. 1987).

[52] *See v. City of Seattle*, 387 U.S. 541, 544-45 (1967).

[53] *See McLaughlin v. A.B. Chance Co.*, 842 F.2d 724 (4th Cir. 1988). The court reasoned that since 29 U.S.C. § 1904.5 requires the annual summary forms to be posted in a place where notices to employees are customarily posted, there can be little expectation of privacy in information that is available to anyone observing the employer's bulletin board. *Id.* at 728.

[54] OSH Act, § 8(b); 29 U.S.C. § 657(b); OSHA *Field Inspection Reference Manual, supra* note 33 at II(A)(c)(3) - Administrative Subpoena.

[55] *See v. City of Seattle*, 387 U.S. at 544-45. The court in *Lone Steer* explained that the employer's defenses do not include the right to insist upon a judicial warrant as a condition precedent to valid administrative subpoena. 464 U.S. at 415.

warrants, OSHA may not obtain enforcement of civil subpoenas through an *ex parte* process. All subpoenas of documents are required to be "sufficiently limited in scope, relevant in purpose, and specific in directive so that compliance will not be unreasonably burdensome."[56]

The Eighth Circuit, in *Donovan v. Union Packing Co.*,[57] upheld OSHA's authority to issue and obtain enforcement of administrative subpoenas for employer records, even independent of an inspection of the premises. The court in *Wollaston Alloys* also held that the Secretary may review forms required under OSHA section 657(c) through a subpoena.[58] The First Circuit noted the specific language of section 657(c), which requires employers to "make, keep, and...make available" records which the Secretary prescribes, and stated that it was "reasonable to assume that an employer would have less of a privacy interest in a document it is required by statute or regulation to maintain than in a document it produces and maintains on its own."[59]

6.4 The Walkaround

Once the inspector has either finished with his / her review of the records, or reached an appropriate transition point, the "walk-around" begins. That is an OSHA term used to describe the process by which the inspector becomes more familiar with the workplace and its employees by simply walking (at a widely varying pace) through it, generally focusing on the places where employees are at work, what they are doing and the processes used by the employer. This process could involve anything from a quick look to a lengthy investigation with extensive photographs, videos, sampling and employee interviews. Frequently, during the walkaround, the inspector makes notes of things to re-inspect or otherwise further investigate.

General programmed inspections (both safety and health inspections) usually involve a walkaround inspection that is fairly broad in scope. The walkaround portion of a health inspection is frequently followed by exposure monitoring (*e.g.*, airborne levels of toxic substances, noise levels, etc.) if the inspector has reason to believe that the workers are exposed to excessive levels in the workplace. The

---

[56] *Donovan v. Lone Steer, Inc.*, 464 U.S. 408, 415 (1984) (quoting *See*, 387 U.S. at 544); *See also, United States v. Morton Salt Co.*, 338 U.S. 632, 652-653 (1950).
[57] *Donovan v. Union Packing*, 714 F.2d 833 (8th Cir. 1983).
[58] *Donovan v. Wollaston Alloys*, 695 F.2d 1, 8 (1st Cir. 1982).
[59] *Id.*

walkaround is used to assist the OSHA compliance officer to determine where and what to monitor.

A more thorough inspection will generally involve the taking of photographs or videos by the inspector, the questioning of some employees, and perhaps sampling for air contaminants and other measurements of exposures to ergonomic (musculoskeletal disorder) risk factors. All of this depends upon the type of inspection (health or safety), the reason why it is being conducted (complaint or routine), the information the OSHA inspector has obtained up to that point, the inspector's schedule and other factors.

6.5 Pictures, Documents and Other Evidence

It is important to remember that an OSHA inspection is an evidence gathering event. The compliance officer is gathering evidence to document and ultimately prove violations. Such documentary support includes: photographs, videotapes, audiotapes, sampling data, statements from the employer and employees, and other company documents (*e.g.*, internal audit reports and the reports of consultants or insurance company loss control representatives).[60] In its recently issued policy on access to employer self-audits, OSHA enunciated a policy of generally not seeking an employer's self-audit reports (whether performed by the employer or an outside consultant on the employer's behalf) in pursuit of a "fishing expedition." However, OSHA indicated that it would seek the pertinent portion of an audit report in circumstances where it had independent evidence of a violation and sought to determine whether the employer was aware of the nature and extent of the hazard.[61]

In addition to the foregoing types of employer documents, evidence may be obtained from trade association articles, minutes from safety meetings, complaints from employees, memoranda and correspondence from safety personnel (especially from plant or corporate safety personnel to management), and notes relating to OSHA activities and industry practice in other companies.[62] Such company documents may be obtained by warrant or subpoena if the inspector has reason to believe they exist, they are not produced voluntarily when requested, and the inspector determines they are important to his investigation.

---

[60] 29 C.F.R. § 1903.7 (b).

[61] Final Policy Concerning the Occupational Safety and Health Administration's Treatment of Voluntary Employer Safety and Health Self-Audits, 65 *Fed. Reg.* 46498 (July 28, 2000).

[62] OSHA Instruction CPL 2.80, Handling of Case to be Proposed for Violation-By-Violation Penalties, Oct. 1, 1990; BNA OSHA Reference File at 21:9649; online: <http://www.osha-slc.gov/OshDoc/Directive_data/CPL_2_80.html>.

Two types of photographs are typically taken by inspectors: (1) those used to obtain a visual representation of the area being inspected; and (2) those used to document specific violations. Compliance officers are increasingly using videotapes to document violations they believe are present during the inspections. While the inspector is gathering evidence, the employer should also be gathering evidence to counter any unfounded allegations that are made by the inspector.

The employer representative may also want to question the OSHA inspector in order to obtain his evaluation of the plant's job safety and health practices and conditions. The responses so obtained could be of help to the company in making improvements and might be useful in a later settlement conference or trial.

Questioning the OSHA inspector may also disclose whether or not he or she is knowledgeable in the industrial processes of the particular business under inspection. That kind of information can also be helpful to an employer—particularly if it later receives a citation that it thinks is unjustified.

6.6 Employee Interviews

Subject to the employer's and employee's Fourth Amendment privacy rights, and the employee's Fifth Amendment right against self-incrimination, Section 8(a)(2) of the Act authorizes the compliance officer to question any employee privately during an OSHA inspection. Moreover, two regulations grant the inspector that same authority.[63] Such interviews, however, are to be conducted within reasonable limits and in a reasonable manner and must be kept as brief as possible.[64] If necessary, the OSH Act[65] authorizes OSHA to issue administrative subpoenas for the purpose of compelling, with the support of a court order, employee interviews.[66]

The importance of employee and employer statements to OSHA inspectors should not be underestimated. An employer should know that an employee's statements made concerning a matter within the scope of his / her employment can

---

[63] 29 C.F.R. §§ 1903.3(a), 1903.7(b).

[64] OSHA *Field Inspection Reference Manual, supra* note 33, at II(A)(4)(e)(1) - Interviews, p. II-15. The purpose of such interviews is to help the compliance officer obtain whatever information he or she deems is necessary or useful in carrying out the inspection effectively. *Id.*

[65] The Act provides further: In making his inspections and investigations under this Act the Secretary may require the attendance and testimony of witnesses and the production of evidence under oath...29 U.S.C. § 657(b).

[66] *Secretary of Labor v. Pete Barrios, et al.,* 1990 O.S.H. Dec. (CCH) § 29052 (Tex. Dist. Ct. 1990).

be used as evidence against the employer in trial.[67] In fact, the most damaging evidence against a company is often such statements made by management personnel to the compliance officer. Frequently, citations are based upon what is said during the inspection by the employer representative (or provided to OSHA through documents the employer voluntarily gives to the inspector) and by employees rather than by what the inspector observes during the walkaround.[68] Accordingly, long before an OSHA inspector arrives, employers are well advised to carefully think this matter through, select the individuals who will serve as the employer representative, develop prudent policies and procedures for handling OSHA inspections, and train and evaluate the designated personnel and procedures by conducting mock inspections.

6.7 Closing Conference

At the completion of the walkaround portion of the inspection, the compliance officer generally conducts a closing conference with both employer and employee representatives.[69] During the closing conference, the inspector will review his / her observations and comments on workplace conditions and practices, and indicate whether citations are likely and, if so, their likely characterization (*e.g.*, willful, serious). The closing conference usually includes further questioning of the employer in an effort to obtain additional support or justification for the inspector's intended course of action. Employers are well advised to be cautious when responding to OSHA questions that are designed to obtain admissions of various conditions of noncompliance. The compliance officer also explains at the closing conference that if the employer receives a citation, a follow-up inspection may be conducted to verify that the citation has been posted as required, and that during the abatement period employees are adequately protected.

Sometimes the closing conference is not held until several days or weeks after the end of the on-site inspection. Delays of this length often occur after on-site inspections that include air sampling for toxic substances. The samples are sent to

---

[67] Federal Rules of Evidence 801(d)(2)(D) Admission by party-opponent. A statement by the party's agent or servant concerning a matter within the scope of the agency or employment, made during the existence of the relationship, is not considered hearsay.

[68] OSHA inspectors do not normally take signed statements, although there is nothing to prevent them from attempting to take such statements. Usually, the inspector makes "field notes" and they record what is said to them in the notes. The notes are usually not verbatim.

[69] While closing conferences are important, like opening conferences, the failure to hold one will not result in the vacating of a citation unless there has been prejudice to the employer. *See Kast Metals Corp.*, 5 O.S.H. Cas. (BNA) 1861 (Rev. Comm'n 1977).

an OSHA laboratory in Salt Lake City for analysis and the inspector must wait for those results before he makes a determination on whether the workplace is in compliance. Delays between the end of the on-site inspection and the closing conference also result from OSHA's internal procedures requiring higher-level approval before a local office can issue "willful" citations or other citations raising important policy issues.

The employer representative must exercise good judgment in deciding whether to attempt to convince an inspector not to issue a citation or to reduce its classification. It may be successful or it may provoke a confrontation, unwittingly reveal information damaging to the company's interest or lead to other adverse consequences (*e.g.*, inform the inspector of a weakness in his / her case which might be cured by further investigation). There are numerous post-citation opportunities for an employer to make its case and, in many situations, those opportunities are much less risky as well as much more likely to be successful.

7.0 General Rules on Managing OSHA Inspections

As noted above, an inspection involves the collection of information on hazards, observations of employee activities, and interviews where appropriate. Based on this information, the inspector will perform an overall evaluation of the employer's safety program. Lastly, a record will be made of all the facts pertinent to any apparent violations, and the inspector's views will be brought to the attention of the employer.

The employer or his representative should stay with the inspector at all times and should not hesitate to ask questions, take his or her own notes and photographs, and express his or her own opinions. If there are any doubts about how to answer, proceed or respond at any time during the process, the employer's representative should advise the inspector that he is going to call a "Time-Out" so he can confer with his supervisor, attorney or whomever else he wishes to consult. Of course, the inspector also will be asking questions of the employer representative. The responses to such questions can and will be used against the employer during a contest or other court proceeding.

8.0 Special Consideration for Construction Sites and Other Multi-Party Worksites

A construction site presents special problems to OSHA and employers because of the existence of multiple and ever-changing employers (*e.g.*, contractors, subcontractors), often for short periods of time. Inspections of such employers are not easily separable into distinct worksites.[70]

There are several key points to remember when dealing with construction sites. First, it is well established that a third party can consent to a search of jointly occupied property, as long as the third party has "common authority" over the premises.[71] A general contractor has "common authority" over the worksite and, therefore, he or she can give a compliance officer permission to investigate the entire construction site.[72] This authority eliminates the expectation of privacy a subcontractor might otherwise assert. Second, even if a compliance officer does not have permission to enter a workplace, the plain view doctrine allows him to cite employers for violations for any conditions that are visible to someone in public areas.

At the opening conference, the general contractor is asked to notify the subcontractors immediately of the inspection and to ask them to assemble in the general contractor's office to discuss the inspection with the compliance officer.[73] A subcontractor who has not been notified of the inspection of his premises will not be included in the inspection, but may nevertheless be subject to citation.[74]

Third, Chapter 3 discusses the continuing evolution of the multi-party worksite rule under which the following construction employers (and possibly employers in other sectors) are liable for an OSH Act violation and subject to citation: 1) an employer who creates a violative condition, 2) an employer who has the right of control over a violative condition, 3) an employer who has the obligation to cure a violative condition, or 4) an employer who exposes its employees to a violative condition, unless the employer demonstrates that it either took whatever steps were reasonable under the circumstances to protect its employees against the hazards or that it lacked the expertise to recognize the condition as hazardous.

---

[70] OSHA *Field Inspection Reference Manual, supra* note 33, at II(B)(4)(c) - Employer Worksite, p.II-25.

[71] *U.S. v. Matlock,* 415 U.S. 164, 171 (1974).

[72] *J.L. Foti Const.* at 715-716.

[73] OSHA *Field Inspection Reference Manual, supra* note 33, at II(A)(3) - Opening Conference, p. II-10.

[74] *Id.*

Chapter 11

Contesting Citations and Penalties

Manesh K. Rath, Esq.
Keller & Heckman, LLP
Washington, D.C.

1.0 Overview

Given the scope and complexity of OSHA's regulations, even the most diligent health and safety programs cannot ensure that an inspector will find no faults, that an accident or incident will not prompt an investigation, and that employees will perfectly implement the health and safety plan. If you do receive a citation, you need to be prepared to respond promptly, preserve your company's rights, and know how to correct any agency errors. Because administrative procedures can affect a company's substantive rights, it is important to be familiar with the applicable rules. Initially, you will be dealing with the inspector and local area office, but if the matter can not be resolved at that level, you must quickly deal with an Administrative Law Judge from the Occupational Safety and Health Review Commission (OSHRC). If you fail to take action within the specified time, the company's right to pursue disputed issues or penalties will be automatically waived.

We start by looking at the citation that OSHA issues. If a settlement agreement cannot be reached with the OSHA Area Director, a Notice of Contest (NOC) must be filed within 15 working days to preserve any right to challenge the citation. Once an employer's NOC is sent to the OSHA Area Director, the Area Director, acting for the Secretary of Labor, must transmit the NOC and all other documents relevant to the contest to the Executive Secretary of the Occupational Safety and Health Review Commission ("Commission") within 15 working days.[1]

The Commission is an autonomous agency created by the OSH Act to perform ad judicatory functions. The Commission serves as a forum for review of the

---

[1] 29 C.F.R. § 2200.33.

Secretary of Labor's citations and proposed penalties that are challenged by an employer. Initially, all cases are heard at the Commission by a single ALJ. An appeal of the ALJ's ruling is reviewed by the OSHRC Commissioners either upon petition filed by a party or at the Commission's own request. Initial or appellate proceedings before the Commission are governed by rules found at 29 C.F.R. Part 2200, *et. seq.*

2.0 The OSHA Citation

Each citation must be in writing and describe with detail the nature of the violation alleged, including a reference to the rule allegedly violated.[2] It will also state the time period OSHA thinks is necessary to correct the alleged violation, also known as the abatement period. Any proposed penalty is usually stated in the citation, but the penalty may be set forth in a separate document.

The citations must be issued promptly. While the OSH Act states no citation may be written six months from the alleged violation, Section 9(a) of the Act mandates reasonable promptness. Therefore, OSHA makes every effort to issue citations within three day of a competed inspection or investigation.

3.0 Classification of OSHA Violations

Citations for workplace hazards are classified by their degree of severity. They may be serious, nonserious, willful, or repeated.[3] These categories of violations are not fixed, and similar conditions can be re-classified into more serious violations as OSHA gains more experience, as is illustrated by citations issued for violations of the Hazard Communication Standard ("HCS"). Between 1988 and 1990, the number of citations issued for serious violations of the HCS increased by a factor of six, from 1,700 to over 10,000. The number of other-than-serious citations dropped from nearly 14,000 to approximately 6,000. While the language of the standard was unchanged, the Agency's interpretation of the standard clearly changed, at least with respect to enforcement and classification of violation severity. For example, more than 60% of the violations in 1990 were categorized as serious, as compared to less than 10% two years earlier.

---

[2] 29 U.S.C. § 658(a).

[3] 29 U.S.C. § 658. OSHA may also designate a violation as *de minimis*; after a citation is issued, OSHA may designate a new citation as failure to abate.

3.1 *De minimis* Violations

De minimis violations are those involving a technical violation of the standard which does not result in a hazard to health or safety. An example of such a violation might be the installation of a stair rail at a height slightly above the maximum permitted under the standard, or the use of a nuisance dust respirator against a non-toxic dust without a respiratory protection program. Such violations are noted, but carry no precedential value. They cannot be used as the basis for a repeat violation. No abatement period is fixed for correction of the condition, and no penalties are assessed. For *de minimis* violations, the Secretary may issue a citation or may elect to issue a notification.[4]

3.2 Other-than-Serious Violations

Other-than-serious (or non-serious) violations are those for which the risk of serious physical harm is small or the likelihood of injury is remote. While a penalty is mandatory for a serious violation, it is discretionary for non-serious violations. Therefore, the Secretary may issue a citation with a zero penalty. Examples of this include failure to comply with various record keeping requirements, or requirements to document safety or health decisions.

3.3 Serious Violations

Serious violations are those which involve the potential for real harm to employees. The penalty for serious violations can be as much as $7,000. A serious violation is found if there is "substantial probability that death or serious physical harm could result from a condition which exists...unless the employer did not and could not with the exercise of reasonable diligence, know of the presence of the violation."[5] Thus, when OSHA issues a citation for a violation of a safety or health standard which did in fact result in an injury or fatality, the citation is almost certain to be for a serious violation.

3.4 Willful Violations

Willful or repeated violations receive the most attention because they carry potential fines of up to $70,000 per violation. A willful violation arises when the employer was aware of the hazard, but either failed to abate the hazard in a timely fashion or

---

[4] 29 C.F.R. § 1903.14(c).
[5] 29 C.F.R. § 17(d).

deliberately ignored the hazard. Thus knowledge must be proved in order for OSHA to establish a willful violation. Such a violation may support criminal charges. However, a misdemeanor conviction is possible and this provision has not been invoked often. In this regard, courts have generally found that state criminal laws are not preempted by OSHA.

3.5 Repeat Violations

OSHA may cite an employer for a repeat violation where the employer has already been notified or cited for s specific violation and is later found to be in violation of the same standard again. As noted above, repeated violations carry potential fines of up to $70,000 per violation. Therefore, while non-serious and serious violation citations carry with them a nominal maximum penalty of up to $7,000, the potential that they could lead to highly burdensome repeat violation citations often makes worthwhile the effort to contest such citations.

3.6 The Egregious Penalty Policy

OSHA has created a directive to circumvent the penalty maximums established by statute. OSHA's policy, known as the "egregious penalty" policy, or the "violation by violation" penalty policy, allows for issuing a separate citation for each instance of a violation. Although OSHA has employed this practice since the mid 1980's, it formalized this policy in a directive in 1990.[6] In some cases, this means OSHA may issue a separate violation for each victim of a violation of a standard. In other cases, such as record keeping violations, OSHA may cite an employer for each failure to record an event. In so doing, OSHA may issue penalties which are potentially millions of dollars for an act or omission which may have otherwise had a maximum of a $70,000 penalty.

Instead of grouping violations for penalty purposes, OSHA separates the violations out and applies a penalty to each violation separately. In general, OSHA applies this method only to violations it deems willful, where OSHA believes that the violation resulted in worker fatalities or a large number of injuries or illnesses, or where the employer has an extensive history of prior violations.[7]

OSHA may only use the egregious penalty policy in instances where it is issuing citations under a particular standard, or when it is issuing a citation under the

---

[6] OSHA Directive, Handling of Cases To Be Proposed for Violation-by-Violation Penalties, CPL 2.80 (October 21, 1990).
[7] OSHA Directive CPL 2.80.

general duty clause for violations where the cited hazard involves more than one violative condition. It may not, however, use the egregious penalty policy to issue multiple citations for a single violation of the general duty clause which affects multiple employees.[8]

3.7 Increasing Size of Penalties

A number of factors have worked together to increase substantially the size of typically OSHA penalties. Prior to 1986, a standard could be violated only once in a particular inspection. But this policy was changed to allow for multiple violations of a single standard when an employer receives a willful violation. Then, as part of the 1990 budget negotiations, Congress increased OSHA's penalty structure by a factor of seven. While this was not totally unreasonable, Congress also sought to turn the Agency into a revenue source by charging it with increasing its revenues from penalties from the then current level of $40 million per year to a total of $900 million per year over five years. In the 20 years of OSHA's existence, total fines have been less that $4 billion, with approximately $90 million in penalties assessed in the last year.

Through increased penalty amounts, multiple penalties, and use of increased severity classifications for violations, OSHA has garnered increased attention from the business community and the public.

3.8 OSHA Penalties in Light of Its Budget

When OSHA issues a citation against an employer for a violation, the penalties can range from very small amounts to significant sums. More often than not, the cost of abatement will exceed the cost of the penalty itself. Yet the cost of abatement must necessarily be factored into a decision as to whether to file a notice of contest.

For non-serious violations, OSHA may issue a penalty of up to $7,000 for each violation.[9] If a citation is determined by OSHA to be for a serious violation, penalties can be assessed up to $7,000 for each violation as well.[10] The difference is that frequently, but not necessarily, the actual amount assessed for a non-serious violation will be less than that for a serious violation. If an employer is cited for a violation which is determined by OSHA to be willful or repeated, OSHA may

---

[8] *See, e.g., Metzlen v. Arcadian Corp.*, (OSHRC Docket No. 93-3270 (September 15, 1995).
[9] 29 U.S.C. § 666(c).
[10] 29 U.S.C. § 666(b).

assess a penalty of up to $70,000, but not less than $5,000 for each violation.[11] If an employer fails to correct a violation for which a citation has been issued, OSHA may assess a penalty of up to $7,000 for each day during which a violation continues.[12] The period in which the employer is deemed to have failed to correct a violation begins to run on the date of final order of the Commission. Thus it is tolled during any contest of a citation.[13] Here again, the employer is faced with a strong incentive to file a notice of contest in a timely fashion. If an employer is cited with a willful violation which results in the death of an employee, the OSH Act provides for a civil penalty of up to $10,000 or imprisonment of up to six months, or both.[14] Likewise, if a person knowingly makes a false statement, that person may face penalties up to $10,000 or imprisonment of up to six months or both.[15]

Issuance of citations and collection of penalties is an important, if not the central, activity of OSHA. OSHA's budget for the fiscal year 2001 dedicated almost 36% towards enforcement activities. From fiscal years 1996 to 2001, OSHA's budget has grown from over $303 million to $426 million, an increase of over 40% in five years. Taking the fiscal year 1998 as an example, OSHA inspected about 3,000 sites and issued about 10,000 citations, from which it collected over $15 million. Its operating budget that year was $336.5 million. Thus, OSHA was able to fund almost five percent of its budgeted revenue through its enforcement activities, a significant percentage.

4.0 Challenging a Citation

4.1 Settlement Agreement with Area Director

An employer may request an informal conference upon receipt of a citation and notice of proposed penalty. In the late 1970's, OSHRC was experiencing a serious backlog of cases and as a result, an action contesting a citation could take several years. As the employer was not required to correct the alleged hazard until the case was resolved, OSHA was concerned about leaving the employees unprotected during this potentially long period. To resolve this dilemma, OSHA developed a policy that allows OSHA area directors to enter into settlement agreements with contesting employers. Under the agreement's terms, OSHA area directors can

---

[11] 29 U.S.C. §666(a).
[12] 29 U.S.C. § 666(d).
[13] 29 U.S.C. § 666(d).
[14] 29 U.S.C. § 666(e).
[15] 29 U.S.C. § 666(g).

reduce an employer's penalties or reclassify the employer's violation, *i.e.*, from serious to nonserious, if the employer agreed to correct the violation immediately.

Under many circumstances, this informal conference with the area director provides the employer with an opportunity to resolve the disputed citation(s) without litigation, a chance to improve its understanding of the applicable standard, and guidance on how to correct the violation(s). Such requests are granted routinely before the end of the fifteen working-day contest period. But, by the end of the fifteen working day period, the employer must file a Notice of Contest ("NOC") or a notice alleging that the time fixed for abatement is unreasonable. Employers should normally file a NOC as a contingency even when negotiating an informal settlement. Once a NOC is filed, the abatement period does not begin to run until the entry of the OSHRC's final order.

4.2 The Notice of Contest

OSHA citations are challenged by the employer filing a Notice of Contest ("NOC"). The NOC must be postmarked within 15 working days of receipt of the citation.[16] A failure to file a timely NOC will convert the citation into a final order not subject to review and will subject the employer to a follow-up inspection to determine whether the violation has been corrected.

The NOC is typically sent to the area director "in writing," and we strongly recommend that it be in writing. But, the Review Commission has found oral Notice of Contest to be valid on rare occasions. No particular form is required for the NOC and the Review Commission is liberal in its interpretation of documents that satisfy the requirement for giving notice. A copy of both the citation and the NOC must be posted at the worksite where it is accessible to affected employees. A NOC may be withdrawn at any stage of the proceedings.

The benefits of contesting a citation range from dismissal or modification of the violation to a reduction of the proposed fines and modification of the abatement date. Employers should be aware that acceptance of certain citations (for example, willful citations) may support criminal charges or increase workers' compensation recoveries in some states. The main disadvantages of contesting a citation are the time that company personnel may need to devote to the hearing process and the cost involved in litigation.

---

[16] Employers are nearly always unsuccessful in seeking extension of the 15-day period. However, an employer may be excused for not filing within the 15-day period if the delay was caused by reliance on erroneous representations, deception, or failure by OSHA to follow proper procedure.

4.3 Extending the Abatement Period

In contrast to challenging whether a violation occurred or the proposed penalties, a Petition for Modification of Abatement (PMA) may be filed when an employer has made a good faith effort to comply, but abatement has not been completed due to factors beyond its control, or when abatement would cause a significant financial hardship to the employer.[17] The PMA must be filed no later than the next working day following the date on which abatement was originally required. It must be in writing and include the following: (1) steps already taken in an effort to comply and the dates those steps were taken; (2) the amount of additional time needed to comply; (3) the reasons the additional time is needed; (4) interim steps being taken to protect employees; and (5) certification that the petition has been posted in a conspicuous place in the worksite. If uncontrollable circumstances prevent the employer from meeting the abatement date and the 15-working-day contest period has expired, a PMA may be filed if accompanied by a statement explaining the circumstances. At that point, the burden of showing the reasonableness in the abatement period is on OSHA.

Once a PMA is filed, the abatement period does not begin to run until the entry of a final order by the Review Commission. However, if the Review Commission finds that the only purpose of the employer's Notice of Contest is to delay or avoid the abatement period, the abatement period begins to run from the date of receipt of the citation.

4.4 Employee Participation

The employer must post the NOC at the worksite so that affected employees are informed and may challenge the reasonableness of the abatement date.

When a citation is issued to an employer, the citation must be posted in a place that is both near the site of the alleged violation and also prominent, or easily seen by affected employees. A good location for an employer to post such a citation is the bulletin board which it may use for other employee notices. Failure to post a citation as required may result in another citation and a civil penalty of up to $1,000.00 for each violation, and even the dismissal of the notice of contest.

Once the employer has received a citation, there is a fifteen working day period during which an employer or an employee representative may request an informal conference with the OSHA Area Director to discuss the citation, penalty or abatement. The parties may not request an informal meeting once the notice of

---

[17] 29 C.F.R. § 1903.14a.

contest has been filed. OSHA has the power to modify its citation, penalty amount or abatement instructions as a result of the informal conference.

Once the employer has filed a notice of contest, it is required to post the notice of contest in the same place as it posted the citation, and to file a certification of posting with the Commission.[18] If an employer fails to meet this requirement, its notice of contest may be dismissed.

Just as an employer may file a notice of contest, affected employees or an employee representative has the right to file a notice contest against a citation alleging that the period of time fixed in the citation for abatement is unreasonable.[19] Once this has been filed, the Secretary must immediately advise the Commission of the notice of contest, and the Commission must set a hearing for the matter. In such matters, although the employee or employee representative has filed the notice of contest, just as with the employer, the Secretary is the complainant because the Secretary filed the citation. Thus the Secretary has the burden of establishing that the abatement period was a reasonable one.

Additionally, an employee representative may plead to the Commission for leave to intervene in a matter involving the Secretary and an employer that has filed a notice of contest.

4.5 Opportunity for Settlement after Filing the Notice of Contest

Settlement is encouraged at all stages of the proceeding as long as it is consistent with the provisions and objectives of the Act. The employer may not withdraw the NOC in favor of settlement unless there is certification of a date on which violations have been or will be abated. Settlement is treated as an admission of liability in the original citation unless a settlement clause provides otherwise. In addition, the settlement agreement may be used in future proceedings concerning subsequent violations to show knowledge of a condition or practice.

Under the rules governing settlements, an Administrative Law Judge (ALJ) is assigned when "there is a reasonable prospect of substantial settlement with assistance of mediation by a Settlement Judge." Affected employees who have party status must receive written notice of a proposed settlement (either by personal service or posting) so they can file objections with the ALJ within 10 days after a settlement has been reached.

---

[18] Commission Rule 7(d).
[19] 29 U.S.C. § 659(c).

5.0 Proceedings before the Administrative Law Judge

Once an employer's NOC is sent to the OSHA Area Director, the Area Director, acting for the Secretary of Labor, must transmit the NOC and all other documents relevant to the contest to the Executive Secretary of the Occupational Safety and Health Review Commission ("Commission") within 15 working days.[20]

Initially, all cases are heard at the Commission by a single ALJ. An appeal of the ALJ's ruling is reviewed by the OSHRC Commissioners either upon petition filed by a party or at the Commission's own request. Initial or appellate proceedings before the Commission are governed by rules found at 29 C.F.R. Part 2200, *et seq.*, and in the absence of a specific provision in the OSHRC procedural rules, proceedings are conducted in accordance with the Federal Rules of Civil Procedure.

5.1 The Complaint

As noted above, the filing of the NOC triggers the Secretary of Labor's obligation to notify the Commission of the contest and to furnish relevant documents and records. Within 20 days after that filing with the Commission, the Secretary is required to file a complaint in the case. 29 C.F.R. § 2200.34.[21] The complaint must set forth all alleged violations and proposed penalties which are contested, stating with particularity: (a) the basis for jurisdiction; (b) the time, location, place, and circumstances of each alleged violation; and (c) OSHA's rationale for the proposed abatement period and penalty for each alleged violation.[22]

5.2 The Answer

Answers, or motions in lieu of an answer, must be filed with the Commission within 20 days after service of the complaint.[23] However, if a motion to dismiss or a motion for a more definite statement has been filed and denied, the answer must be filed within the time frame specified by the judge in the order denying the motion.

---

[20] 29 C.F.R. § 2200.33.

[21] As of December 10, 1992, the Commission returned to notice pleading (from fact pleading). Nevertheless, an employer may move for a more definite statement of the Secretary's allegations before filing an answer upon showing that it cannot frame a responsive answer to the allegations contained in the complaint.

[22] 29 U.S.C. § 659(a).

[23] 29 C.F.R. § 2200.34(b)(1).

Though pleadings are liberally construed and easily amended,[24] an employer is advised to raise all available defenses in the answer, or as early as possible in an amended pleading, because the employer may run the risk that the omitted defenses will otherwise be deemed waived and unavailable to him at a later time.[25]

5.3 Defenses

Defenses recognized by the Commission and the courts can be categorized in several ways, but a common approach is to distinguish between 'procedural' and 'substantive' defenses. We highlight here some of the defenses that are frequently involved in or unique to challenges to an OSHA citation. But this is not an exhaustive list and should not be treated as such. For example, an employer may disagree with OSHA's conclusion that a particular activity or practice violated a rule, that a particular rule applies to the employer's operations, or that a General Duty Clause violated is adequately demonstrated. These types of defenses must all be evaluated on a case-by-case basis.

5.3.1 Procedural Defenses

The 'procedural' group involves challenges focusing on the validity of the Secretary's enforcement procedures and the procedures involved in the adjudication by the Commission of the contested case. This group includes defenses involving the failure of the Secretary to conform to constitutional and statutory requirements in conducting inspections of worksites and improper service of citation.

Six Month Statute of Limitations: The Act states that no citation may be issued under the Act after the expiration of six months following the occurrence of any violation.[26] Thus, an employer may raise the defense that a citation is barred by the six month statute of limitations set out in section 9(c) of the Act.[27] However, even this seemingly clear language is open to interpretation. The Secretary may cite violations which occurred within six months preceding the citation's issuance.[28] Where a violation remains a violation which exposes employees to a zone of danger, a citation is properly issued if the last day on which employees are exposed to a zone of danger was within the six month period, even if most of the time in which the

---

[24] *See Fed. Reg. Civ. P.* 15(a).
[25] 29 C.F.R. § 2200.34(b)(4).
[26] 29 U.S.C. § 658(c).
[27] 29 U.S.C. § 658(c).
[28] *Secretary v. Dayton Tire*, OSHRC Docket No. 94-1374, 1994 WL 913343 (O.S.H.R.C. 1994)) at *2.

violation existed was outside of the six month period.[29] If a violation occurred from January 1 until March 30 of a particular year, and a citation was issued on September 30, or six months after the last date on which the violation occurred, then the citation would have been timely issued.[30]

Promptness: An employer may raise the defense that the Secretary failed to issue a citation with reasonable promptness. This is not to be confused with the Secretary's burden to issue a citation within the six month period of limitations. Under the reasonable promptness standard, OSHA is expected to make every effort reasonable to issue citations within seventy-two (72) hours of the completion of an inspection or investigation.[31] This expectation, however, is not obligatory, and OSHA's failure to observe this 72 hour period is not fatal to its citation.[32]

5.3.2 Substantive Defenses

Substantive defenses more directly address the employers ability to comply with a regulation or whether a violation actually occurred within the meaning of the OSH Act and OSHA regulations. These defenses must be proved by a preponderance of the evidence. The "substantive defenses" available to an employer are numerous and can range from challenges to the Secretary's *prima facie* case,[33] to arguments that the standard at issue does not apply to the employer. In addition to these basic defenses, however, a number of well-recognized independent substantive defenses have evolved from past litigation.

Impossibility of compliance is often raised by employers who cannot challenge the Secretary's *prima facie* case. This defense permits an employer to avoid liability for non-compliance by showing that compliance would prevent the employee from performing its required work.[34] To successfully assert this defense, the employer

---

[29] *Secretary v. Southern Nuclear Operating Co., Inc.*, OSHRC Docket No. 97-1450, 1999 WL 569136 (O.S.H.R.C. 1997).

[30] *Secretary v. Southern Nuclear Operating Co., Inc.*, OSHRC Docket No. 97-1450, 1999 WL 569136 (O.S.H.R.C. 1997).

[31] See OSHA *Field Operating Manual*, Chapter 10.

[32] *Chicago Bridge & Iron Co.*, 1 OSHC 1485 (1974).

[33] Judges rarely vacate the Secretary's citation on this ground, but it does occasionally happen. When a citation is vacated for failure to make out a prima facie case, it usually occurs because the Secretary has either filed to demonstrate that the cited standard applies to the facts asserted as being in violation or the Secretary simply fails to prove the facts necessary to support the citation. *See, e.g. Secretary v. Scafar Contracting, Inc.*, OSHRC Docket No. 97-0960 (2000); *Secretary v. Nova Fedrick / AJV*, OSHRC Docket No. 99-1511 (2000).

[34] *See, e.g., Cleveland Electric Illuminating Co.*, 13 OSHC 2209, 1989 OSHD 28,494 (Rev. Comm'n 1989).

must demonstrate not only that compliance was economically infeasible or impossible, but that alternative means of protection either were being used or were unavailable.[35] In most cases in which the defense is raised, the employer's problem is more one of difficulty of compliance rather than impossibility. Thus, the Commission has rarely concluded the defense to be proven.[36]

The *greater hazard defense* is a well-established doctrine that on some occasions allows employers to escape sanctions for violations of otherwise applicable safety regulations. The employer must establish that the act of abating a violation would itself pose an even greater threat to the safety and health of their employees.[37] There is a triple burden on an employer who seeks to assert the greater hazard defense. The employer must demonstrate: (a) that the hazards of compliance are greater than the hazards of noncompliance, (b) that alternative means of protecting employees are unavailable, and (c) that a variance is unavailable or inappropriate.[38]

The defense of *unpreventable employee misconduct* is one of several defenses relating to an employer's lack of control. In essence these defenses assert that the employer lacked control to correct or prevent an employee's behavior or to identify or correct a safety or health hazard. To establish the unpreventable employee misconduct defense, the employer must show that: (a) it has established work rules designed to prevent the violation; (b) it has adequately communicated these rules to its employees; (c) it has taken steps to diligently discover violations; and (d) it has effectively enforced the rules when violations have been discovered.[39]

While often raised, the employee misconduct defense is difficult for an employer to establish. To be successful, the employer must show that at every stage, it exerted its utmost ability to achieve safety or health compliance. Evidence that an

---

[35] *Sec. of Labor v. Dun-Par Engineered Form Co.*, 843 F.2d 1135, 1138 (8th Cir. 1988); *M.J. Lee Constr. Co.*, 7 OSHC 1140 (Rev. Comm'n 1979); see also *A/C Electric Co. v. OSHRC*, 956 F.2d 530 (6th Cir. 1991) (Employer's failure to raise "impossibility" defense in answer waives it).

[36] *Tube-Lok Prods.*, 9 OSHC 1369 (Rev. Comm'n 1981).

[37] *Pennsylvania Power & Light Co., v. OSHRC*, 737 F.2d 350 (3rd Cir. 1984); *Ocean Electric Corp. v. Sec. of Labor*, 594 F.2d 396 (4th Cir. 1979); *Brennan v. OSHRC*, 511 F.2d 1139 (9th Cir. 1975); *Capital Electric Line Builders of Kansas, Inc. v. Mar*shall, 678 F.2d 128 (10th Cir. 1982).

[38] *Modern Drop Forge Co. v. Sec. of Labor*, 683 F.2d 1105, 1116 (7th Cir. 1982). This aspect of the defense contemplates a recurring or continuous situation for which a variance application may be suitable. A variance allows an employer to use compliance means different than the means dictated by a standard.

[39] *P. Gioioso & Sons, Inc. v. OSHRC*, 115 F.3d 100 (1st Cir. 1997); *New York State Electric & Gas Corp. v. Sec. of Labor and OSHRC*, 88 F.3d 98 (2nd Cir. 1996); *Jensen Construction.*, 7 OSHC 1477, 1979 OSHD 23,664 (Rev. Comm'n 1979).

employer failed to discharge or otherwise adequately discipline an employee for violations of the employer's safety or health rules would be substantial evidence that an employer failed to enforce those rules.[40]

The *borrowed servant* defense arises in the instance where an employer leases employees to another employer. If the staffing agency has totally relinquished control over the employee to another employer, then it may be able to raise the "borrowed servant" defense and avoid liability for OSHA violations. The element of control over the employee's conduct is the critical issue in this defense.

The *multi-employer worksite doctrine* is another defense involving the employer's lack of ability to abate a safety or health hazard. In the early days of the Act's enforcement, OSHA determined that penalizing individual subcontractors on construction job sites was an inefficient method of leveraging its enforcement activities, and that, often, the employer who was held liable was not responsible for creating or changing the conditions that violated the standard. Indeed, with strict labor rules on many construction sites, it was a violation of the contract for one tradesman to perform a task that was reserved to another trade under the terms of the labor agreement. For example, it was impossible for a painter to correct an electrical hazard. In response, OSHA developed the multi-employer work site doctrine, under which any of three employers can be liable for and OSHA citation: the controlling employer, the employer who creates the hazard, and the employer whose employees are exposed to the hazard. An employer defense to such a charge is that it took all reasonable steps to correct the hazard.

The *IBP* case addressed the question of how far the host employer's liability extends in general industry.[41] There, IBP hired a contractor to clean its meat-packing plant, and installed its own inspectors on site for quality control. A member of the cleaning staff, employed by the contractor, failed to lockout a machine while cleaning it and was killed. The accident occurred in spite of clear and repeated instructions to the contractor from IBP to conform to the lockout / tagout rule, and IBP's inspectors had observed clear lockout / tagout violations on a number of prior occasions. The Occupational Safety and Health Review Commission (OSHRC) ruled that IBP had not taken all reasonable steps to control the contractors compliance behavior, since it could have canceled the contract. The appellate court rejected that approach, saying that contract termination was an extreme and unrealistic expectation. The court held that IBP had taken reasonable steps.

---

[40] *Asplundh Tree Expert Co.*, 6 OSHC 1951 (1978).
[41] *IBP, Inc. v. Hermann*, 144 F.3d 861 (D.C. Cir., 1998), 18 O.S.H. Cas. (BNA) 1353, 1998 O.S.H.D. § 31,577; *Secretary v. IBP, Inc.*, OSHRC Docket No. 93-3059.

The foregoing are not the only affirmative defenses available to employers. Other, more traditional legal defenses such as *res judicata*[42] or failure to prove the employer's knowledge of the underlying violation[43] are also available to employers.

An employer contemplating a challenge to an OSHA citation is urged to research the ever-growing body of case law on this subject and make use of the best affirmative defenses available under the circumstances of its case.

5.4 Discovery

At any time after the filing of the answer or the first responsive pleading or motion that delays the filing of an answer (*e.g.*, a motion to dismiss), a party may initiate discovery, that is, investigate the facts of the case or determine where OSHA and the employer agree or disagree on the facts. Discovery includes: (a) production of documents or request for permission to enter upon land or other property for inspection and other purposes; (b) requests for admissions, and (c) interrogatories.[44] Discovery must be initiated early enough to permit completion of discovery at least 7 days prior to the hearing date, unless otherwise ordered by the judge.[45] Discovery requests may focus on any matter that is not privileged and that is relevant to the subject matter of the pending case.[46]

A judge or the Commission may limit the frequency or extent of discovery methods if it determines: (a) that the discovery sought is unreasonably cumulative or duplicative or is obtainable from some other source that is more convenient, less burdensome or less expensive; (b) the party seeking discovery has had an ample opportunity to obtain the information sought by discovery in the action, or the discovery is unduly burdensome or expensive, taking into account the needs of the case, limitations on the party's resources, and the importance of the issues and litigation.[47]

On a showing of good cause (to protect a party or person from, *e.g.*, annoyance, embarrassment, oppression or undue burden or expense in discovery procedures),

---

[42] *See, e.g., Continental Can Co. v. Marshall*, 603 F.2d 590 (7th Cir. 1979((upholding the District Court's issuance of an injunction prohibiting the issuance of citations to an employer with several plants where prior citation were vacated on the basis of company-wide economic infeasibility); *Kaiser Engineers, Inc.*, 6 OSHC 1845, 1978 OSHD 22,845 (ALJ 1978).

[43] *Brennan v. OSHRC*, 511 F.2d 1130, 2 OSHC 1646 (9th Cir. 1975).

[44] 29 C.F.R. § 200.52 *et seq.*

[45] 29 C.F.R. § 2200.52(a)(2).

[46] 29 C.F.R. § 2200.52(b).

[47] 29 C.F.R. § 2200.52(c).

the judge or the Commission may also issue a protective order that discovery not be had, that it be limited or that it be permitted only under specified terms and conditions. 29 C.F.R. § 2200.52(d).

When a party refuses or obstructs discovery, a judge may issue an order compelling discovery.[48] A failure to comply may lead to sanctions, which include the following:

a. an order that designated facts shall be taken to be established for purposes of the case in accordance with the claim of the party obtaining the order;

b. an order refusing to permit the disobedient party to support or to oppose designated claims or defenses or prohibiting it from introducing designated matters in evidence;

c. an order striking out pleadings or parts thereof or staying further proceedings until the order is obeyed; and

d. an order dismissing the action or proceeding or any part thereof or rendering a judgment by default against the disobedient party.[49]

The Commission does not have the power to enforce its orders under the threat of contempt proceedings. That approach requires a supporting order of a federal district court.

5.4.1 Production of Documents

Parties may serve on any other party requests to permit the inspection and copying of designated documents or to inspect and copy, test or sample any tangible things which are in the possession, custody or control of the party on whom the request is served.[50] This includes permission to enter upon designated land or other property in the possession or control of the party who receives the request for the purpose of inspection and measuring, surveying, photographing, testing or sampling.[51] Such requests must be specific as to the described items and categories, including specification of a reasonable time, place and manner of making the inspection and performing the related actions. The party upon whom the request is served, shall serve a written response within 30 days after service of the request, unless the requesting party allows a longer time.[52] Absent an objection to a request to inspect

---

[48] 29 C.F.R. § 2200.52(c).
[49] 29 C.F.R. § 2200.52(e).
[50] 29 C.F.R. § 2200.53.
[51] 29 C.F.R. § 2200.53.
[52] 29 C.F.R. § 2200.53.

and copy documents, it is customary, but not required, for parties to provide copies of requested documents instead of actually providing access to inspect and an opportunity to copy.

5.4.2 Requests for Admissions

Parties may serve upon any other party written requests for admissions for purposes of the pending action of the genuineness and authenticity of any document described or attached to the request, or the truth of any specified matter of fact.[53] The number of requested admissions is limited to 25 (including subparts) without an order from a judge or the Commission.[54] Parties seeking to serve more than 25 requested admissions have the burden of persuasion in establishing that the complexity of the case or number of citation items necessitates a greater number of requested admissions.

Within 30 days after service of the requests, the party to whom requests are directed must respond or each matter is deemed to be admitted. Responding parties must serve a written response specifically admitting or denying the matter involved in whole or in part with respect to each area of inquiry or asserting that it cannot be truthfully admitted or denied and setting forth in detail the reasons why this is so, or parties must state an objection, describing in detail the reasons for that response.[55] Matters admitted are deemed to be conclusively established unless the judge or the Commission on motion permits the withdrawal or modification of the admission.[56]

5.4.3 Interrogatories

Parties are permitted to serve interrogatories on any other party. Interrogatories are written questions that the receiving party must respond to in writing. In the OSHRC context, interrogatories may not exceed 25 questions (including subparts) without an order of a judge or the Commission.[57] As with requests for admission, parties seeking to serve a greater number of questions have the burden of establishing that a greater number of interrogatories is required due to the complexity of the case or number of citation items. Responses to interrogatories must be made in

---

[53] 29 C.F.R. § 2200.54.
[54] 29 C.F.R. § 2200.54.
[55] Id.
[56] Id.
[57] 29 C.F.R. § 2200.55.

good faith and as completely as the answering party's information permits.[58] A responding party cannot give lack of information as a response or as a reason for failure to respond unless the party states that it has made reasonable inquiry and that information known or readily obtainable is insufficient to enable an answer to the substance of the interrogatory. Each interrogatory must be answered separately and fully under oath or affirmation. If an interrogatory is objected to, the objection should be stated in lieu of the response. Such responses or objections must be served upon the propounding party within 30 days after service of the interrogatories unless the judge allows a shorter or a longer period of time.[59]

5.4.4 Depositions

Depositions are sworn testimony of a person in the presence of a court reporter and counsel for all the parties. Depositions of parties, intervenors or witnesses are only allowed by agreement of all of the parties or on order of a judge or the Commission after the filing of a motion which states good cause and just reasons.[60] Depositions allowed by a judge or the Commission may be taken after 10 days written notice to the other party or parties.[61] Depositions taken in this fashion may be used for discovery, to contradict or impeach the testimony of a deponent as a witness or for other purposes permitted by the Federal Rules of Evidence and the Federal Rules of Civil Procedure. Provision is made in the rules for telephonic and video depositions as alternatives to the standard in-person stenographic deposition.[62]

5.4.5 Subpoenas

Upon application of any party, the judge may issue to the applying party subpoenas that require the attendance and testimony of witnesses and the production of any evidence, including relevant books, records, correspondence or documents in his possession or under his / her control. The party to whom the subpoena is issued is responsible for its service.[63] Applications for subpoenas are made in an *ex parte* fashion, and the subpoena must show on its face the name and address of the party at whose request the subpoena was issued. Persons served with a subpoena may, within five days after service of the subpoena, move in writing to revoke or modify

---

[58] 29 C.F.R. § 2200.55.
[59] 29 C.F.R. § 2200.55.
[60] 29 C.F.R. § 2200.56.
[61] 29 C.F.R. § 2200.56.
[62] 29 C.F.R. § 2200.56.
[63] 29 C.F.R. § 2200.57(a).

the subpoena if he / she does not intend to comply. Motions to revoke or modify are to be served on the party at whose request the subpoena was issued. Subpoenas may be revoked or modified by a judge or the Commission if the evidence required by the subpoena does not relate to any matter under investigation or in question in the proceeding, or if the subpoena does not describe the evidence required by the subpoena with sufficient particularity or if the subpoena is otherwise invalid. If any person fails to comply with a subpoena issued upon the request of a party, the Commission's counsel is obligated to initiate enforcement proceedings in the appropriate federal district court.[64] However, neither the Commission nor its counsel thereby assumes responsibility for the effective prosecution of the proceeding before the court to enforce the subpoena.[65]

5.4.6 Exhibits

Exhibits offered into evidence at a hearing should be marked for identification with the case docket number, a designation of the entity offering the exhibit and numbered consecutively. The admission of exhibits may be denied, in the judge's discretion, because of their excessive size, weight or other problematic characteristics (although parties may offer photographs, models or representations of such exhibits).

Parties may request (by motion) the return of a physical exhibit within 30 days after the time has expired for filing a petition for review of a Commission final order in a federal appellate court or within 30 days after the completion of proceedings by the Court of Appeals. Otherwise, if the OSHRC Executive Secretary determines that it is no longer necessary for use in any Commission proceeding, the exhibit may be disposed of. Moreover, the Commission's rules permit any person to request custody of a physical exhibit for use in any court or tribunal.[66]

5.4.7 Trade Secrets and Protection of Confidential Information

All claims of privilege (including trade secrets and other matters the confidentiality of which is protected by law), must be asserted in writing or (if during a hearing) on the record. Such claims must identify the information that would be disclosed and for which a privilege is claimed, and specifically allege those facts which show that the information is privileged. Such claims are to be supported by affidavits, depositions or testimony and must specify the relief sought. Assertions of privilege may be

---

[64] 29 C.F.R. § 2200.57(e).
[65] 29 C.F.R. § 2200.57.
[66] 29 C.F.R. § 2200.70(f).

accompanied by a motion for a protective order or by a motion that the allegedly privileged information be received and that the claim be ruled on in an *ex parte* or *in camera* fashion. Parties wishing to make a response to a claim of privilege must do so in a timely fashion, as specified in the Commission's Rules.[67]

5.5 Hearing

Under Section 10(c) of the OSH Act, an employer that contests a citation or notification of penalty for failure to abate must be afforded a formal hearing in accordance with the principles of the Administrative Procedure Act ("APA").[68] Notice of a hearing and its location will be given to the parties, under Commission rules, "at least thirty days in advance of the hearing," unless the hearing has been previously postponed, in which case the notice will be given 10 days in advance.[69] Employers are required to serve the hearing notice on affected employees. A hearing may be postponed by the judge on his own initiative or for good cause shown upon the motion of a party. This motion must be received at least seven days prior to the hearing, unless good cause is shown for the late filing.[70] Stays of proceedings, on the other hand, are "not favored" by the commission,[71] though they are occasionally granted. The applicant's motion must state the position of the other parties, and set forth the reasons a stay is sought and the length of stay requested.[72]

As discussed earlier, proceedings before the ALJ are governed by the Federal Rules of Evidence and of Procedure. Accordingly, Commission hearing practice is much like a regular civil trial, but without a jury. Witnesses are examined under oath and cross-examined.[73] Documentary evidence can be received, subject to the procedural strictures of Commission Rule 70.

The judge who presides at the trial will not, as a rule, make his decision from the bench. When the trial is over, he will give the parties time to obtain copies of the transcript of the proceedings and prepare briefs setting forth their views on the evidence and the applicable law. The briefs are mailed to the judge at a date he will

---

[67] 29 C.F.R. § 2200.11.
[68] 29 U.S.C. § 659(c).
[69] 29 C.F.R. § 2200.60.
[70] 29 C.F.R. § 2200.62.
[71] 29 C.F.R. § 2200.63(a).
[72] *Id.*
[73] 29 C.F.R. § 2200.69.

establish.[74] Usually this date is a couple of months after the trial ends. The judge will then take a couple of additional months before deciding the case.

5.6 Simplified Proceedings

Cases are eligible for simplified proceedings under "E-Z Trial" procedures.[75] E-Z Trial includes the following features: (1) pleadings generally are not required or permitted; (2) discovery generally is not permitted; (3) the Federal Rules of Evidence do not apply, and (4) interlocutory appeals are not permitted. Any party may request an E-Z Trial within 20 days of the notice of docketing, and the judge must act upon the request within 15 days after the request has been filed. At any time, a party may move to discontinue application of the simplified rules to the case.

6.0 Review by the Commissioners

6.1 Interlocutory Review

Normally, challenges to a ruling by the administrative law judge occur after the ALJ files the final decision. Under limited circumstances, a party may seek review while the proceeding before the ALJ are in progress. These are known as interlocutory appeals or review. A party may seek the Commission's interlocutory review of a judge's ruling, which is at OSHRC's discretion. A petition seeking such review must assert (and the Commission must find) that the review involves an important question of law or policy on which there is substantial ground for difference of opinion and that immediate review will materially expedite the final disposition of the proceedings. Alternatively, the Commission must find that the judge's ruling will result in the disclosure of information that is alleged to be privileged before the Commission can review the judge's report.[76]

A petition for interlocutory review may be filed within five days after receipt of the ruling from which review is sought. Unless granted within 30 days of its receipt by the Commission's Executive Secretary, the petition is denied. However, denial of a petition for interlocutory review does not preclude a party from raising an objection to the judge's interlocutory ruling in a petition for discretionary review.

---

[74] 29 C.F.R. § 2200.74.
[75] 29 C.F.R. §2200.202.
[76] 29 C.F.R. §2200.73.

When a petition for interlocutory review deals with a judge's ruling concerning an alleged trade secret, its filing stays the effect of the judge's ruling until the Commission denies the petition or rules on its merits.

6.2 Final Decisions and Reports of Judges

Judges must issue written decisions in pending proceedings and, on the 21st day after transmitting a decision to the parties, file a report with the Commission's Executive Secretary.[77] The report includes the decision, the record in the proceeding, petitions for discretionary review, and statements in opposition to such petitions. If no Commissioner directs review of a report by the 30th day after the docketing of the report, the Judge's decision becomes a final order of the Commission.[78]

6.3 Discretionary Review

Review by the Commission is not a matter of right. It may be granted by a Commissioner as a matter of discretion on his / her own motion[79] or on the petition of a party.[80] Petitions for OSHRC review must be filed with the judge within 20 days after transmittal of the decision to the parties or by filing such a petition directly with the Executive Secretary within 20 days of the docketing of the Judge's report.[81] Parties may file cross-petitions for discretionary review, which may be conditional, i.e., limiting their request to circumstances in which an opposing party's petition for review is granted. Cross-petitions must be filed with the judge within 20 days of transmittal of the decision to the parties, or with the Executive Secretary within 27 days after the docketing of the Judge's report.[82]

6.4 Commission Review

Unless otherwise specified, when the Commission agrees to review a case, it has jurisdiction to review the entire case.[83] Ordinarily, the issues to be decided will be those stated in the direction for review, those raised in the parties' petitions or those

---

[77] 29 C.F.R. § 2200.90(b)(2).
[78] 29 C.F.R. § 2200.90(d).
[79] 29 C.F.R. § 2200.92(b).
[80] 29 C.F.R. § 2200.91.
[81] 29 C.F.R. § 2200.91(b).
[82] 29 C.F.R. § 2200.91(c).
[83] 29 C.F.R. § 2200.92(a).

stated in any later order. Review can be directed by a Commissioner on his / her own motion, within 30 days after the docketing of the judge's report. Typically, such instances would be limited to matters which raise novel questions of law or policy, or where conflicts exist between rulings of different judges.[84] The Commission ordinarily will request the parties to file briefs on issues before the Commission. Rules governing the filing of briefs are found at 29 C.F.R. § 2200.93.

6.5 Stay of Commission Order

Parties aggrieved by a final order of the Commission may file a motion for a stay while the matter is within the jurisdiction of the Commission.[85] A motion for a stay of the final order must state the reason the stay is sought and the length of time requested for a stay.[86] The Commission may order a stay for the period requested or for any length of time which the Commission determines is appropriate.[87]

---

[84] 29 C.F.R. § 2200.92(b).
[85] 29 C.F.R. § 2200.94.
[86] 29 C.F.R. § 2200.94(b).
[87] 29 C.F.R. § 2200.94(c).

Chapter 12

Criminal Enforcement of Violations

Michael C. Formica, Esq.
Marshall Lee Miller, Esq.[1]
Baise and Miller, P.C.
Washington, D.C.

1.0 Overview

It is well known that employers who are charged with violating OSHA laws and regulations face substantial civil penalties. Less well known is that they may also be subject to criminal sanctions. For other environmental laws, criminal penalties are threatened for a wide variety of offenses. In the OSHA area, the applicability of criminal charges is restricted to narrow circumstances: (1) if an employer's willful violation of an OSHA standard results in the death of an employee,[2] (2) if an individual gives advanced notice of an inspection,[3] or (3) if an individual or employer knowingly makes any false statement, representation, or certification under the OSH Act.[4] Unsurprisingly, the most stringent measures occur when violations result in the death of an employee. That means, however, that if a willful violation "only" causes paralysis, brain damage, or a comatose state without leading to death,[5] the OSH Act treats them as less serious and authorizes only civil penalties.

Moreover, unlike civil complaints, which OSHA prosecutes with its own staff, criminal charges are only prosecuted by the United States Department of Justice (DOJ), following referral by OSHA. The DOJ prosecutions, however, have been fairly limited. As a result, in an ironic reversal of roles, pressure has built on the

---

[1] Mr. Formica is an associate and Mr. Miller is a partner with the Washington, D.C., law firm of Baise & Miller, P.C.; <www.dclaw.net>.
[2] 29 U.S.C. § 666(e).
[3] 29 U.S.C. § 666(f).
[4] 29 U.S.C. § 666(g).
[5] All of these, of course, are situations that many victims might consider worse than death.

states to reassert their authority to regulate workplace conditions and increase criminal enforcement for workplace safety violations.

While these attempts were originally successful in State courts, some argue that the OSH Act preempts any state efforts to regulate the workplace through criminal enforcement actions. The Supreme Court apparently agrees, finding in *Gade v. National Solid Waste Management Association*[6] that a state's ability to regulate occupational safety and health standards is pre-empted by the OSH Act, unless approved by OSHA.[7] This, and the requirement of death prior to criminal prosecution, increased the pressure on the United States Environmental Protection Agency (EPA) to prosecute for workplace violations under the various environmental statutes, including the Clean Water Act, Clean Air Act, and Resource Conservation and Recovery Act. There have also been calls on the national political scene in the last year to strengthen criminal enforcement provisions under the OSH Act.

2.0 Federal Prosecution

OSHA does not handle criminal enforcement under the OSH Act. Instead, OSHA simply refers potential cases to the DOJ for review. The actual process of referral, however, can be quite drawn out and complicated.[8] In most situations, the OSHA Area Director refers any potential actions to the appropriate Regional Administrator. He will then hand the case to an OSHA Solicitor, either Regional or National, who makes a final decision to refer the case for DOJ prosecution. Following DOJ acceptance, and assignment of a U.S. Attorney, a grand jury is convened to set forth an indictment. As complex as the referral process is, OSHA spent 20 years devising an internal system to effectively keep track of referrals.[9]

2.1   Definition of "Employee"

In this age of temporary workers, independent contractors, and subcontracting, debate over who is an employee under the OSH Act has grown. Traditionally, an employee is someone directly under control of the employer. Under the common law of agency, independent contractors are not necessarily agents and therefore not

---

[6] 505 US 88 (1992).
[7] About half of the States currently have OSHA-approved state plans.
[8] Of course, the process may always be shortened or expedited if the facts are such that media coverage and political pressure make swift prosecution necessary.
[9] OSHA Interpretation Memo, Leo Carey to Regional Administrators, "Procedures For Tracking Criminal Referrals," May 31, 1990.

necessarily considered employees. Neither are the employees of a subcontractor or temporary agency, especially in relation to many mandated employee benefits.

However, under the "Multi-Employer Doctrine," any employer who creates a safety hazard, and willfully violates OSHA standards, on a multi-employer work site, may be criminally liable for the death of a worker, whether their own or that of another employer.[10] In *Pitt-Des Moines,* an accident resulting from a deviation from industry standards and inadequate training caused a structure to collapse, killing both a Pitt-Des Moines employee and a subcontractor's employee. Pitt-Des Moines was found guilty of a willful violation due to their failure to follow standard industry procedures in securing steel beams and fined $1 million. On appeal, the 7th Circuit Court of Appeals upheld application of the "multi-employer doctrine."

The doctrine, however, is not without controversy. In addition to complaints concerning its migration away from the traditional law of agency, in the case of *IPB, Inc. v. Herman,* the Court of Appeals for the D.C. Circuit has criticized its broad expansion of liability outside the express language of the OSH Act.[11]

2.2 Willful Violations Causing Death to Employee

The most serious penalties follow from the willful violation of an OSHA standard or rule when that violation results in death to an employee. While for first-time offenders the penalty is identical to that for making false statements, $10,000 and six months imprisonment, maximum penalties for a repeat offender double to $20,000 and one year in prison.[12]

A "willful violation" is a knowing violation.[13] An employer must know, prior to the fatal accident, the essential facts and legal requirements.[14] However, to prosecute for a willful violation, the employer needs not to have exhibited "intentional disregard," but just "plain indifference," towards the safety requirements promulgated under the OSH Act.

The 1998 conviction of Roy G. Stoops is an example. Mr. Stoops, the owner of C&S Erectors of Nolesville, Indiana, failed to correct safety hazards and provide fall

---

[10] *US v. Pitt-Des Moines, Inc,* 168 F.3d 976, (7th Cir. 1999).
[11] *IPB, Inc. v. Herman,* 144 F.3d 861 (D.C. Cir. 1998).
[12] As discussed below, penalties are commonly increased by tying OSH Act prosecutions to additional prosecutions under existing environmental laws or raising more common criminal violations, such as conspiracy, at the same time.
[13] *US v. Ladish Malting,* 135 F.3d 484 (7th Cir. 1998).
[14] *Id.*

protection at a job site, even after receiving complaints from contractors. Following the fatal fall of an employee, an Indiana court found that Mr. Stoops' failure to act and listen to concerns amounted to a willful act, that taken together with C&S's substantial history of OSHA violations led the court to sentence Mr. Stoops to four months in prison.[15]

"Knowing" has also been interpreted as "know or *should* have known." In other words, a person cannot insulate himself from liability by either willful blindness or negligent conduct. As a strict liability statute, the normal criminal elements of malice or specific intent are removed.[16] While an omission or failure to act may be willful if voluntary or intentional,[17] the exact nature and extent of the employers' knowledge and understanding is an issue to be factored and examined by a jury considering criminal penalties.[18]

2.3 False Statements and Advance Notice

Perjury, or lying under oath, as well as false statements to federal officers are illegal under federal criminal law.[19] In making official filings required under the OSH Act, the signatory is affirming, under oath, the truth of any statements and documents. Knowing violation of this oath is a crime punishable by up to six months in prison and/or a $10,000 fine. In other words, since perjury is already a crime, with more substantial punishment under U.S. criminal statutes, there is nothing exceptional in specifying that this law also applies to workplace situations. In addition, the OSH Act also makes the unauthorized disclosure of forthcoming inspections punishable by a fine of $1000 and imprisonment for up to six months.[20]

---

[15] "Construction Worker's Death Results In Jail Time For Indiana Employer," OSHA National News Release, USDL 98-421, October 15, 1998.

[16] *Ensign-Bockford Corporation v. OSHRC*, 717 F.2d 1419 (D.C. Cir. 1983).

[17] *Daniel Intern Corporation v. Donovan*, 705 F.2d 382 (10th Cir. 1983).

[18] *U.S. v. Ladish Malting, supra.*

[19] 18 U.S.C. § 1621 permits a fine and/or imprisonment of not more than five years for knowingly making any false material declaration or using any other information, including any book, paper, document, record, recording, or other material, knowing the same to contain any false material declaration, in a proceeding before or ancillary to any court or grand jury of the United States.

[20] 29 U.S.C. § 666(f).

3.0 State Enforcement

In light of what some see as the restricted criminal deterrence of the OSH Act, states were increasingly bringing their own actions against employers whose conduct resulted in death or serious injury to employees. For the most part, these efforts were successful. Starting with *People v. Chicago Magnet Wire Corp.*,[21] several state supreme courts have found that OSHA did not preempt criminal prosecution of corporate officials. In *Chicago Wire*, an Illinois manufacturer was indicted for reckless endangerment by failing to provide adequate safety precautions to prevent exposure to the chemicals used in manufacturing wire. Likewise, in the leading case *People v. Pymm Thermometer*, the New York Court of Appeals affirmed a decision that criminal prosecution for reckless endangerment associated with employee exposure to mercury was not preempted by the federal OSH Act.[22]

The state courts, looking at the OSH Act's definition of standards, have generally found criminal laws are not standards. Instead, they find that criminal laws serve a general purpose separate from regulation of the workplace. In *Pymm*, the court held that unlike the OSH Act, which established standards to prevent death or injury from occurring, New York's criminal laws were reactive, designed for punishment of acts already committed, and this serves a purpose separate from regulation of the workplace.

In *Gade*, the U.S. Supreme Court looked at whether OSHA preempted an Illinois licensing standard applicable to hazardous waste workers. The Court held that where a federal standard is in effect, the unauthorized state regulation of occupational, safety, and health issues is in conflict with the purposes and objectives of the OSH Act and is therefore impliedly preempted.[23] Justice O'Conner, writing for the Court, found the statement in § 18(b), that a State "shall" submit a plan if it wishes to "assume responsibility" for development of standards, indicated Congress' intent to preempt state law. As a result, a State may not enforce its own standards without federal approval.[24]

The Court found that those dual-impact regulations, that embody several purposes in addition to regulation of occupational, safety, and health, cannot avoid preemption.[25] However, the Court acknowledged that general applicability statutes,

---

[21] *People v. Chicago Magnet Wire Corp.*, 534 NE.2d 962 (Ill. 1989).
[22] *People v. Pymm Thermometer Company*, 561 NYS.2d 687 (NY 1991).
[23] *Gade* at 98.
[24] *Id.*
[25] *Id.* at 107.

such as fire and traffic safety laws, that do not conflict with OSHA regulations and regulate the conduct of workers and non-workers alike, are not generally preempted.[26] To the extent that State prosecution is carried out under already existing general criminal and safety provisions, they should be allowed.

4.0 Prosecution under Environmental Statutes

Because of the limitations on criminal prosecution under OSHA and the concomitant problems under state and common law doctrines, some prosecutors have sought to deal with workplace situations by finding violations under the more flexible environmental laws.

At the federal level, the EPA is not constrained by the fatality requirement of the OSH Act. It has therefore been able to apply criminal laws in a wide range of circumstances including some that touch on occupational situations. The major environmental statutes provide a host of criminal violations and penalties, and almost every environmental law provides some criminal liability. In most cases, violations can exist for both "knowing" and "negligent" conduct. It is not uncommon to find penalties of over $25,000 a day per violation and prison terms in excess of 15 years.

In light of this, EPA and OSHA have entered into Memorandums of Understanding (MOUs) whereby the two agency's work together to enforce both environmental and health and safety regulations in the workplace. These MOUs provide for joint inspections by EPA and OSHA investigators, a system for referrals of violations between the agencies, exchanges of data and other evidence uncovered during investigations, and cross training.

For example, the Resource Conservation and Recovery Act (RCRA) provides "cradle to grave" controls and requirements on generators and transporters, and treatment, storage, and disposal facilities of hazardous wastes. RCRA imposes criminal penalties of up to $50,000 and two years imprisonment for knowing violations of the act. It also has put in place a system of heightened criminal liability for knowingly placing persons in danger of imminent death or serious bodily injury. Fines can run as high as $250,000 for an individual and $1,000,000 for a Corporation, in addition to prison terms of up to 15 years.

---

[26] *Id.*

5.0 Recent Legislation

In light of the inability of the OSH Act to provide criminal penalties for violations causing serious bodily harm, in addition to a belief by some that the criminal penalties provided are inadequate to serve a deterrence effect, calls have been made to toughen the OSH Act by expanding the range of activities that result in criminal liability as well as increase the penalties provided under the Act. So far, these demands have been on the political periphery. Consumer activist Ralph Nader made criminal enforcement of workers deaths a major point in his third party campaign for the presidency of the United States, and liberal Senator Paul Wellstone (D-MN) introduced legislation in 1999 that expanded the coverage of the OSH Act to federal employees;[27] significantly increased the penalties for willful violations causing death to an employee by providing prison terms up to 10 years for initial violation and 20 years for subsequent violations while at the same time removing the limits on the amount of financial penalty imposed;[28] and strengthened the whistleblower protections.[29] With Senator Ted Kennedy (D-MA), an original sponsor of Senator Wellstone's bills, ascending to chairmanship of the influential Senate Health, Education, Labor and Pensions Committee, new versions of these bills can be expected along with renewed efforts to overhaul OSHA and the OSH Act generally. However, even most supporters of a stronger OSH Act have not pushed for expanding the scope of criminal penalties under the OSH Act. That will probably come, if it ever does, only after some highly publicized accident.

---

[27] Federal Employee Safety Act S. 650 106th Congress-1st Session.
[28] Wrongful Death Accountability Act S. 651 106th Congress-1st Session.
[29] Safety and Health Whistleblower Protection Act S. 652 106th Congress-1st Session.

Chapter 13

Judicial Review of Enforcement Actions

Stanley M. Spracker, Esq.
John B. O'Loughlin, Jr., Esq.
Weil, Gotshal & Manges, LLP
Washington, D.C.

1.0 Overview

Parties that are unsatisfied with a final order of the Occupational Safety and Health Review Commission (OSHRC) may seek judicial review of the decision. Because the primary adjudicative function under the Occupational Safety and Health Act (OSH Act)[1] is vested in the Labor Department's administrative law judges (ALJs) and the OSHRC, the role of the federal circuit courts of appeal in enforcement cases is limited to appellate review and enforcement of final Commission orders.[2] As discussed in Chapter 2, the federal circuit courts also provide pre-enforcement judicial review of health and safety standards promulgated under the OSH Act if an adversely affected party petitions for judicial review within 60 days after the standard is promulgated.

The role of the federal courts under the OSH Act is limited. First, parties may not bring actions in the federal courts of appeal without first exhausting their administrative remedies. Second, the jurisdiction of the federal district courts is

---

[1] 29 U.S.C. § 660. See Chapter 1 for a comprehensive summary of the OSH Act.

[2] For purposes of this chapter, a final order of the "Commission" can refer to (1) decisions of the Occupational Safety and Health Review Commission (OSHRC) or (2) unappealed decisions of Administrative Law Judges (ALJs) who conduct full evidentiary hearings on behalf of the Commission. Decisions of ALJs become final orders of the Commission if the employer or the Secretary of Labor does not seek review or if OSHRC declines a petition for discretionary review. For a fuller discussion, see Chapter 11.

limited to ancillary procedural matters.[3] Finally, the OSH Act does not create a federal cause of action for violations of health and safety standards, so employees may not seek direct enforcement against employers in the federal courts.

Any party who is adversely affected by a final Commission order may file an appeal in the appropriate U.S. circuit court of appeal. The circuit courts have exclusive jurisdiction to hear appeals of final orders of the Commission. Review by a circuit court panel is generally restricted to the issues preserved for appeal and is decided on the basis of the written record of the proceedings below, the briefs submitted by the parties, and oral argument. As with a review of any final action of an administrative agency, the circuit courts generally give a high level of deference to decisions of the Commission. Orders will only be overturned if the Commission's factual findings are not supported by substantial evidence in the record or if the Commission's legal conclusions are arbitrary and capricious, an abuse of discretion, or otherwise not in accordance with law. A decision of a circuit court is final unless the court remands the case back to the Commission for further proceedings or the U.S. Supreme Court agrees to hear a discretionary appeal of the case.

2.0 Jurisdiction

2.1 Parties That Have Standing to Bring an Appeal

Section 11 of the OSH Act authorizes federal courts to review final orders of the Commission. Under section 11(a), "[a]ny person adversely affected or aggrieved by an order of the Commission...may obtain a review of such order."[4] This provision provides direct access for employers to seek judicial review of an adverse Commission decision. Section 11(b) of the Act also authorizes the Secretary of Labor to seek judicial review of a final Commission order.[5] Other parties have the right to seek judicial review as well. For example, employees or their representatives (including unions) may file a notice with the Department of Labor seeking to become a party to a case and may file a Notice of Contest (NOC) to challenge as unreasonable the

---

[3] The jurisdiction of the district courts in matters under the OSH Act is limited to: (1) actions by the Secretary of Labor to enforce administrative subpoenas, 29 U.S.C. § 657(b); (2) actions by the Secretary to enforce the antidiscrimination provisions of the OSH Act, § 660(c)(2); (3) actions by the Secretary to restrain imminent dangers, § 660(c)(2); and (4) actions on behalf of the United States to recover civil penalties, § 662(a), (d).

[4] 29 U.S.C. § 660(a).

[5] 29 U.S.C. § 660(b).

period set for an employer to abate a violation.[6] Any employee (or employee representative) who elected party status before the Commission and who is adversely affected by the Commission's final decision may seek judicial review of the decision.[7]

On the other hand, most courts do not interpret section 11(b) as granting the Commission standing to defend its actions in federal court.[8] The basis for the majority view that the Commission has no standing to appeal is that Congress intended the Commission to act like a court. The parties to the dispute are the Secretary of Labor (the plaintiff) and the employer charged with the violation (the defendant). Accordingly, the Commission is an adjudicative entity with no stake in the outcome of the litigation.[9] The Fourth and Fifth Circuit Courts of Appeal, however, have concluded that the Commission may participate in judicial proceedings on grounds that it serves an administrative function as well as an adjudicative one.[10] In that respect, the Commission's functions are similar to that of the National Labor Relations Board or the Federal Trade Commission.

The courts have also concluded that manufacturers do not have standing to appeal a Commission decision on an OSHA violation involving the manufacturer's machinery. In *R.T. Vanderbilt Co. v. OSHRC*,[11] for example, although the Commission had allowed an equipment manufacturer to intervene in the administrative proceeding against the employer, the Sixth Circuit Court of Appeals held that the manufacturer did not have standing to seek judicial review of the Commission's decision because the manufacturer was not in the "zone of interest" covered by the Act.

To bring an appeal, a party must establish that it has been injured by a Commission order. In reality, this is a fairly easy standard to meet because any employer

---

[6] 29 C.F.R. §§ 2200.20-.22.

[7] 29 U.S.C. § 660(a).

[8] *See Oil, Chem. & Atomic Workers v. OSHRC*, 671 F.2d 643, 651 (D.C. Cir. 1982); *General Electric Co. v. OSHRC*, 583 F.2d 61, 63 n.3 (2d Cir. 1978); *Marshall v. Sun Petroleum Prods.*, 622 F.2d 1176 (3d Cir. 1980); *Marshall v. OSHRC*, 635 F.2d 544, 547 (6th Cir. 1980); *Dale Madden Constr., Inc. v. Hodgson*, 502 F.2d 278, 280-81 (9th Cir. 1974); *Brennan v. OSHRC*, 505 F.2d 869, 871 (10th Cir. 1974).

[9] *See Oil, Chem. & Atomic Workers v. OSHRC*, 671 F.2d 643, 652 (D.C. Cir. 1982).

[10] *See Brennan v. Gilles & Cotting, Inc.*, 504 F.2d 1255, 1267 (4th Cir. 1974)(holding that the Commission may appear in federal court to defend its enforcement policies); *Diamond Roofing Co. v. OSHRC*, 528 F.2d 645, 648 n.8 (5th Cir. 1976)(holding that the Commission, as any administrative agency, is a proper party in suits brought by the Secretary of Labor or a third party seeking review of its decisions).

[11] 728 F.2d 815 (6th Cir. 1984).

ordered to pay a civil penalty or take action to abate a violation has been aggrieved.[12]

2.2 Courts That Have Jurisdiction over Appeals

Section 11 of the OSH Act states that final Commission orders may be appealed in the U.S. circuit courts of appeal. In particular, an employer may bring an appeal in the circuit where the violation is alleged to have occurred, in the circuit where the employer has its principal office, or in the U.S. Court of Appeals for the District of Columbia Circuit.[13] The Secretary of Labor also may appeal a final order in the circuit where the violation is alleged to have occurred or in the circuit where the employer has its principal office.[14] The jurisdiction of the U.S. circuit courts of appeal to review decisions of the Commission is exclusive, and the judgment of a circuit court is final unless the U.S. Supreme Court agrees to a discretionary review.[15]

3.0 Timing

3.1 Final Commission Orders

Any party with standing, including the Secretary of Labor, may appeal a final Commission order by filing a petition within 60 days of the final order with an appropriate U.S. circuit court of appeals. Under the Federal Rules of Appellate Procedure, appellants must adhere strictly to this deadline or the court may dismiss the petition for lack of jurisdiction.[16] An employer seeking an appeal may also submit a petition for a stay of a final Commission order because an appeal, by itself, does not delay enforcement of the order, and the Secretary of Labor may seek penalties or cite the employer for failure to abate a violation during the pendency of

---

[12] *See RSR Corp. v. Donovan*, 733 F.2d 1142, 1145-46 (5th Cir. 1984)(remanding case to administrative law judge for further findings, but holding that employer had established standing because OSHA had imposed a civil penalty). On the other hand, decisions on procedural issues are not reviewable. *CH2M Hill Central, Inc. v. OSHRC*, 131 F.3d 1244 (7th Cir. 1997).

[13] 29 U.S.C. § 660(a).

[14] 29 U.S.C. § 660(b).

[15] 29 U.S.C. § 660(a).

[16] 29 U.S.C. § 660(a); *Fed. Reg.* App. P. 15. *See Consolidated Andy, Inc. v. Donovan*, 642 F.2d 778, 779 (5th Cir. 1981).

the appeal.[17] If no timely petition for appeal is received by the clerk of the court, the Commission's findings of fact, conclusions of law, and order are deemed conclusive. Because Commission orders are not self-enforcing (see Chapter 11), the Secretary of Labor must petition the court to enforce an order if the employer does not comply.[18]

3.2 Exhaustion of Administrative Remedies

A prerequisite to obtaining judicial review of a Commission order is that the order must be final. This means that the party (typically an employer) must have first exhausted all available remedies. As discussed in Chapter 11, the employer must file a Notice of Contest (NOC) of the citation or penalty with the Secretary of Labor within 15 working days after receiving notice. If the party does not contest the citation or penalty, the citation will become a final Commission order that is not subject to judicial review.[19] The NOC must indicate whether the employer is contesting all or part of the citation, the civil penalty, or the abatement. If the employer contests only part of the order, the uncontested part of the order becomes binding upon the employer upon the expiration of the contest period.[20] Upon the filing of a timely NOC, an employer's challenge is subject to a full evidentiary proceeding before an Administrative Law Judge (ALJ). At the conclusion of the trial, an employer that is not satisfied with the ALJ's decision may file a Petition for Discretionary Review (PDR) of the ALJ's decision with the Review Commission.[21] (See Chapter 11 for a complete discussion of the procedures for trials before ALJs and appeals before the Review Commission.) If the Commission declines to review the ALJ's decision, or if the employer is not satisfied with the Commission's decision after review, the employer may seek judicial review in an appropriate U.S. circuit court of appeal.

An employer may not seek judicial review if it has not followed the required administrative procedures. In particular, an employer may not seek judicial review

---

[17] 29 U.S.C. § 660(a). *Fed. Reg.* App. P. 18. The Court of Appeals has the power to grant temporary relief or an injunction during the pendency of an appeal. *Brennan v. Winters Battery Mfg. Co.*, 531 F.2d 317 (6th Cir. 1976). In addition, the Commission itself has the discretion to stay its own orders pending an appeal. *Lance Roofing v. Hodgson*, 343 F. Supp. 685 (N.D. Ga.), *aff'd*, 409 U.S. 1070 (1972).

[18] 29 U.S.C. § 660(b).

[19] OSH Act § 10(a); 29 U.S.C. § 659(a).

[20] *Penn-Dixie Steel Corp. v. OSHRC*, 533 F.2d 1078, 1079-81 (7th Cir. 1977).

[21] 29 C.F.R. § 2200.91.

of a citation if it did not first submit an NOC. Similarly, it may not seek judicial review of an ALJ's decision if it did not first submit a PDR to the Commission irrespective of whether the Commission agreed to review the case.[22] Furthermore, an employer may not raise issues before the court that were not identified in the PDR.[23] For example, in *P. Gioioso & Sons Inc., v. OSHRC*,[24] the employer attempted to raise issues on appeal that were argued before the ALJ but not expressly included in the PDR submitted to the Commission. The court held that it lacked jurisdiction over the issues because the petitioner had failed to preserve them for appeal.[25] In addition, in the event the Commission seeks on its own initiative to review an ALJ decision that was not appealed by the employer,[26] the employer must follow the Commission's review procedures to completion before seeking a judicial challenge. Finally, a decision of the Commission must be a final agency action in order to be reviewable in court.[27] Thus, a Commission order remanding a case back to an ALJ for final disposition is not reviewable, and intermediate decisions of the Commission on procedural questions (*e.g.*, evidence, discovery, or procedure) are not directly reviewable.[28]

Section 6(f) of the OSH Act provides that any party adversely affected by an OSHA standard may seek pre-enforcement judicial review by filing a petition prior to 60 days after promulgation of the final rule.[29] The Act is silent, however, on the question of whether a safety standard upon which a violation is based can be challenged during judicial review of the enforcement order under section 11(a).[30] The courts are in agreement that substantive challenges to the validity of a standard may be raised either during pre-enforcement review or during an enforcement

---

[22] 29 C.F.R. § 2200.91(f). *See, e.g., Keystone Roofing Co. v. OSHRC*, 539 F.2d 960, 962 (3rd Cir. 1976)(order to employer which sought direct judicial review of ALJ decision was a final Commission order, but was unreviewable by the court of appeals because employer had failed to petition the Commission for discretionary review).

[23] *Globe Contractors, Inc. v. Herman*, 132 F.3d 367 (7th Cir. 1997)(issues not reviewable if not included in PDR).

[24] 115 F.3d 100, 107 (1st Cir. 1997)

[25] Similarly, the Commission will ordinarily not review issues not first raised before the ALJ. 29 C.F.R. § 2200.92(c).

[26] 29 C.F.R. § 2200.92(b).

[27] *See, e.g., Northeast Erectors Ass'n v. Secretary of Labor*, 62 F.3d 37, 40 (1st Cir. 1995)(change in OSHA enforcement policy concerning an OSHA standard was not subject to challenge by trade association until such time as an employer is cited for violation of the standard).

[28] *CH2M Hill Central, Inc. v. OSHRC*, 131 F.3d 1244, 1246-47 (7th Cir. 1997).

[29] 29 U.S.C. § 655(f).

[30] 29 U.S.C. § 660(a).

review. The courts are not in agreement, however, on whether a party may challenge the procedural aspects of a rule during an enforcement review. For example, in *Deering Milliken, Inc., Unity Plant v. OSHRC*,[31] the court held that the pre-enforcement review provision in section 6(f) of the Act did not bar a procedural attack on the validity of the underlying standard during a proceeding challenging an enforcement order. On the other hand, in *National Industrial Contractors, Inc. v. OSHRC*,[32] the Court of Appeals for the Eighth Circuit held that the procedural validity of OSHA standards may be challenged only during pre-enforcement review pursuant to section 6(f). The Eighth Circuit distinguished between procedural challenges, which it said were restricted to pre-enforcement review under section 6(f), and substantive challenges, which could be brought either during pre-enforcement review or during a challenge to an enforcement order under section 11(a). Other circuits, however, have declined to draw the distinction between procedural and substantive challenges.[33] In addition, the burden of proof in a legal challenge to a final OSHA rule during an enforcement action is different from what is required during pre-enforcement review of a final OSHA rule. In a proceeding under section 6(f), the Secretary of Labor has an affirmative obligation to demonstrate that the rule as adopted is "reasonable and consistent" with its purpose. In a challenge to an enforcement action under section 11(a), however, the petitioner must produce evidence showing why the standard, as applied to the petitioner, is arbitrary, capricious, unreasonable, or contrary to law.[34]

3.3 Constitutional Challenges

A court of appeals has jurisdiction to consider the constitutionality of the OSH Act during an appeal of a final Commission order.[35] Although the Commission is not empowered to rule on challenges to the constitutionality of a particular action,[36] most courts have concluded that such fundamental issues should be raised at the

---

[31] 630 F.2d 1094 (5th Cir. 1980).

[32] 583 F.2d 1048 (8th Cir. 1978).

[33] *Deering Milliken, Inc., Unity Plant v. OSHRC*, 630 F.2d 1094 (5th Cir. 1980) (discussing the difference between challenges to OSHA's compliance with procedural requirements in promulgating a rule and substantive attacks on the rule itself, but declining to adopt a restriction on post-enforcement review of either type of challenge). *See also Atlantic & Gulf Stevedores, Inc. v. OSHRC*, 534 F.2d 541 (3rd Cir. 1976); *Noblecraft Indus., Inc. v. Secretary of Labor*, 614 F.2d 199 (9th Cir. 1980).

[34] *Atlantic & Gulf Stevedores, Inc. v. OSHRC*, 534 F.2d 541 (3rd Cir. 1976).

[35] *Mohawk Excavating, Inc. v. OSHRC*, 549 F.2d 859 (2d Cir. 1977).

[36] *See Buckeye Indus., Inc. v. Secretary of Labor*, 587 F.2d 231, 235 (5th Cir. 1979).

administrative level in order to preserve the matter for judicial appeal.[37] In a few reported cases, courts have allowed fundamental challenges to Commission decisions that were not raised below to proceed on grounds that they are "extraordinary conditions" within the meaning of section 11(a) of the OSH Act.[38] As a general matter, however, the cautious approach for any practitioner would be to raise formally all issues at the administrative review level.

4.0 Scope of Judicial Review

4.1 Procedural Matters

The operating procedures for enforcement hearings before an ALJ incorporate the Federal Rules of Civil Procedure, which prescribe practice in the federal district courts.[39] An ALJ will rule on evidentiary matters in accordance with the Federal Rules of Evidence.[40] A comprehensive discussion of the Commission's rules of procedure and evidence is provided in Chapter 11. When a Commission decision is appealed to a circuit court, the Federal Rules of Appellate Procedure apply. In particular, Rules 15-20 set forth the rules governing review and enforcement of decisions of administrative agencies and commissions.[41] In addition, each circuit court has local rules that supplement the Federal Rules. The rules are quite detailed in establishing deadlines for the submission of transcripts and briefs, page limits on the length of briefs, the number of copies that must be filed and in what format, and setting forth the manner in which papers must be served. Because the rules can vary between jurisdictions, practitioners must pay close attention to the details of the circuit in which the appeal is filed.[42]

---

[37] *See, e.g., In re* Establishment Inspection of Kohler Co., 935 F.2d 810 (7th Cir. 1991)(holding that Fourth Amendment challenge to inspection warrant should have been raised below).

[38] "No objection that has not been urged before the Commission shall be considered by the court, unless the failure or neglect to urge such objection shall be excused because of extraordinary circumstances." 29 U.S.C. § 660(a). *See, e.g., McGowan v. Marshall*, 604 F.2d 885, 892 (5th Cir. 1979)(holding that constitutional challenge of OSH Act before Commission would be "fruitless").

[39] 29 C.F.R. § 2200.2.

[40] 29 C.F.R. § 2200.71.

[41] *Fed. Reg.* App. P. 15-20.

[42] For example, the Seventh Circuit requires litigants to file 15 copies of briefs with the court along with a disk copy, and sets explicit limits on type size, page counts, and even the color of brief covers. The Seventh Circuit also imposes requirements for the content of briefs, such as the jurisdictional statement and appendices. *See,* Circuit Rules of the United States Court of

(continued on the following page)

There is no evidentiary hearing on an appeal to the circuit court. Normally, a three-judge panel decides the appeal on the basis of the written record, the briefs filed by the parties, and the oral argument before the circuit court panel. Oral argument is usually limited to 15 minutes per side. The appeal is narrow in scope and, as discussed above, the issues are limited to those raised before the Commission in the PDR. In rare cases, a petitioner may seek permission from the circuit court to submit additional evidence and, upon reviewing such additional evidence, the Commission may revise its decision.[43]

4.2 Standard of Review for Conclusions of Law

The OSH Act does not include standards for judicial review of conclusions of law. Accordingly, the generic provisions in section 10(e) of the Administrative Procedure Act (APA) apply.[44] Under the APA, courts will not set aside an agency decision unless it is:

- arbitrary, capricious, an abuse of discretion, or otherwise not in accordance with law;

- contrary to constitutional right, power, privilege, or immunity;

- in excess of statutory jurisdiction, authority, or limitations;

- short of statutory right;

- without observation of procedure required by law;

- unsupported by substantial evidence; and

- unwarranted by the facts to the extent that the facts are subject to trial *de novo* by the reviewing court.[45]

Appeals for the Seventh Circuit (Sept. 1999) and Practitioner's Handbook for Appeals to the United States Court of Appeals for the Seventh Circuit (1999 Edition).

[43] "If any party shall apply to the court for leave to adduce additional evidence and shall show to the satisfaction of the court that such additional evidence is material and that there were reasonable grounds for the failure to adduce such evidence in the hearing before the Commission, the court may order such additional evidence to be taken before the Commission and to be made a part of the record." 29 U.S.C. § 660(a).

[44] 5 U.S.C. § 706(2).

[45] 5 U.S.C. § 706(2). Under the Administrative Procedure Act, in addition to the "abuse of discretion" and "arbitrary and capricious" standards of review, a court can also overturn a decision on grounds that it is "otherwise not in accordance with law." Where, as here, the agency's statute does not provide a specific standard of review for legal conclusions, any of

(continued on the following page)

As a practical matter, this level of review is highly deferential to the Commission. For example, in *Reich v. Simpson, Gumpertz & Heger, Inc.*,[46] the court held that the Commission's legal conclusions must not be disturbed unless they are "arbitrary, capricious, an abuse of discretion, or otherwise not in accordance with law."

Courts give the Commission significant latitude in interpreting the OSH Act and OSHA safety standards unless a decision is arbitrary and capricious or an abuse of discretion. The U.S. Supreme Court has held that OSHA's construction of its own regulations is entitled to substantial deference "because applying regulations to complex or changing circumstances calls upon the agency's unique expertise and policymaking prerogatives."[47] In some cases, the Commission's interpretation of a safety standard may differ from that of OSHA. The Supreme Court has held that, in such cases, the Commission must defer to OSHA's reasonable interpretation of a safety standard because OSHA, and not the Commission, is vested with promulgating and enforcing safety standards and is, therefore, in a better position to interpret such standards.[48]

4.3 Standard of Review for Findings of Fact

The Commission's factual findings, and the inferences derived from them, must be supported by substantial evidence on the record considered as a whole.[49] The term "substantial evidence" has been defined by courts as that which a reasonable person would accept as supporting a conclusion.[50] Although this standard appears to be somewhat stricter than the abuse of discretion standard applied to questions of law, as discussed above, courts nevertheless accord substantial deference to the determinations of an ALJ concerning evidentiary and factual matters because the ALJ has had the opportunity to observe witnesses first hand, whereas the Commission and the courts have no such opportunity. So long as the Commission sets forth a

---

these overlapping standards of the APA can be grounds for overturning a decision of the Commission.

[46] 3 F.3d 1, 2 (1st Cir. 1993).
[47] *Martin v. OSHRC*, 499 U.S. 144, 150-51 (1991)(internal citations omitted).
[48] *Id.* at 152-153.
[49] 29 U.S.C. § 660(a). See *Reich v. Simpson, Gumpertz & Heger, Inc.* 3 F.3d 1, 2 (1st Cir. 1993)(Commission's findings of fact must be based on substantial evidence in the record).
[50] *Martin Painting & Coating Co. v. Marshall*, 629 F.2d 437, 438 (6th Cir. 1980).

rational argument connecting the facts of the case with its decision, therefore, courts are unlikely to overturn a decision based on the facts.[51]

Not all courts apply the substantial evidence standard in the same manner. For example, in *Austin Road Co. v. OSHRC*,[52] the court overturned a Commission decision on grounds that the decision was speculative and not supported by substantial evidence in the record. The court held that the Secretary failed to prove that the defendant's activities "affected interstate commerce," an essential prerequisite to establishing OSHA's jurisdiction.[53] As a result, OSHA failed to demonstrate that the defendant was an "employer" within the meaning of the OSH Act, and the court declined to enforce the citation.

In other cases, courts have remanded Commission orders where the record was insufficient to provide a "substantial basis" for the Commission's decision. For example, in *Builders Steel Co. v. Marshall*,[54] the Eighth Circuit Court of Appeals held that the Commission did not provide an adequate factual foundation for the court to find that the Commission's order was reasonable. In *Builders Steel*, the employer was cited for failing to provide adequate fall protection for welders working approximately 29 feet above a concrete floor. The Commission concluded that the warehouse on which the employees were working was a single-story building subject to the regulation found at 29 C.F.R. § 1926.105(a), which required fall protection for workers operating higher than 25 feet. The employer contended that the building was a multi-tiered structure subject to a different regulation found at § 1926.750(b)(2)(i), which required fall protection for workers operating at heights of greater than 30 feet. After finding that there was insufficient evidence in the record to support the Commission's conclusion that the building was a single-story structure subject to section 105(a), the court vacated the order and remanded the case for the Commission to supplement the record to show which standard applied to the building.[55]

Penalty assessments are evaluated by a court using the highly deferential "abuse of discretion" standard. So long as a penalty is within OSHA's statutory authority

---

[51] *Brennan v. OSHRC*, 501 F.2d 1196 (7th Cir. 1974).

[52] 683 F.2d 905 (5th Cir. 1982).

[53] *Id.* at 907. Under the Commerce Clause of the U.S. Constitution, the authority of Congress to regulate commerce is limited to interstate commerce. A long series of cases have included within that power commercial activities that merely "affect" interstate commerce even if they are conducted entirely intrastate. *See, e.g., Wickard v. Filburn*, 317 U.S. 111 (1942).

[54] 575 F.2d 663 (8th Cir. 1987).

[55] *Id.* at 667.

and is based on some findings set forth in the record, a court is unlikely to vacate a penalty approved by the Commission.[56]

4.4 Precedential Effect of Judicial Decisions

The Commission is not bound by decisions of federal circuit courts even if they are in conflict with the Commission's interpretation of the OSH Act or OSHA standards. Unless the U.S. Supreme Court overturns one of its decisions,[57] the Commission acts in accordance with its own precedent. One circuit court has concluded that the Commission must respect the decisions of a circuit court at least with respect to cases arising in that circuit.[58] Moreover, when a case is remanded to the Commission by a circuit court, the Commission is bound to follow the instructions of the appellate court.[59] In addition, the Commission may not depart arbitrarily from its own precedent without a reasoned explanation in the record.[60]

5.0 Conclusion

The U.S. circuit courts of appeal have exclusive jurisdiction to hear appeals of final orders of the Commission. Prior to obtaining judicial review, parties must first exhaust their administrative remedies or the Commission must decline to hear a discretionary appeal. Judicial review is not a full evidentiary hearing and is restricted to the issues preserved for appeal. Courts generally give a high level of deference to decisions of the Commission, and decisions will only be overturned if the Commission's factual findings are not supported by substantial evidence in the record or if the Commission's legal conclusions are arbitrary and capricious, an abuse of discretion, or otherwise not in accordance with law. A decision of a circuit court is final unless the court remands the case back to the Commission for further proceedings or the U.S. Supreme Court agrees to hear a discretionary appeal of the case.

---

[56] *See, e.g., D&S Grading Co. v. Secretary of Labor*, 899 F.2d 1145 (11th Cir. 1990).

[57] A petition for review of Circuit Court decisions concerning OSHA enforcement can be submitted to the Supreme Court in accordance with 28 U.S.C. § 1254. 29 U.S.C. 660(a).

[58] *See Smith Steel Casting Co. v. Donovan*, 725 F.2d 1032 (5th Cir. 1984)(stating that holding by court of appeal on legal question is binding on Review Commission in all cases arising within that circuit until and unless court of appeals or Supreme Court overturns that holding).

[59] *Butler Lime & Cement Co. v. OSHRC*, 658 F.2d 544 (7th Cir. 1981)(holding that failure to follow Circuit Court's instruction was reversible error).

[60] *See Graphic Communications International Union v. Salem Gravure Division of World Color Press, Inc.*, 843 F.2d 1490 (D.C. Cir. 1988)(holding that Commission's unsupported departure from its own precedent was arbitrary and capricious).

Chapter 14

Imminent Danger Inspections

Kathryn McMahon-Lohrer, Esq.
Jennifer E. McCadney, Esq.
Collier Shannon Scott, LLP
Washington, D.C.

1.0 Overview

In October 1993, an employee at a tire manufacturing plant in Oklahoma City sustained fatal injuries when his head was crushed by a piece of equipment that unexpectedly started in violation of OSHA standards.[1] Prompted by this tragic accident, OSHA inspected the plant and imposed a record-setting fine of $7.5 million on the company. Labor Secretary Robert Reich personally delivered the citations that alleged nearly 100 violations of the lockout / tagout rule.[2] The agency posted an imminent danger notice for other employees at the plant and then sought an injunction requiring Dayton Tire, the manufacturing plant, to immediately abate the alleged life-threatening violations. The federal district court refused to award the injunction because the agency could not prove a real risk of future injury if the machines at issue were not locked out.

This recount is typical of circumstances surrounding imminent danger cases, aside from the high-profile nature of OSHA's service of its citation. It is a case from which one can make several observations that tend to be typical of imminent danger notice cases. First, imminent danger cases are controversial. The agency cannot force an employer to cease and desist dangerous activity without going to court—thus it will carefully deliberate before seeking an injunction. Furthermore, courts are careful to balance threats to worker safety against concerns over employers' productivity. Second, in almost all cases, imminent danger inspections are triggered by on-the-job fatalities or catastrophic accidents brought to OSHA's attention

---

[1] *See* U.S. Judge Denies Immediate Abatement Request, Finding No Imminent Danger at Dayton Tire Plant, 23 O.S.H. (BNA) 1811 (May 25, 1944).
[2] *Id.*

either by employee notices or the evening news. Finally, much to the chagrin of employers, imminent dangers typically become headline news.

While imminent danger cases tend to generate publicity, and therefore appear to occur with some frequency, OSHA actually rarely initiates them. Given their high-profile and urgent nature, however, these cases require immediate attention by employers. Infrequent as imminent danger inspections are, an understanding of the notice, response and remedial process is crucial for any OSHA practitioner. This chapter synthesizes the existing data and information on imminent dangers and outlines the investigation and notice procedures associated with imminent dangers from multiple perspectives—those of a practitioner, employer, employee and OSHA investigator, and discusses competing rights and responsibilities. Finally, it analyzes the potential impact of proposed amendments to the provisions of the Occupational Safety and Health Act.

1.1 Background on Imminent Danger Inspections

A primary purpose of OSHA is to encourage employers and employees to reduce workplace hazards. The agency carries out this duty primarily through the development of occupational safety and health standards. Where OSHA has not promulgated specific standards, employers must comply with the Act's general duty clause.[3] Workplace inspections are the principal means by which OSHA enforces these mandatory job safety standards and the general duty clause.

OSHA has broad authority to conduct workplace inspections to determine whether establishments covered by the Act are complying with the agency's promulgated standards. In order to regulate its response to inspection-triggering events or requests, OSHA has developed a hierarchy of inspection priorities. Thus, while multiple actions or thresholds exist triggering an OSHA inspection, the agency will prioritize its inspection as follows: imminent dangers, catastrophes and fatal accidents, employee complaints, programmed high-hazard inspections, and follow-up inspections.[4] Although OSHA inspectors follow the same inspection procedure regardless of the inspection-triggering event, an employer will want to note the type and purpose of the inspection when contesting a citation or proposed penalty – there are slight, yet important differences between regular and imminent danger inspections.

---

[3] The general duty clause mandates that each employer, "shall furnish...a place of employment which is free from recognized hazards that are causing or are likely to cause death or serious physical harm to his employees." OSH Act § 5(a); 29 U.S.C § 654(a).

[4] *See* OSHA Field Inspection Reference Manual (FIRM) 2.103 (I)(B)(3)(a).

What sets imminent danger inspections apart from regular inspections is their priority status within the agency and the agency's authority to require (with court approval) abatement of the hazard *before* the employer can litigate his position. As the name suggests, imminent dangers are treated seriously, and with deliberate urgency. Typically, upon the completion of a workplace inspection an OSHA inspector will issue a citation, which levies an allegedly appropriate penalty and sets a date by which the employer must abate the alleged hazard. The employer has 15 working days from the date of the citation to contest the citation, the time set for hazard abatement, or the proposed penalty. Normally an employer's contest of a citation stays the requirement that he abate the alleged hazard. However, if the employer contests a case involving an imminent danger, the Secretary of Labor may seek to enjoin the dangerous practice or condition that poses an immediate threat of serious harm to employees. Thus, in an imminent danger situation, the Secretary may act to protect employees and compel the employer to fix workplace dangers prior to the contest hearing. The act also empowers employees to help themselves. If the Secretary of Labor fails to request an injunction to abate an imminent hazard, employees may bring a writ of mandamus compelling the Secretary to take action.

Imminent danger determinations are not classifications of a particular type of violation, rather they categorize the practice or hazard at the workplace temporally. The classification of an alleged violation measures the degree of harm to an employee and the culpability of the employer. Imminent dangers, on the other hand, notify employees of grave risks that are certain to happen in the proximate future. For instance, a citation is classified as a "serious violation" where there is "substantial probability that death or serious physical harm could result and that the employer knew, or should have known, of the hazard."[5] Imminent dangers, by definition, exist where there is reasonable certainty that a danger exists that can be expected to cause death or serious physical harm immediately.[6] Thus, a workplace hazard or practice may pose both an imminent danger and constitute a serious violation.

In practice, imminent danger inspections and notices are rare. Even more rare are requests for injunction initiated by the Secretary of Labor to compel abatement of a hazard. For instance, OSHA's New England region reported that over a five-year period beginning in 1985, it posted an average of one imminent danger notice a year.[7] Because of the high-profile nature of these citations, however, and because of the need to act quickly if OSHA requires immediate abatement based on an

---

[5] *See* OSH Act § 17(k); 29 U.S.C. § 666(k); OSHA FIRM 2.103 (III)(C)(2)(b).

[6] *See* OSH Act § 13(a); 29 U.S.C. § 662(a).

[7] OSHA Posts Imminent Danger Notice Citing Hazards at Massachusetts Plant, 21 O.S.H. (BNA) 492 (Oct. 2, 1991).

imminent danger, it is important to understand how the imminent danger process works.

2.0 Statutory Framework and Regulatory Guidance

The rules and procedures governing the agency's response to imminent dangers are covered in Sections 8 and 13 of the OSH Act. Section 8 discusses OSHA inspections, investigations and record keeping. In particular, Section 8(f)(1) outlines the procedure by which an employee may request an inspection when faced with imminent danger. Section 13(a) contains procedures authorizing the agency to act to counteract imminent dangers, including petitioning a district court for injunctive relief to eliminate the danger.

Additionally, OSHA has developed regulatory guidance that outlines the inspection procedures to be followed when dealing with imminent danger inspections.[8]

2.1 Imminent Dangers Defined

Imminent dangers are "...any conditions or practices in any place of employment which are such that a danger exists which could reasonably be expected to cause death or serious physical harm immediately or before the imminence of such danger can be eliminated through the enforcement procedures otherwise provided by this Act...."[9]

The Act does not separately define "conditions or practices." However, the plain meaning of the terms suggest that an employer may affirmatively create imminent dangers through the crafting of practices or procedures that are likely to cause occupational hazards. The terms also suggest that a dangerous situation could arise indirectly, based on a combination of work conditions that may or may not be within the employer's control. The sparse case law that exists relating to imminent dangers suggests that the most important elements for the agency to establish are "imminency" and "danger" within the meaning of the statute.

OSHA's Field Inspection Manual requires the satisfaction of two requirements before a hazard attains the status of imminent danger. First, there must be sense of urgency calling for immediate action. In other words, there must be a reasonable likelihood that the accident will occur immediately, or if not immediately, then

---

[8] *See* OSHA FIRM Chapters I and II.
[9] OSH Act § 13(a); 29 U.S.C. 662(a).

before abatement would otherwise occur.[10] Second, the "harm threatened must be death or serious physical harm."[11] The threat of death is a clear requirement. Serious physical harm, on the other hand, is typically limited to situations where an individual is in danger of losing a limb, digit or use of a bodily function, such as sight, hearing or mobility. The FIRM also categorizes dangers to human health as another form of "serious physical harm." This is especially so when an employee is exposed to a toxic substance or other health hazard that causes "...harm to such a degree as to shorten life or cause substantial reduction in physical or mental efficiency even though the resulting harm may not manifest itself immediately."[12]

3.0 Inspection Procedure

3.1 OSHA Investigation

There are several sources from which OSHA learns of imminent dangers. Commonly investigations are prompted by outside tips, most of which come from other safety or regulatory entities, such as local fire departments or the National Institute for Occupational Safety and Health. Sometimes an imminent danger investigation ensues from evidence gathered during a regularly-scheduled OSHA inspection. Often, inspections follow publically-reported accidents or catastrophes. Indeed, accidents involving significant publicity may be considered as either a complaint or referral.[13] Inspections almost always ensue following accidents involving the death or serious injury of a worker.

Employees may also notify OSHA of imminent dangers they encounter in the course of their work. An employee or representative of an employee may request an inspection by giving notice to the Secretary or his authorized representative if he believes that a "violation of a safety or health standard exists and threatens physical harm, or that an imminent danger exists.[14] Employees have the option of filing a formal or nonformal complaint.[15] While, OSHA inspections of all formal serious complaints are mandatory, whereas inspections of nonformal complaints are discretionary,[16] OSHA must immediately make onsite investigations of all reports or

---

[10] *See* OSHA FIRM 2.103 (II)(B)(3)(b)(1).
[11] OSHA FIRM 2.103 (II)(B)(3)(b)(2).
[12] *See* OSHA FIRM 2.103 (II)(B)(3)(b)(2).
[13] OSHA FIRM 2.103 (I)(c)(u).
[14] OSHA Act § 8(f)(1) 20 U.S.C. § 657(f)(1).
[15] OSHA FIRM 2.103 (I)(C)(2)(c)and (d).
[16] OSHA FIRM 2.103 (I)(C)(6).

complaints of imminent dangers once the Area Director determines that there is a reasonable basis for the allegation.

OSHA facilitates the employee complaint process by providing its OSHA-7 Form, Notice of Alleged Safety or Health Hazards, on its website.[17] Employees may file the complaint online or download the form and then fax or mail it to their local OSHA office. Employees may also submit a request for an onsite inspection without using the OSHA-7 as long as it is in writing and establishes with reasonable particularity the grounds for the notice.[18] The person sending notification, other than OSHA-7, must attach his signature and provide a copy of the notice to his employer no later than at the time of inspection.[19] In order to protect privacy, the person giving notice may request that his name and the names of individual employees referred to do not appear in the employer's copy or on any copies that will be made available publicly.[20]

Employees may also file nonformal, oral or unsigned complaints with OSHA. In fact, because of their urgent nature, OSHA encourages employees to report imminent dangers by telephone by calling its imminent danger hotline. As stated above, OSHA must respond to any legitimate report of an imminent danger with an onsite investigation. This differs from a response to a typical nonformal complaint, that does not involve an allegation of imminent danger, where the Area Director has discretion in determining whether to initiate an inspection based on his or her assessment of the gravity and seriousness of the hazard.[21]

3.2 Secretary of Labor's Response

Upon the receipt of any imminent danger allegation from an employee or employee representative, the Area Director of an OSHA office must immediately determine whether there is a reasonable basis for the allegation.[22] If the Area Director determines there are no reasonable grounds to believe that a violation or danger exists, he must inform the employee or employee representative in writing of such determina-

---

[17] See www.osha.gov. The OSHA website is a valuable resource for employees who wish to notify their local OSHA office of a potential imminent danger. The agency has posted guidance and hotline numbers to assist employees with any specific questions they may have pertaining to OSHA's complaint procedure.

[18] OSH Act § 8(f)(1); 20 U.S.C. § 657(f)(1).

[19] OSH Act § 8(f)(1); 20 U.S.C. § 657(f)(1).

[20] OSH Act § 8(f)(1); 20 U.S.C. § 657(f)(1).

[21] OSHA FIRM 2.103 (I)(C)(7)(d)(2).

[22] See OSH Act § 8(f)(1); 29 U.S.C. § 657(f)(1); OSHA FIRM 2.103 (I)(B)(3)(a).

tion.[23] However, in the event the inspector affirms that a reasonable basis exists, the OSH Act and agency regulations require an inspection of the incident, according it the highest priority with respect to other inspections.[24] If an immediate inspection cannot be made, the Area Director (AD) or Compliance Safety and Health Officer ("CSHO") must immediately contact the employer and when known, the employee representative in order to obtain as many pertinent details as possible concerning the situation and attempt to have any employees affected by imminent danger voluntarily removed.[25] Agency regulations permit the CSHO to consider expanding the scope of investigation based on the information available during the inspection process.[26]

If the inspection reveals conditions or practices that constitute an imminent danger exist, the AD or CSHO must advise the employer of the determination. The AD or CSHO must then request the employer to notify its employees and remove them from exposure to the imminent danger.[27] Furthermore, the OSHA official should encourage the employer to do whatever is possible to take voluntary measures to promptly eliminate the danger. Often, the imminent danger notice will include an order and proposed plan of action that upon compliance will result in the removal of the notice. Orders issued pursuant to a finding of imminent danger under § 13(a) may require an employer to take steps necessary to avoid, correct or remove the danger.[28] They also may prohibit the presence of individuals in locations or under conditions where an imminent danger exists. This restriction does not apply to individuals whose presence is necessary to remedy the imminent danger or maintain capacity so that the facility or site can return to normal business operations. It is important to note, however, that an OSHA inspector does not have unilateral authority to force the employer to comply with the order. The inspector also cannot order a shutdown of the facility without first obtaining a court order. If the AD or CSHO petitions the Secretary for relief, he must inform the affected employees and employer.[29]

---

[23] OSH Act § 13(c); 29 U.S.C. § 662(c).
[24] See OSHA FIRM 2.103 (I)(B)(3)(a).
[25] See OSHA FIRM 2.103 (I)(C)(5)(c).
[26] See OSHA FIRM 2.103 (II)(B)(3)(c)(1).
[27] See OSHA FIRM 2.103 (II)(B)(3)(c)(2).
[28] OSH Act § 13(a); 29 U.S.C. 622(a).
[29] OSH Act § 13(c); 29 U.S.C. 622(c).

3.3 Employer Response

The Act anticipates that an employer's preferred response will be to voluntarily comply with an inspector's request to remove work hazards. If an employer voluntarily and permanently eliminates an imminent danger, the OSHA inspector need not institute an imminent danger proceeding, nor complete a Notice of Alleged Imminent Danger.[30] However, the inspector must still issue a citation and penalty.[31] If the employer cannot or does not take voluntary action, OSHA has authority to petition a U.S. district court to force the employer to cease operations and eliminate the dangerous conditions.[32] The desired remedy is either an injunction or temporary restraining order ("TRO").[33] In such instances, it is important to note that only the AD and Regional Solicitor may order the closing of an operation pursuant to a TRO. A CSHO has no such authority and may not direct employees to leave the workplace.[34] The CSHO may only post an OSHA-8 form and then contact the AD, who makes the ultimate decision to begin the process for obtaining a TRO. The Regional Administrator must be notified of all TRO proceedings.[35]

4.0 Remedying Imminent Dangers – Injunction Proceedings

The standard under which a district court orders a preliminary injunction pursuant to an imminent danger determination differs slightly from the usual injunction test. Generally, a court will order a preliminary injunction if the movant establishes that: (1) there is a substantial likelihood of prevailing on the merits; (2) the movant will suffer irreparable injury in the absence of relief; (3) the threatened injury to the

---

[30] *See* OSHA FIRM 2.103 (II)(B)(3)(c)(2)(a).

[31] *See* OSHA FIRM 2.103 (II)(B)(3)(c)(2)(a).

[32] OSH Act § 13(a); 29 U.S.C. 662(a).

[33] "Upon the filing of any such petition the district court shall have jurisdiction to grant such injunctive relief or temporary restraining order pending the outcome of an enforcement proceeding pursuant to this Act. The proceeding shall be as provided by Rule 65 of the Federal Rules, Civil Procedure, except that no temporary restraining order issued without notice shall be effective for a period longer than five days." OSH Act § 13(b); 29 U.S.C. 662(b).

[34] *See Able Contractors, Inc.* 1978 OSAHRC Lexis 337 at *17 (June 14, 1978); 6 OSHC (BNA) 2135 (stating the limit of an inspector's authority is to inform the employees of the danger and recommend to his area director that injunctive releif be sought). Notwithstanding the statement in the previously cited case, the Commission refused to vacate a citation where the inspector told a worker not to return to a dangerous work condition. It acknowledged that the inspector had unlawfully shut down the job, but did not feel a vacation was appropriate when the employer, at the time of the inspection, was preparing to shut down the job for the day.

[35] *See* OSHA FIRM 2.103 (II)(B)(3)(c)(2)(b)(1).

movant outweighs any damage the relief would cause to the opposing party; (4) the injunction is not adverse to public interest.[36]

When faced with a TRO petition based on the existence of an imminent danger, a court must be able to say that employees are in fact in imminent danger. The court makes this determination by assessing whether conditions are such that a danger exists which could reasonably be expected to cause death or serious harm either (1) immediately or (2) before the imminence of such danger can be eliminated through regular enforcement procedures.[37] Furthermore, a reviewing court must conclude that there is "more than a *mere possibility* that an employee *may* be injured."[38] Thus, a court must affirmatively conclude, from the viewpoint of a reasonable person, that a real risk of injury of death exists. Typically, where the Secretary decides to seek an injunction it is because an employee is faced with immediate death or serious harm.

In the Dayton Tire case, for instance, Secretary Reich was criticized for his novel use of the imminent danger clause in seeking a TRO.[39] Until that time, most injunctions sought under an imminent danger determination were based on what is known as the "first prong" of OSH §13(a), which gives the Secretary authority to "restrain any conditions or practices...which could reasonably be expected to cause death or serious physical harm immediately...."[40] Secretary Reich instead based the Department's injunction request on the "second prong" which permits the Secretary to restrain conditions or practices that could cause death or serious physical harm before the danger can be eliminated through normal administrative law enforcement procedures.[41] Dayton Tire opposed the use of the injunction clause in this way, arguing that it affords the Secretary too much discretion.[42] Namely, the second prong permits an injunction when "some employees may suffer some kind of injury at some time in the future."[43]

---

[36] *See Reich v. Dayton Tire*, 853 F.Supp. 376, 379 (W.D. Okla. 1994) (citing *Lundgrin v. Claytor*, 619 F.2d 61, 63 (10th Cir. 1980).

[37] *See Reich v. Dayton Tire*, 853 F.Supp 376 (W.D. Okla. 1994).

[38] *See Reich v. Dayton Tire*, 853 F.Supp 376 (W.D. Okla. 1994) (Emphasis added).

[39] *See* Firestone Case Represents 'Novel' Use of Imminent Danger Provision, Dear Says, 23 O.S.H. Rep. (BNA) 1660 (Apr. 27, 1994).

[40] OSH Act § 13(a); 29 U.S.C. 662(a).

[41] OSH Act § 13(a); 29 U.S.C. 662(a).

[42] U.S. Judge Denies Immediate Abatement Request, Finding No Imminent Danger at Dayton Tire Plant, 23 O.S.H. Rep. (BNA) 1811 (May 25, 1994).

[43] U.S. Judge Denies Immediate Abatement Request, Finding No Imminent Danger at Dayton Tire Plant, 23 O.S.H. Rep. (BNA) 1811 (May 25, 1994).

Ultimately, the district court denied injunctive relief under OSH § 13(a) in the Dayton Tire proceeding because it could not conclude that employees were in imminent danger. There was not sufficient evidence presented to establish a prima facie case for injunction according to the court. The testimony did not demonstrate that the injuries were the result of noncompliance with the cited standard, no witnesses testified to the presence of any future risk, no employees indicated a belief of imminent danger, and only one of four inspectors believed the violations posed an imminent danger.[44] The court's opinion did not address the Secretary's novel use of the second prong, and there have been no subsequent legal challenges against its use.

Accordingly, the use of the "second" prong of Section 13(a) as a valid basis for obtaining an injunction remains unresolved. Given the infrequency of such petitions, it is unlikely to re-appear before a court anytime soon. Nevertheless, the language remains and OSHA has stated that it will use all tools at its disposal to combat imminent dangers.[45]

5.0 Employee Rights

When faced with an imminent danger, employees may seek protection by asserting a combination of statutory and judicially recognized rights.[46] Employee rights fall within two general categories—the right to seek agency protection and the right to invoke self protection. The former is statutory, and the later regulatory. An employer may not discriminate against an employee for exercising any rights granted under the Act.[47] Knowledge of the scope and limitations on these rights are necessary when assessing the appropriate response to an imminent danger from an employee's perspective.

5.1 Right to Inform and Assist With Inspection

As previously noted, the Act permits employees to notify OSHA of an imminently dangerous workplace, condition or practice and request an inspection.[48] Employees

---

[44] *See Reich v. Dayton Tire*, 853 F.Supp 376, 380 (W.D.Okla. 1994).

[45] *See* Firestone Case Represents 'Novel' Use of Imminent Danger Provision, Dear Says, 23 O.S.H. Rep. (BNA) 1660 (Apr. 27, 1994) where OSHA Administrator Joseph A. Dear said OSHA will use "what works" to compel employers to provide safe conditions for their workers.

[46] *See Whirlpool Corp. v. Marshall*, 445 U.S. 1, 10 (1980).

[47] OSH Act § 11(c); 29 U.S.C. 660(c).

[48] OSH Act § 8(f)(1); 29 U.S.C. § 657(f)(1).

also have a limited right to assist with the workplace inspection[49] and aid the court in determining whether a risk of imminent danger exists.[50] Finally, an employee may compel the Secretary of Labor to seek injunctive relief if he believes that the Secretary has wrongfully declined to do so.[51] Subsection 5.2 discusses an employee's right to seek a writ of mandamus in such situations in more detail.

5.2 Requesting a Writ of Mandamus

In the event the Secretary of Labor arbitrarily or capriciously fails to seek relief for an imminent danger, any employee or representative of an employee who may be injured may bring an action in a U.S. district court and "compel the Secretary to seek such an order and any further relief as may be appropriate".[52]

Procedurally, an employee may file a writ of mandamus action in a district in which the imminent danger is alleged to exist, a district where the employer has its principal office or the District of Columbia.

The Act limits an employee's relief request to mandating that the Secretary of Labor take some action. The clause, " ...and any further relief as may be appropriate," does not encompass a private suit for damages alleging negligent failure of the Secretary to respond to imminently dangerous conditions.[53] Furthermore, the right to seek a writ is a remedy provided for the exclusive use of *employees*. The Act does not deprive an *employer* of procedural due process if the Secretary of Labor fails to formally initiate imminent danger proceedings after posting an imminent danger notice. One company unsuccessfully argued that in such a situation it did not have an opportunity to challenge the Secretary's interpretation of standards in the imminent danger proceeding.[54] The court found no violation because the Secretary did not have to take action under the Act and the employer could challenge the citations.[55] Thus, failure to take action when the Act so requires exposes the Secretary of Labor to suit by an employee only.

---

[49] OSH Act §§ 8(a)(2), (e), and (f)(2); 29 U.S.C. §§ 657(a)(2), (e), and (f)(2).

[50] *See* OSH Act § 11(c)(1); 29 U.S.C. § 660(c)(1).

[51] OSH Act § 13(d); 29 U.S.C. § 662(d).

[52] OSH Act § 13(d); 29 U.S.C. 662(d).

[53] *Caldwell v. U.S.,* 6 O.S.H. Cas. (BNA) 1410-12 (D.D.C. 1978) (holding that Congress considered and rejected a cause of action for employees against the United States for failure to seek appropriate relief under the Act.).

[54] *Secretary of Labor v. Catapano,* 17 O.S.H. Cas. (BNA) 1776, 1779 (O.S.H. Rev. Commission 1996).

[55] *Secretary of Labor v. Catapano,* 17 O.S.H. Cas. (BNA) 1776, 1779 (O.S.H. Rev. Commission 1996).

5.3 Refusing Dangerous Work

There is no general right under the OSH Act for employees to refuse work. In fact, refusing work may result in disciplinary action by an employer. However, employees do have the right to refuse a job if they believe in good faith that they are exposed to an imminent danger.[56] The Supreme Court addressed this very issue in *Whirlpool Corp. v. Marshall* and upheld the corresponding regulation promulgated by the Secretary of Labor as consistent with the Act.

The Court found this implied right to refuse to be exposed to an imminent danger to be an essential component in carrying out the Act's goal of assuring "safe and healthful working conditions." The right was narrowly carved out to permit an employee to take protective action in the unusual situation where an employer refuses to voluntarily eliminate an imminent danger and provisions providing for prompt judicial relief are ineffective. For instance, if an employee were ordered to immediately perform a task in the face of danger, there would be no time to notify an OSHA official, file a notice of imminent danger or seek a court injunction or TRO. The Department of Labor has no authority to shut down a worksite without district court approval, or otherwise unilaterally provide immediate protection to an employee.[57] Furthermore, the employee has no authority to order the employer to correct the hazardous condition. In these rare situations, an employee is permitted to protect himself by refusing to work, without discrimination and without pay, on a good faith belief that completing the assigned task would put him in grave danger of death or serious injury.

The "good faith" element means that the employee must reasonably believe that such dangerous conditions exist. The employee may not refuse work merely to harass the employer or disrupt business. Thus, if an imminent danger is later not found to exist, the employee will not be penalized if his belief was reasonable. It is important to note that there is no corresponding right to walk off the job because of mere unsafe conditions. Employees must stay on the job until the problem can be resolved.[58]

OSHA establishes the following conditions for refusing to work:[59]

---

[56] 29 C.F.R. § 1977.12(b)(2) (1999).

[57] Legislators did contemplate giving the Secretary to take more protective action, such as ordering a shutdown of a no more than five days, but all bills with such language did not pass. See H.R. 16785 § 12(a) 91st Cong., 2d Sess. (1970).

[58] See Refusing To Work Because Conditions are Dangerous, available at <http://www.osha.gov/as/opa/worker/refuse.html>.

[59] See Refusing To Work Because Conditions are Dangerous, available at <http://www.osha.gov/as/opa/worker/refuse.html>.

- where possible, the employee has asked the employer to eliminate the danger and the employer has failed to do so;
- the employee refuses the work in good faith;
- a reasonable person would agree that there is a real danger of death or serious injury; and
- there is imminency, or not enough time due to the urgency of the hazard to have it corrected through regular enforcement (such as requesting an OSHA inspection).[60]

If all conditions are present, OSHA recommends that the employee should then (1) ask the employer to correct the hazard; (2) ask the employer for other work; (3) tell the employer that he will not perform the work until the hazard is corrected; and (4) remain at the worksite until the employer orders otherwise. At this point, an employee should contact OSHA if the employer takes discriminatory action against him for refusing the work. It should be noted that the Act contains no "strike with pay" provision. An employer is only prevented from discriminating against workers who exercise their rights under the Act and need not pay for the time an employee refuses to work.[61]

6.0 Penalties

The OSH Act does not provide for the assessment of special penalties associated with the posting of imminent danger notices or the failure or refusal to correct a hazard or remove employees from conditions that constitute an imminent danger. However, the agency may issue citations and corresponding penalties for serious, willful or repeat violations just as it does for non-imminent danger inspections. Thus, there is no specific penalty for violating an imminent danger order. This does not mean, however, that the existence of an imminent danger is not considered when assessing a penalty. In fact, in some cases, the presence of an imminent danger notice provides sufficient evidence to support a finding of a more serious violation than otherwise would have been found.

In *American Bridge*, an administrative law judge of the Occupational Safety and Health Review Commission denied the respondent's challenge of an OSHA-issued willful violation.[62] The judge noted the respondent's conduct following the posting

---

[60] See also, 29 C.F.R. § 1977.12 (1999).
[61] See generally, Whirlpool Corp. v. Marshall, 445 U.S. 1 (1980).
[62] See American Bridge, 1979 OSAHRC Lexis 81 (O.S.H. Rev. Commission Nov. 19, 1979); 1979 O.S.H. Dec. (CCH) § 24,124 (O.S.H. Rev. Commission Nov. 19, 1979).

of an alleged imminent danger notice as evidence supporting his decision.[63] The judge stated that although the notice "...does not constitute a citation of alleged violations or a notice of proposed penalties,"[64] it established that the respondent knew that it needed to take protective measures for the safety of its employees. Furthermore, because of the notice, the respondent had knowledge of the specific standards that were allegedly violated.[65] Thus, by proceeding with normal operations notwithstanding the notice, the respondent's actions established conscious, deliberate noncompliance and a sufficient basis to support a willful violation.

In 1991 Congress proposed legislation that would permit OSHA to assess a daily civil penalty for the failure of an employer to immediately correct a hazard or remove employees from imminent danger.[66] The bill failed to gain sufficient support and to date no new legislation calling for the creation of specific imminent danger penalties has been introduced. This chapter more fully addresses this bill and other proposed legislation affecting imminent dangers in Subsection 9.0.

7.0 Imminent Danger Inspections in Practice

Imminent danger postings may be triggered by concern over ongoing injuries at facilities. A series of accidents within a short period of time will usually catch OSHA's attention. For instance, within days of an OSHA investigation of a door manufacturing plant where two workers suffered finger amputations, another employee suffered crushing injuries to both hands. OSHA posted imminent danger signs on the two machines where the employees were injured. The signs were removed when the company eliminated the hazard by providing guards for the machines.[67] The company was assessed a total fine of $1,124,500 – it is unclear whether the imminent danger posting increased the sum. However, of the total fine, $210,000 was proposed for three alleged willful violations of machine guarding requirements and failure to implement a lockout / tagout program.

In one case, OSHA posted an imminent danger notice after receiving a tip from a local fire department that a manufacturing plant refused to correct a number of

---

[63] *See American Bridge*, 1979 OSAHRC Lexis 81 at *21 (O.S.H. Rev. Commission Nov. 19, 1979); 1979 O.S.H. Dec. (CCH) § 24,124 (O.S.H. Rev. Commission Nov. 19, 1979).

[64] *See American Bridge*, 1979 OSAHRC Lexis 81 at *21 (O.S.H. Rev. Commission Nov. 19, 1979); 1979 O.S.H. Dec. (CCH) § 24,124 (O.S.H. Rev. Commission Nov. 19, 1979).

[65] *Id.*

[66] The Comprehensive Occupational Safety and Health Reform Act, H.R. 3160, 102nd Cong. (1991).

[67] *See* Texas Manufacturer Faces $1.1 Million Fine for Safety and Health Violations, OSHA National News Release USDL 00-310, Oct. 24, 2000.

fire code violations.⁶⁸ The dangerous conditions included blocked isles, missing fire doors, dead bolted exit doors, and combustible materials improperly stored. After posting the notice, the employer shut down the plant and sent home all employees. The company later agreed to cooperate with the agency to correct the violations.

Imminent danger proceedings also have been used, albeit infrequently, to remedy health hazards.⁶⁹ For a health hazard to be considered an imminent danger, an OSHA inspector must conclude that "toxic substances or health hazards are present and that exposures to them will cause irreversible harm to such a degree as to shorten life or cause reduction in physical or mental efficiency even though the resulting irreversible harm may not manifest itself immediately."⁷⁰ OSHA has exercised this authority in one case by removing thirteen workers with high blood-lead levels from a battery breaking plant.⁷¹ The notice of imminent danger was filed upon the completion of a National Institute for Occupational Safety and Health ("NIOSH") health hazard evaluation which was carried out at the request of OSHA officials. According to the NIOSH evaluation, 13 of the company's 15 employees had average blood-lead levels that exceeded the agency's lead standard. On the same day of the posting, and notwithstanding the fact that the company removed its employees from the workplace, OSHA sought and was granted a TRO. Approximately 5 days later, the company entered into a consent agreement to a preliminary injunction to ensure the preservation of employee rights.⁷²

In another reported case,⁷³ the Sixth Circuit upheld an OSHA subpoena compelling a company to produce documents establishing that it was taking precautions to protect employees from lead exposure. OSHA issued the subpoena primarily to determine whether employees were at risk of an imminent danger of death or

---

⁶⁸ *See* OSHA Posts Imminent Danger Notice Citing Hazards at Massachusetts Plant, 21 O.S.H. (BNA) 492 (Oct. 2, 1991).

⁶⁹ *See Secretary of Labor v. Gene T. Jones Tire & Battery Distributors, N. Ala.* CV-91-P-1654-S (July 19, 1991). *Secretary of Labor v. Master Metals Inc., N.Ohio* 190-CV-1010 (June 1, 1990) (awarding TRO where employees exposed to excessive levels of airborne lead and employer refused to remove workers, and employees themselves declined to leave). *See also*, Labor Department Gets Restraining Order Against Cleveland Firm Over Lead Exposures, 20 O.S.H. Rep. (BNA) 3 (June, 6, 1990).

⁷⁰ OSHA FIRM 2.103 (II)(B)(3)(b)(2).

⁷¹ *See Secretary of Labor v. Gene T. Jones Tire & Battery Distributors, N. Ala.* CV-91-P-1654-S (July 19, 1991).

⁷² *See* OSHA Gets Restraining Order to Remove Alabama Workers from Excessive Lead Exposure, 21 O.S.H. Rep. (BNA) 227 (July 24, 1991).

⁷³ *See Secretary of Labor v. Manganas*, 70 F.3d 434 (6th Cir. 1995). *See also*, Sixth Circuit Upholds OSHA Subpoena, 25 O.S.H. Rep. (BNA) 876 (Nov. 29, 1995); Scope of OSHA's Subpoena Power Form 'Novel" Appeal in Manganas Case, 24 O.S.H. Rep. (BNA) 1154 (Oct. 26, 1994).

physical injury. It could not seek an injunction because work had temporarily ceased due to winter weather. The appellant argued that since it filed a Notice of Contest, the Secretary of Labor could not act, via a subpoena, to fulfill continuing information-gathering and inspection responsibilities under the Act.[74] Specifically the appellant argued that 29 U.S.C. § 659(b) limits the Secretary's right to obtain such information because the period permitted for a correction of a cited violation does not begin to run until a final order by the Commission.[75] Thus, OSHA could not take any action until a final decision was rendered on the administrative challenge to the earlier alleged violations.[76] The Court upheld the subpoena stating that the OSH Act empowers the Secretary to require the protection of witnesses and evidence under oath. Furthermore, the court found that administrative subpoenas should generally be enforced if the requested information is within the agency's authority. The court noted that the Secretary must petition the courts to prevent imminent dangers to workers and may be subject to legal action by employees if it fails to abide by OSH provisions.[77]

8.0 Comparison of Federal Standards with MSHA and State OSHA Programs

OSHA is not the only federal agency with authority to conduct workplace inspections and post imminent danger notices. The Mine Safety and Health Agency ("MSHA") has jurisdiction to regulate the health and safety of mine workers according to its own standards.[78] Furthermore, OSHA shares concurrent jurisdiction with state agencies over certain statutory-designated workplaces.[79] OSHA monitors these state plan programs on an annual basis, and also conducts a set of one-time and follow-up evaluations to identify deficiencies and affirm corrections.[80] Both MSHA and state health and safety plans accord their inspectors differing levels of authority in the imminent dangers area.

When compared to the scope and authority of MSHA and some state inspectors, an OSHA inspector has significantly less authority to take immediate action without first seeking court approval when threatened with an imminent danger.

---

[74] 70 F.3d 434, 436.
[75] 70 F.3d 436.
[76] *Id.* at 437.
[77] 70 F.3d 434, 437.
[78] Federal Mine Safety and Health Act of 1977, 30 U.S.C. § 801 et seq.
[79] OSH Act § 18; 29 U.S.C. 667.
[80] OSH Act § 18(f); 29 U.S.C. 667(f).

Further, the options available to MSHA and some states officials accord them more flexibility to counteract imminent dangers when the employee refuses to take voluntary remedial steps. OSHA does not possess similar authority.[81] These differences are most stark when it comes to shut-down authority and the assessment of special penalties for imminent dangers.

For instance, MSHA inspectors have the authority to issue an imminent danger/withdrawal order requiring the immediate shutdown of a facility.[82] A withdrawal order can restrict employee access to a piece of equipment, section or area of a mine or an entire mine. Furthermore, MSHA may impose criminal penalties up to $25,000 for failure of refusal to comply with a withdrawal order.[83] The employer may face imprisonment for not more than one year.[84]

Some states also permit their health and safety officials to unilaterally authorize a shutdown of operations in the face of an imminent danger.[85] Others, rather than permitting the closure of an entire facility, permit inspectors to post notices prohibiting employees from gaining access to areas that present an imminent danger. A state official may take all the aforementioned steps without first obtaining court approval. Nevertheless, all states have some type of appeals process that permits employers to contest the agency's decision.[86]

In Minnesota, Maryland, Michigan, Oregon and Washington, state officials may issue orders of immediate restraint and can "red tag" machines and equipment that violate state standards.[87] Notice of a red tag prohibits the use of such machines until the employer eliminates the hazard. Similarly, state health and safety officials in Vermont may close a workplace or portion of a workplace where an imminent danger exists without obtaining a TRO.[88] There is a $5,000 per day penalty for disobeying an order.[89] Thus, depending on the state, employees exposed to

---

[81] See Occupational Safety and Health: Changes Needed in the Combined Federal-State Approach, GAO/HEHS-94-10; See also, Federal OSHA Has Less Authority Than Mine Agency, Some States, GAO Says, 21 O.S.H. Rep. (BNA) 1669 (May 20, 1992).

[82] Federal Mine Safety and Health Act of 1977, 30 U.S.C. sec. 801 et. seq. MSHA § 107(a).

[83] Federal Mine Safety and Health Act of 1977, 30 U.S.C. sec. 801 et. seq. MSHA § 110(d).

[84] Federal Mine Safety and Health Act of 1977, 30 U.S.C. sec. 801 et. seq. MSHA § 110(d).

[85] See Occupational Safety and Health: Changes Needed in the Combined Federal-State Approach, GAO/HEHS-94-10; see also, Federal OSHA Has Less Authority Than Mine Agency, Some States, GAO Says, 21 O.S.H. Rep. (BNA) 1669 (May 20, 1992).

[86] See Occupational Safety and Health: Options to Improve Hazard-Abatement Procedures in the Workplace, GAO/HRD-92-105 at 17.

[87] See Imminent Danger Restraint available at <www.osha.slc.gov/fso/osp/oshpa/>.

[88] Id.

[89] See Imminent Danger Restraint available at <www.osha.slc.gov/fso/osp/oshpa/>.

imminent dangers have various degrees of protection afforded them and employers face various types of enforcement actions.

9.0 Legislation

Congress has introduced very little legislation that would have any major impact on imminent danger procedures. All proposed bills address what some believe to be perceived weaknesses in OSHA authority to adequately deal with imminent dangers. Most all amendments that have been introduced propose broadening OSHA's inspector authority or imposing specific penalties associated with imminent danger notices.

H.R. 3160, introduced in the early 1990s, had it won Congressional approval, would have represented the most significant overhaul of OSHA's imminent danger authority.[90] The bill would have allowed the imposition of substantial civil penalties – $10,000 minimum and $50,000 maximum – if the employer refused to correct a hazard or remove employees from imminent danger situations. It would also have permitted the Secretary of Labor to post a notice to employees using a tag or other device identifying the source of imminent danger. The bill also would have permitted OSHA to require that employers abate serious hazards while a citation is being contested and allow employees an expedited appeal of OSHA's decision. Finally, it would have made an employee's right to refuse dangerous work a statutory right, rather than implied through agency regulation.

Even if these changes had been implemented, OSHA, nevertheless, would continue to have less authority to immediately address imminent dangers than MSHA and certain other state safety and health plans provide. OSHA officials would still need to obtain a court order before removing all employees from work areas where they are exposed to imminent dangers.

There have been other, less dramatic, proposals to the OSH Act that would alter the way the agency deals with imminent dangers. One proposal would have granted employees the right to refuse to perform a duty identified as the source of an imminent danger, and also prohibited employer discrimination for such refusal. The bill also would have provided for civil penalties against the employer for each day during which an employee endured exposure to harm.[91] Another less significant, yet interesting proposed amendment to the OSH Act, would have provided an exemption from citation to employers of individuals who perform rescues of

---

[90] Comprehensive Occupational Safety and Health Reform Act, H.R. 3160, 102nd Cong. (1991).
[91] Comprehensive Occupational Safety and Health Reform Act, H.R.1280, 103rd Cong. § 510 (1993).

individuals exposed to imminent dangers as a result of a life threatening accident.[92] Finally, a bill introduced in 2000 would require the Secretary of Labor to issue regulations specifying the application of the OSH Act to workplaces in employee residences. The bill would require OSHA to promulgate regulations that would specify the action to be taken when a complaint or referral indicates that a violation of a safety or health standard exists or exposes an employee to an imminent danger at an employee residence worksite.[93]

It is extremely unlikely that Congress will introduce any legislation during the current administration that will significantly overhaul the OSH Act or expand the authority of OSH inspectors in imminent danger scenarios. In fact, one may expect just the opposite. The new Bush Administration is more likely to subject OSHA inspectors citations to exacting scrutiny and impose tighter reign over their inspection authority. House Republicans were critical of OSHA's position in Dayton Tire and complained that the agency prematurely issued its notice. The outcome merely supported their belief that OSHA imposes unreasonable burdens on the regulated community.[94] If the fate of the Clinton administration ergonomics standards are any indication, there will be no broadening of OSHA inspection authority in the new administration. As for the Bush administration's use of imminent danger authority, an conservative prediction for the next four years would have to be that it would be reserved for situations where dangerous conditions are glaringly obvious.

---

[92] Heroic Efforts to Rescue Others Act, S.AMDT 1497 to S. 4, 103rd Cong. § 3 (1994).
[93] Home Office Worker Protection Act of 2000, H.R. 4098 106th Cong. (2000).
[94] *See* Reich Defends OSHA Enforcement Activities As Republicans on Appropriations Panel Attack, 24 O.S.H. (BNA) 1724 (Jan. 25, 1995).

Chapter 15

State Plans

Lesa L. Byrum, Esq.
Keller and Heckman, LLP
Washington, D.C.

1.0 Overview

In passing the OSH Act in 1970, Congress recognized that many states had effective programs for worker protection. Congress included provisions in the OSH Act permitting states (1) to exercise jurisdiction over any occupational safety and health issue where no federal standard exists; (2) to assume responsibility for development and enforcement of state rules equivalent to federal standards; (3) to operate an occupational safety and health program comparable in funding, authority, and staffing to the federal program; and (4) to enforce more stringent standards where compelled by local conditions and where the standards would not unduly burden interstate commerce.

2.0 Approval of State Plans

2.1 Introduction

Section 18 of the OSH Act requires a state to submit a plan to OSHA if the state desires to assume responsibility for the development and enforcement of standards "relating to" any occupational safety or health issue with respect to which a federal standard has been issued. When a state submits a plan for approval, a notice is published in the *Federal Register* summarizing the plan's contents, inviting public comment, and stating where the plan may be reviewed by the public.[1]

OSHA may elect to hold a hearing, formal or informal, on the proposed plan's approval; however, a hearing is not required.[2] If OSHA proposes to reject a plan,

---

[1] 29 C.F.R. § 1902.11.
[2] 29 C.F.R. § 1902.11(f).

the state must be provided an opportunity for a formal hearing.[3] After evaluation of the proposed plan, consideration of public comments, and hearing (if any), OSHA's decision to approve or reject the plan is published in the Federal Register.[4]

2.2 Program Content

Subsection (c) of § 18 lists the criteria a state program must meet before it can be approved by OSHA. A state program must:

1. designate a state agency or agencies as responsible for administering the plan in the state;

2. provide for the development and enforcement of safety and health standards relating to one or more safety or health issues which are or which will be at least as effective as the federal standards in providing safe and healthful employment and places of employment;

3. provide for a right of entry and inspection of all workplaces which is at least as effective as the corresponding federal right and which includes a prohibition on advance notice of inspections;

4. provide satisfactory assurances that the agency or agencies responsible have or will have the legal authority and qualified personnel necessary for the enforcement of the standards;

5. provide satisfactory assurances that the state will devote adequate funds to the administration and enforcement of the standards;

6. provide satisfactory assurances that the state will establish and maintain a comparable occupational safety and health program for the state's public sector employees;

7. require employers to make reports to the Secretary of Labor in the same manner and extent as if the plan were not in effect; and

8. provide that the state will make reports to the Secretary of Labor providing information required by the Secretary.

---

[3] 29 C.F.R. § 1902.19.
[4] 29 C.F.R. § 1902.20(b) and 1902.23.

2.3 Steps for Approval

The first step in the state plan process is obtaining approval for a "developmental plan."[5] To obtain this approval, a state must assure OSHA that within three years it will have in place all the structural elements necessary for an effective occupational safety and health program as set forth in the regulations.[6] The plan must include the specific actions the state proposes to take and a schedule for their accomplishment, not to exceed three years.[7] If necessary changes require state legislative action, a copy of the legislation or draft must be submitted along with a statement of the governor's support for the legislation and a legal opinion that the legislation will meet the requirements of the OSH Act consistent with the state's law.[8]

During the developmental period, federal OSHA and the state have concurrent enforcement authority in the state. The *Federal Register* notice indicating plan approval will identify the exact level of federal enforcement in the state.[9]

Once a state has completed and documented all its developmental steps, it is eligible for operational certification. This certification makes no judgment as to the performance of the state plan, but merely attests to the structural completeness of the plan. Under the regulations, operational status is achieved when the state's enabling legislation has been enacted, approved standards adopted, a sufficient number of qualified personnel are enforcing the standards, and provisions for appealing citations and penalties are in effect.[10] OSHA cannot grant final approval and withdraw federal operations until having observed the state program for at least three years, or for one year of actual operations in the case of developmental approvals.[11] The state is required to make periodic reports to the Secretary concerning its program. "Operational status" commits OSHA to suspend the exercise of discretionary federal enforcement in the activities covered by the state plan and emphasize monitoring activities.[12] The state is then responsible for conducting enforcement activity, including inspections in response to employee complaints.[13]

---

[5] 29 C.F.R. § 1902.2(b).
[6] 29 C.F.R. § 1902.2(b).
[7] 29 C.F.R. § 1902.2(b).
[8] 29 C.F.R. § 1902.2(b).
[9] 29 C.F.R. § 1902,20(b)(iii).
[10] 29 C.F.R. § 1954.3(b).
[11] 29 C.F.R. § 1954.3(a)(1).
[12] 29 C.F.R. § 1954.3(b).
[13] 29 C.F.R. § 1954.3(b).

After at least one year following operational certification, the state becomes eligible for final approval if OSHA determines that the state is providing worker protection at least as effective as the protection provided under the federal regime.[14] In addition, to obtain final approval, the state must meet the fifteen factors set forth in 29 C.F.R. § 1902.37. Some of these factors include establishing a sufficient number and training of enforcement personnel, timely adoption of federal standards, and participation in OSHA's computerized inspection data system. This determination is in effect the final approval for a state plan. Upon final approval, OSHA relinquishes all direct enforcement authority, except for issues not covered by the state plan.[15]

Even after final approval, OSHA retains responsibility for monitoring the state programs, and approval may be withdrawn when: (1) a state fails to complete its developmental plan within three years; (2) it is no longer reasonable to expect that the plan will meet the criteria for the completion of the developmental plan within three years; (3) when the state has failed to comply substantially with the state plan; or (4) when the state voluntarily withdraws its plan. The OSH Act provides for notice and an opportunity for the state to be heard on an action by OSHA to withdraw approval, as well as a right to judicial review.[16] Once approval is withdrawn, federal OSHA's authority and duty to enforce the OSH Act in the state resumes.

3.0 Complaints against State Plans

Anyone finding inadequacies or other problems with a state plan program may file a complaint about State Program Administration (CASPA) with the appropriate OSHA regional administrator.[17] OSHA, through a regional administrator, investigates all such complaints based upon factors enumerated in the regulations, and where a complaint is found to be valid, requires the state to take appropriate corrective action.[18] The regional administrator considers the extent to which the complaint affects a substantial number of people, the number of complaints received about the same or similar issues, whether the complainant has exhausted state remedies, and the extent to which the complaint relates to effective federal

---

[14] 29 C.F.R. §§ 1902.35 and 1902.39.
[15] 29 C.F.R. § 1902.42(c).
[16] 29 C.F.R. §§ 1902.47 and 1902.49.
[17] 29 C.F.R. § 1954.20.
[18] 29 C.F.R. §§ 1954.20 and 1954.21.

policy.[19] So-called CASPA complaints could include delays in processing cases, inadequate inspections or deviation from mandates in the enforcement of the plan.

States with approved plans are required to inform the public of the right to file complaints with OSHA about the manner in which the state plan is being administered.[20] The names of complainants are to remain confidential and may not be released to the state or the public.[21]

4.0 Approved State Plans

Twenty-one states and two territories have state plans approved by the Secretary of Labor under the Act.[22] Although the law provides for withdrawal of approval by the Secretary after notice and hearing,[23] no state plans have had their approval involuntarily withdrawn. In 1991, as a result of allegations of ineffective operation of the state's plan, OSHA proposed to withdraw approval of North Carolina's program. The state vigorously opposed OSHA's action, and succeeded in retaining approval after making significant changes in funding, resources and enforcement.

Congress provided for funding of up to fifty percent of the cost of administering and enforcing programs in the states, and up to ninety percent to improve state capabilities in this area.[24] Operation grants are renewed each fiscal year.

States without OSHA-approved plans are preempted by the OSH Act from exercising authority over any area of occupational safety and health covered by federal OSH standards, even if the state standard would be more strict than the federal standard.[25] However, a non-plan state may exercise authority over any hazard for which there is no corresponding federal standard in effect.[26] Non-plan states may also conduct non-enforcement activities such as providing information and consultative services to employers and employees.

---

[19] 29 C.F.R. § 1954.20(c)(2).

[20] 29 C.F.R. § 1954.22.

[21] 29 C.F.R. § 1954.21(a).

[22] Connecticut, New York and New Jersey plans cover public sector employment only. Alaska, Arizona, California, Hawaii, Indiana, Iowa, Kentucky, Maryland, Michigan, Minnesota, Nevada, New Mexico, North Carolina, Oregon, Puerto Rico, South Carolina, Tennessee, Utah, Vermont, Virgin Islands, Virginia, Washington, and Wyoming operate complete state plans covering both private sector employees and public sector employees. *See* Table 1.

[23] 29 C.F.R. § 1955.

[24] OSH Act § 23.

[25] OSH Act § 18(b).

[26] OSH Act § 18(a).

5.0 State Standards

The states with approved state plans promulgate and enforce their own standards. Generally, the states have six months after promulgation of a new federal standard to adopt an equivalent standard under their own rules.[27] The majority of states simply adopt the federal standards. States adopting standards identical to the federal standard must submit copies of the state standard along with a certification that they are identical to the federal standard. States adopting standards that are not identical to the federal standard must submit a comparison between the state and federal standard, as well as a statement as to why the state standard should be deemed as effective as the corresponding federal standard.

In some instances, the states have acted in advance of the federal government because of perceived needs that have not been addressed. In several instances, the states have adopted standards which are significantly different from the federal standards. For example, in 1983, Maryland adopted provisions limiting exposure to lead in the construction industry, while the federal standard has yet to be adopted. More recently, California adopted a standard addressing Process Safety Management that applies to a smaller group of industries than does the federal rule, while covering a larger list of chemicals at lower quantities.

Approximately thirteen of the states with state plans have adopted or retained state standards that are more restrictive than the federal standards, or which address areas in which there are no comparable federal standards.[28] Many of these standards were simply carried over from state health and safety programs that existed prior to the passage of the OSH Act. Some examples include standards on elevators, boiler and pressure vessels, confined spaces, and exterior window cleaning on skyscrapers. A number of states have more expansive hazard communication standards as well.[29] In 1978, Minnesota adopted specific regulations based upon its severe winter climate. Minnesota determined that due to the cold climate, working outdoors at certain times required extra safeguards and required employers to provide heated privies and appropriate shelter for eating lunch and changing clothes.

Most states have laws providing for misdemeanor criminal penalties where willful or knowing violations can be proven, particularly in the event of an employee's death. Sanctions range from several thousands of dollars up to one hundred thousand dollars. In some cases, the criminal sanctions provided under state law can

---

[27] 29 C.F.R. § 1953.23.
[28] California, Washington, Kentucky, Virginia, Nevada, Oregon, North Carolina, Utah, Maryland, Minnesota, Michigan, Vermont, and Alaska.
[29] Tennessee, Minnesota, Alaska, Michigan, Iowa, and California.

be assessed for violating "general responsibilities" under state law as well as for violation of individual standards.

Seven states require employers to establish a safety and health program.[30] These programs emphasize worksite analysis to identity actual and potential hazards, technical and administrative control of hazards, and training for all personnel, including supervisors and managers.

6.0 Preemption

The state plan provisions of the OSH Act and the vitality of the federal rules were given new life in 1992 in a United States Supreme Court decision which invalidated a state law requiring heavy equipment operators on hazardous waste sites to have additional training and qualifications in excess of OSHA requirements. In Gade v. National Solid Wastes Management Association,[31] the Court held that the OSH Act and the Hazardous Waste Operations and Emergency Response regulations (HAZWOPER)[32] preempt a state occupational safety and health law even though the state law purports to have a "dual" purpose or impact; that is, that it protects public as well as worker safety or health. Such a dual purpose state law may stand only if OSHA has explicitly approved it as part of a state plan established pursuant to § 18 of the OSH Act.

Four justices found that the OSH Act "impliedly" preempted the state law because the state law stands as an obstacle to the accomplishment and execution of the OSH Act's full purposes, while a fifth justice found "express" preemption. The *Gade* plurality found that state occupational safety and health standards regulating an issue on which a federal standard exists conflict with Congress' intent to subject employers and employees to only one set of regulations. Laws which regulate workers simply as members of the general public, the court cautioned, cannot fairly be characterized as occupational standards. The Illinois statute under challenge in *Gade*, however, was found to be directed at workplace safety, and thus not saved from preemption by its dual purpose.

The decision has important implications for other state laws that have been enacted ostensibly to protect the public, but which also have an effect on worker safety and health. Prior to *Gade*, three federal circuits had ruled that state statutes which impact public safety as well as worker safety are not necessarily preempted by the OSH Act. The Court's decision in *Gade* calls into question the continuing force

---

[30] California, Hawaii, Minnesota, Nevada, North Carolina, Oregon and Washington.
[31] 505 U. S. 73, 112 S. Ct. 2374 (1992).
[32] 29 C.F.R. § 1910.120.

of these decisions. In particular, state statutes that regulate Right-to-Know through labeling and hazard communication are suspect under the *Gade* court's rationale.

Table 1 Status of Approved State Plans

State	Initial Approval	Final Approval	Operational Status Agreement 1	Standards Different 2	Standards Maritime 3
Alaska	July 1973	Sept. 1984		*	
Arizona	Oct. 1974	June 1985			
California	Apr. 1973		*	*	*
Connecticut [4]	Oct. 1973				
Hawaii	Dec. 1973	Apr. 1984		*	
Indiana	Feb. 1974	Sept. 1986			
Iowa	July 1973	July 1985			
Kentucky	July 1973	June 1985			
Maryland	June 1973	July 1985			
Michigan	Sep. 1973		*	*	
Minnesota	May 1973	July 1985			*
Nevada	Dec. 1973		*		
New Jersey [4]	Jan. 2001				
New Mexico	Dec. 1975		*		
New York [4]	June 1984				
North Carolina	Jan. 1973		*		
Oregon	Dec. 1972		*	*	*
Puerto Rico	Aug. 1977		*		
South Carolina	Nov. 1972	Dec. 1987			
Tennessee	June 1973	July 1985			
Utah	Jan. 1973	July 1985			
Vermont	Oct. 1973		*		*
Virginia	Sep. 1976	Nov. 1988			
Virgin Islands	Aug. 1973	5			
Washington	Jan. 1973		*	*	*
Wyoming	Apr. 1974	June 1985			

1 Concurrent federal jurisdiction suspended

2 Standards frequently not identical to federal

3 Plan covers on-shore maritime employment

4 Plan covers only state and local government employees

5 Final approval granted Apr. 1984, suspended Nov. 1992.

Source: OSHA Office of State Programs.

Chapter 15 Appendix

Alaska Department of Labor and Workforce Development
P.O. Box 21149
1111 W. 8th Street, Room 306
Juneau, Alaska 99802-1149
Ed Flanagan, Commissioner
(907) 465-2700
Fax: (907) 465-2784
Richard Mastriano, Program Director
(907) 269-4904
Fax: (907) 269-4915

Industrial Commission of Arizona
800 W. Washington
Phoenix, Arizona 85007-2922
Larry Etchechury, Director, ICA
(602) 542-4411
Fax: (602) 542-1614
Darin Perkins, Program Director
(602) 542-5795
Fax: (602) 542-1614

California Department of Industrial Relations
455 Golden Gate Avenue - 10th Floor
San Francisco, California 94102
Steve Smith, Director
(415) 703-5050
Fax:(415) 703-5114
Dr. John Howard, Chief
(415) 703-5100
Fax: (415) 703-5114
Vernita Davidson, Manager, Cal / OSHA Program Office
(415) 703-5177
Fax: (415) 703-5114

Connecticut Department of Labor
200 Folly Brook Boulevard
Wethersfield, Connecticut 06109
Shaun Cashman, Commissioner
(860) 566-5123
Fax: (860) 566-1520
Conn-OSHA
38 Wolcott Hill Road
Wethersfield, Connecticut 06109
Donald Heckler, Director (860) 566-4550
Fax: (860) 566-6916

Hawaii Department of Labor and Industrial Relations
830 Punchbowl Street
Honolulu, Hawaii 96813
Leonard Agor, Director (808) 586-8844
Fax: (808) 586-9099
Jennifer Shishido, Administrator (808) 586-9116 Fax: (808) 586-9104

Indiana Department of Labor
State Office Building
402 West Washington Street, Room W195
Indianapolis, Indiana 46204-2751
John Griffin, Commissioner (317) 232-2378
Fax: (317) 233-3790
John Jones, Deputy Commissioner
(317) 232-3325 Fax: (317) 233-3790

Iowa Division of Labor
1000 E. Grand Avenue
Des Moines, Iowa 50319-0209
Byron K. Orton, Commissioner (515) 281-6432
Fax: (515) 281-4698
Mary L. Bryant, Administrator (515) 281-3469
Fax: (515) 281-7995

Kentucky Labor Cabinet
1047 U.S. Highway 127 South,
Suite 4
Frankfort, Kentucky 40601
Joe Norsworthy, Secretary (502) 564-3070
Fax: (502) 564-5387
William Ralston, Federal\State Coordinator
(502) 564-3070 ext.240
Fax: (502) 564-1682

Maryland Division of Labor and Industry, Department of Labor, Licensing and Regulation
1100 North Eutaw Street, Room 613
Baltimore, Maryland 21201-2206
Kenneth P. Reichard, Commissioner
(410) 767-2999 Fax: (410) 767-2300
Ileana O'Brien, Deputy Commissioner
(410) 767-2992
Fax: 767-2003
Keith Goddard, Assistant Commissioner, MOSH (410) 767-2215
Fax: 767-2003

Michigan Department of Consumer and Industry Services
Kathleen M. Wilbur, Director
Bureau of Safety and Regulation
P.O. Box 30643
Lansing, MI 48909-8143
Douglas R. Earle, Director
(517) 322-1814
Fax: (517)322-1775

Minnesota Department of Labor and Industry
443 Lafayette Road
St. Paul, Minnesota 55155
Gretchen B. Maglich, Commissioner
(651) 296-2342 Fax: (651) 282-5405
Rosyln Wade, Assistant Commissioner
(651) 296-6529
Fax: (651) 282-5293
Administrative Director, OSHA Management Team
(651) 282-5772
Fax: (651) 297-2527

Nevada Division of Industrial Relations
400 West King Street, Suite 400
Carson City, Nevada 89703
Roger Bremmer, Administrator (775) 687-3032
Fax: (775) 687-6305
Occupational Safety and Health Enforcement Section (OSHES)
1301 N. Green Valley Parkway
Henderson, Nevada 89014
Tom Czehowski, Chief Administrative Officer
(702) 486-9044
Fax:(702) 990-0358
[Las Vegas (702) 687-5240]

New Jersey Department of Labor
John Fitch Plaza - Labor Building
Market and Warren Streets
P.O. Box 110
Trenton, New Jersey 08625-0110
Mark B. Boyd, Commssioner (609) 292-2975
Fax: (609) 633-9271
Leonard Katz, Assistant Commissioner
(609) 292-2313
Fax: (609) 1314
Louis J. Lento, Program Director, PEOSH
(609) 292-3923
Fax: (609) 292-4409

New Mexico Environment Department
1190 St. Francis Drive
P.O. Box 26110
Santa Fe, New Mexico 87502
Peter Maggiore, Secretary (505) 827-2850
Fax: (505) 827-2836
Sam A. Rogers, Chief (505) 827-4230
Fax: (505) 827-4422

New York Department of Labor
W. Averell Harriman State Office Building - 12, Room 500
Albany, NY 12240
James Dillon, Acting Commissioner (518) 457-2741 Fax: (518) 457-6908
Richard Cucolo, Division Director (518) 457-3518 Fax: (518) 457-6908

North Carolina Department of Labor
4 West Edenton Street
Raleigh, North Carolina 27601-1092
Cherie Berry, Commissioner (919) 807-2900
Fax: (919) 807-2855
John Johnson, Deputy Commissioner, OSH Director (919) 807-2861
Fax: (919) 807-2855
Kevin Beauregard, OSH Assistant Director
(919) 807-2863
Fax:(919) 807-2856

Oregon Occupational Safety and Health Division
Department of Consumer & Business Services
350 Winter Street, NE, Room 430
Salem, Oregon 97310-0220
Peter DeLuca, Administrator (503) 378-3272
Fax: (503) 947-7461
David Sparks, Deputy Administrator for Policy
(503) 378-3272
Fax: (503) 947-7461
Michele Patterson, Deputy Administrator for Operations (503) 378-3272
Fax: (503) 947-7461

Puerto Rico Department of Labor and Human Resources
Prudencio Rivera Martinez Building
505 Munoz Rivera Avenue
Hato Rey, Puerto Rico 00918
Victor Rivera Hernandez, Acting Secretary
(787) 754-2119
Fax: (787) 753-9550
Myrna, Velez, Acting Assistant Secretary
(787) 754-2119 / 2171
Fax: (787) 767-6051

South Carolina Department of Labor, Licensing, and Regulation
Koger Office Park
Kingstree Building
110 Centerview Drive
PO Box 11329
Columbia, South Carolina 29211
Rita McKinney, Director (803) 896-4300
Fax: (803) 896-4393
William Lybrand, Program Director
(803) 734-9644 Fax: (803) 734-9772

Tennessee Department of Labor
710 James Robertson Parkway
Nashville, Tennessee 37243-0659
Michael E. Magill, Commissioner
(615) 741-2582 Fax: (615) 741-5078
John Winkler, Acting Program Director
(615) 741-2793
Fax: (615) 741-3325

Utah Labor Commission
160 East 300 South, 3rd Floor
PO Box 146650
Salt Lake City, Utah 84114-6650
R. Lee Ellertson, Commissioner (801) 530-6901
Fax: (801) 530-7906
Jay W. Bagley, Administrator (801) 530-6898
Fax: (801) 530-6390

Vermont Department of Labor and Industry
National Life Building - Drawer 20
Montpelier, Vermont 05620-3401
Tasha Wallis, Commissioner (802) 828-2288
Fax: (802) 828-2748
Robert McLeod, Project Manager
(802) 828-2765 Fax: (802) 828-2195

Virgin Islands Department of Labor
2203 Church Street
Christiansted, St. Croix, Virgin Islands 00820-4660
John Sheen, Commissioner (340) 773-1990
Fax: (340) 773-1858
Marcelle Heywood, Program Director
(340) 772-1315
Fax: (340) 772-4323

Virginia Department of Labor and Industry
Powers-Taylor Building
13 South 13th Street
Richmond, Virginia 23219
Jeffrey Brown, Commissioner (804) 786-2377
Fax: (804) 371-6524
Jay Withrow, Director, Office of Legal Support
(804) 786-9873
Fax: (804) 786-8418

Washington Department of Labor and Industries
General Administration Building
PO Box 44001
Olympia, Washington 98504-4001
Gary Moore, Director (360) 902-4200
Fax: (360) 902-4202
Michael Silverstein, Assistant Director [PO Box 44600]
(360) 902-5495
Fax: (360) 902-5529
Steve Cant, Program Manager, Federal-State Operations [PO Box 44600]
(360) 902-5430
Fax: (360) 902-5529

Wyoming Department of Employment
Workers' Safety and Compensation Division
Herschler Building, 2nd Floor East
122 West 25th Street
Cheyenne, Wyoming 82002
Stephan R. Foster, Safety Administrator
(307) 777-7786
Fax: (307) 777-3646

**Updated through February 2001
Source: OSHA Office of State Programs**

Appendix

OCCUPATIONAL SAFETY AND HEALTH ACT

as amended[1]

29 U.S.C. 651 et seq.

Be it enacted by the Senate and House of Representatives of the United States of America in Congress assembled, That this Act may be cited as the "Occupational Safety and Health Act of 1970."

CONGRESSIONAL FINDINGS AND PURPOSE

29 USC 651
Sec. 2

(a) The Congress finds that personal injuries and illnesses arising out of work situations impose a substantial burden upon, and are a hindrance to, interstate commerce in terms of lost production, wage loss, medical expenses, and disability compensation payments.

(b) The Congress declares it to be its purpose and policy, through the exercise of its powers to regulate commerce among the several States and with foreign nations and to provide for the general welfare, to assure so far as possible every working may and woman in the Nation safe and healthful working conditions and to preserve out human resources—

   (1) by encouraging employers and employees in their efforts to reduce the number of occupational safety and health hazards at their places of employment, and to stimulate employers and employees to institute new and to perfect existing programs for providing safe and healthful working conditions;

   (2) by providing that employers and employees have separate but dependent responsibilities and rights with respect to achieving safe and healthful working conditions;

   (3) by authorizing the Secretary of Labor to set mandatory occupational safety and health standards applicable to businesses affecting interstate commerce, and by creating an Occupational Safety and Health Review Commission for carrying out adjudicatory functions under the Act;

   (4) by building upon advances already made through employer and employee initiative for providing safe and healthful working conditions;

   (5) by providing for research in the field of occupational safety and health, including the psychological factors involved, and by developing innovative methods, techniques, and approaches for dealing with occupational safety and health problems;

   (6) by exploring ways to discover latent diseases, establishing causal connections between diseases and work in environmental conditions, and conducting other research relating to health problems, in recognition of the fact that occupational health standards present problems often different from those involved in occupational safety;

   (7) by providing medical criteria which will assure insofar as practicable that no employee will suffer diminished health, functional capacity, or life expectancy as a result of his work experience;

   (8) by providing for training programs to increase the number and competence of personnel engaged in the field of occupational safety and health;

---

[1] PL 91-596, December 29, 1970, as amended by PL 93-237, January 2, 1974; PL 95-251, March 27, 1978; PL 97-375, December 21, 1982; PL 98-139, October 31, 1983; PL 98-620, November 8, 1984; PL 99-499, October 17, 1986; PL 100-202, December 22, 1987; PL 101-508, November 5, 1990; PL 102-550, October 28, 1992; PL 104-1, January 23, 1995; PL 104-208, September 30, 1996; PL 105-78, November 13, 1997; PL 105-197, July 16, 1998; PL 105-198, July 16, 1998; PL 105-241, and September 28, 1998.

(9) by providing for the development and promulgation of occupational safety and health standards;
(10) by providing an effective enforcement program which shall include a prohibition against giving advance notice of any inspection and sanctions for any individual violating this prohibition;
(11) by encouraging the States to assume the fullest responsibility for the administration and enforcement of their occupational safety and health laws by providing grants to the States to assist in identifying their needs and responsibilities in the area of occupational safety and health, to develop plans in accordance with the provisions of this Act, to improve the administration and enforcement of State occupational safety and health laws, and to conduct experimental and demonstration projects in connection therewith;
(12) by providing for appropriate reporting procedures with respect to occupational safety and health which procedures will help achieve the objectives of this Act and accurately describe the nature of the occupational safety and health problem;
(13) by encouraging joint labor-management efforts to reduce injuries and disease arising out of employment.

SHORT TITLE

29 USC 651 Note

This Act may be cited as the "Occupational Safety and Health Administration Compliance Assistance Authorization Act of 1998".

DEFINITIONS

29 USC 652
Sec. 3

For the purposes of this Act–
(1) The term "Secretary" means the Secretary of Labor.
(2) The term "Commission" means the Occupational Safety and Health Review Commission established under this Act.
(3) The term "commerce" means trade, traffic, commerce, transportation, or communication among the several State, or between a State and any place outside thereof, or within the District of Columbia, or a possession of the United States (other than the Trust Territory of the Pacific Islands), or between points in the same State but through a point outside thereof.
(4) The term "person" means one or more individuals, partnerships, associations, corporations, business trusts, legal representatives, or any organized group of persons.
(5) The term "employer" means a person engaged in a business affecting commerce who has employees, but does not include the United States (not including the United States Postal Service) or any State or political subdivision of a State.
(6) The term "employee" means an employee of an employer who is employed in a business of his employer which affects commerce.
(7) The term "State" includes a State of this United State, the District of Columbia, Puerto Rico, the Virgin Islands, American Samoa, Guam, and the Trust Territory of the Pacific Islands.
(8) The term "occupational safety and health standard" means a standard which requires conditions, or the adoption or use of one or more practices, means, methods, operations, or processes, reasonably necessary or appropriate to provide safe or healthful employment and places of employment.
(9) The term "national consensus standard" means any occupational safety and health standard or modification thereof which (1) has been adopted and promulgated by a nationally recognized standards-producing organization under procedures whereby it

can be determined by the Secretary that persons interested and affected by the scope or provisions of the standard have reached substantial agreement on its adoption, (2) was formulated in a manner which afforded an opportunity for diverse views to be considered and (3) has been designated as such a standard by the Secretary, after consultation with other appropriate Federal agencies.

(10) The term "established Federal standard" means any operative occupational safety and health standard established by an agency of the United States and presently in effect, or contained in any Act of Congress in force on the date of enactment of this Act.

(11) The term "Committee" means the National Advisory Committee on Occupational Safety and Health established under this Act.

(12) The term "Director" means the Director of the National Institute for Occupational Safety and Health.

(13) The term "Institute" means the National Institute for Occupational Safety and Health established under this Act.

(14) The term "Workmen's Compensation Commission" means the National Commission on State Workmen's Compensation Laws established under this Act.

APPLICABILITY OF THIS ACT

29 USC 653
Sec. 4

(a) This Act shall apply with respect to employment performed in a work place in a State, the District of Columbia, the Commonwealth of Puerto Rico, the Virgin Islands, American Samoa, Guam, the Trust Territory of the Pacific Islands, Wake Island, Outer Continental Shelf lands defined in the Outer Continental Shelf Lands Act, Johnston Island, and the Canal Zone. The Secretary of the Interior shall, by regulation, provide for judicial enforcement of this Act by the courts established for areas in which there are no United States district courts having jurisdiction.

(b) (1) Nothing in this Act shall apply to working conditions of employees with respect to which other Federal agencies, and State agencies acting under section 274 of the Atomic Energy Act of 1954, as amended (42 U.S.C. 2021), exercise statutory authority to prescribe or enforce standards or regulations affecting occupational safety or health.

(2) The safety and health standards promulgated under the Act of June 30, 1936, commonly known as the Walsh-Healy Act (41 U.S.C. 35 et seq.), the Service Contract Act of 1965 (41 U.S.C. 351 et seq.), Public Law 91-54, Act of August 9, 1969 (40 U.S.C. 333), Public Law 85-742, August 23, 1958 (33 U.S.C. 941), and the National Foundation on Arts and Humanities Act (20 U.S.C. 951 et seq.) are superseded on the effective date of corresponding standards, promulgated under this Act, which are determined by the Secretary to be more effective. Standards issued under the laws listed in this paragraph and in effect on or after the effective date of this Act shall be deemed to be occupational safety and health standards issued under this Act, as well as under such other Acts.

(3) The Secretary shall, within three years after the effective date of this Act, report to the Congress his recommendations for legislation to avoid unnecessary duplication and to achieve coordination between this Act and other Federal laws.

(4) Nothing in this Act shall be construed to supersede or in any manner affect any workmen's compensation law or to enlarge or diminish or affect in any other manner the common law or statutory rights, duties, or liabilities of employers and employees under any law with respect to injuries, diseases, or death of employees arising out of, or in the course of, employment.

DUTIES

29 USC 654
Sec. 5

(a) Each employer–
  (1) shall furnish to each of his employees employment and a place of employment which are free from recognized hazards that are causing or are likely to cause death or serious physical harm to his employees;
  (2) shall comply with occupational safety and health standards promulgated under this Act.

(b) Each employee shall comply with occupational safety and health standards and all rules, regulations, and orders issued pursuant to this Act which are applicable to his own actions and conduct.

OCCUPATIONAL SAFETY AND HEALTH STANDARDS

29 USC 655
Sec. 6

(a) Without regard to chapter 5 of title 5, United States Code, or to the other subsections of this section, the Secretary shall, as soon as practicable during the period beginning with the effective date of this Act and ending two years after such date, by rule promulgate as an occupational safety or health standard any national consensus standard, and any established Federal standard, unless he determines that the promulgation of such a standard would not result in improved safety or health for specifically designated employees. In the event of conflict among any such standards, the Secretary shall promulgate the standard which assures the greatest protection of the safety or health of the affected employees.

(b) The Secretary may by rule promulgate, modify, or revoke any occupational safety or health standard in the following manner:
  (1) Whenever the Secretary, upon the basis of information submitted to him in writing by an interested person, a representative of any organization of employers or employees, a nationally recognized standards-producing organization, the Secretary of Health and Human Services, the National Institute for Occupational Safety and Health, or a State or political subdivision, or on the basis of information developed by the Secretary or otherwise available to him, determines that a rule should be promulgated in order to serve the objectives of this Act, the Secretary may request the recommendations of an advisory committee appointed under section 7 of this Act. The Secretary shall provide such an advisory committee with any proposals of his own or of the Secretary of Health and Human Services, together with all pertinent factual information developed by the Secretary or the Secretary of Health and Human Services, or otherwise available, including the results of research, demonstrations, and experiments. An advisory committee shall submit to the Secretary its recommendations regarding the rule to be promulgated within ninety days from the date of its appointment or within such longer or shorter period as may be prescribed by the Secretary, but in no event for a period which is longer than two hundred and seventy days.
  (2) The Secretary shall publish a proposed rule promulgating, modifying, or revoking an occupational safety or health standard in the Federal Register and shall afford interested persons a period of thirty days after publication to submit written data or comments. Where an advisory committee is appointed and the Secretary deter-mines that a rule should be issued, he shall publish the proposed rule within sixty days after the submission of the advisory committee's recommendations or the expiration of the period prescribed by the Secretary for such submission.

(3) On or before the last day of the period provided for the submission of written data or comments under paragraph (2), any interested person may file with the Secretary written objections to the proposed rule, stating the grounds therefor and requesting a public hearing on such objections. Within thirty days after the last day for filing such objections, the Secretary shall publish in the Federal Register a notice specifying the occupational safety or health standard to which objections have been filed and a hearing requested, and specifying a time and place for such hearing.

(4) Within sixty days after the expiration of the period provided for the submission of written data or comments under paragraph (2), or within sixty days after the completion of any hearing held under paragraph (3), the Secretary shall issue a rule promulgating, modifying, or revoking an occupational safety or health standard or make a determination that a rule should not be issued. Such a rule may contain a provision delaying its effective date for such period (not in excess of ninety days) as the Secretary determines may be necessary to insure that affected employers and employees will be informed of the existence of the standard and of its terms and that employers affected are given an opportunity to familiarize themselves and their employees with the existence of the requirements of the standard.

(5) The Secretary, in promulgation standards dealing with toxic materials or harmful physical agents under this subsection, shall set the standard which most adequately assures, to the extent feasible, on the basis of the best available evidence, that no employee will suffer material impairment of health or functional capacity even if such employee has regular exposure to the hazard dealt with by such standard for the period of his working life. Development of standards under this subsection shall be based upon research, demonstrations, experiments, and such other information as may be appropriate. In addition to the attainment of the highest degree of health and safety protection for the employee, other considerations shall be the latest available scientific data in the field, the feasibility of the standards, and experience gained under this and other health and safety laws. Whenever practicable, the standard promulgated shall be expressed in terms of objective criteria and of the performance desired.

(6) (A) Any employer may apply to the Secretary for a temporary order granting a variance from a standard or a provision thereof promulgated under this section. Such temporary order shall be granted only if the employer files an application which meets the requirements of clause (B) and establishes that (i) he is unable to comply with a standard by its effective date because of unavailability of professional or technical personnel or of materials and equipment needed to come into compliance with the standard or because necessary construction or alteration of facilities cannot be completed by the effective date, (ii) he is taking all available steps to safeguard his employees against the hazards covered by the standard, and (iii) he has an effective program for coming into compliance with the standard as quickly as practicable. Any temporary order issued under this paragraph shall prescribe the practices, means, methods, operations, and processes which the employer must adopt and use while the order is in effect and state in detail his program for coming into compliance with the standard. Such a temporary order may be granted only after notice to employees and an opportunity for a hearing: Provided, That the Secretary may issue one interim order to be effective until a decision is made on the basis of the hearing. No temporary order may be in effect for longer than the period needed by the employer to achieve compliance with the standard or one year, whichever is shorter, except that such an order may be renewed not more than twice (I) so long as the requirements of this paragraph are met and (II) if an application for renewal is filed at least 90 days prior to the expiration date of this order. No interim renewal of an order may remain in effect for longer than 180 days.

(B) An application for a temporary order under this paragraph (6) shall contain:
(i) a specification of the standard or portion thereof from which the employer seeks a variance,

(ii) a representation by the employer, supported by representations from qualified persons having firsthand knowledge of the facts represented, that he is unable to comply with the standard or portion thereof and a detailed statement of the reasons therefor.
(iii) a statement of the steps he has taken and will take (with specific dates) to protect employees against the hazard covered by the standard,
(iv) a statement of when he expects to be able to comply with the standard and what steps he has taken and what steps he will take (with dates specified) to come into compliance with the standard, and
(v) a certification that he has informed his employees of the application by giving a copy thereof to their authorized representative, posting a statement giving a summary of the application and specifying where a copy may be examined at the place or places where notices to employees are normally posted, and by other appropriate means.

A description of how employees have been informed shall be contained in the certification. The information to employees shall also inform them of their right to petition the Secretary for a hearing.

(C) The Secretary is authorized to grant a variance from any standard or portion thereof whenever he determines, or the Secretary of Health and Human Services certifies, that such variance is necessary to permit an employer to participate in an experiment approved by him or the Secretary of Health and Human Services designed to demonstrate or validate new and improved techniques to safeguard the health or safety of workers.

(7) Any standard promulgated under this subsection shall prescribe the use of labels or other appropriate forms of warning as are necessary to insure that employees are apprised of all hazards to which they are exposed, relevant symptoms and appropriate emergency treatment, and proper conditions and precautions of safe use or exposure. Where appropriate, such standard shall also prescribe suitable protective equipment and control of technological procedures to be used in connection with such hazards and shall provide for monitoring or measuring employee exposure at such locations and intervals, and in such manner as may be necessary for the protection of employees. In addition, where appropriate, any such standard shall prescribe the type and frequency of medical examinations or other tests which shall be made available, by the employer or at his cost, to employees exposed to such hazards in order to most effectively determine whether the health of such employees is adversely affected by such exposure. In the event such medical examinations are in the nature of research, as determined by the Secretary of Health and Human Services, such examinations may be furnished at the expense of the Secretary of Health and Human Services. The results of such examinations or tests shall be furnished only to the Secretary or the Secretary of Health and Human Services, and, at the request of the employee, to his physician. The Secretary, in consultation with the Secretary of Health and Human Services, may by rule promulgated pursuant to section 553 of title 5, United States Code, make appropriate modifications in the foregoing requirements relating to the use of labels or other forms of warning, monitoring, or measuring, and medical examinations, as may be warranted by experience, information, or medical or technological developments acquired subsequent to the promulgation of the relevant standard.

(8) Whenever a rule promulgated by the Secretary differs substantially from an existing national consensus standard, the Secretary shall, at the same time, publish in the Federal Register a statement of the reasons why the rule as adopted will better effectuate the purposes of this Act than the national consensus standard.

(c) (1) The Secretary shall provide, without regard to the requirements of chapter 5, title 5, United States Code, for an emergency temporary standard to take immediate effect upon publication in the Federal Register if he determines (A) that employees are exposed to grave danger from exposure to substances or agents determined to be toxic or physically harmful or from new hazards, and (B) that such emergency standard is necessary to protect employees from such danger.

(2) Such standard shall be effective until superseded by standard promulgated in accordance with the procedures prescribed in paragraph (3) of this subsection.

(3) Upon publication of such standard in the Federal Register the Secretary shall commence a proceeding in accordance with section 6(b) of this Act, and the standard as published shall also serve as a proposed rule for the proceeding. The Secretary shall promulgate a standard under this paragraph no later than six months after publication of the emergency standard as provided in paragraph (2) of this subsection.

(d) Any affected employer may apply to the Secretary for a rule or order for a variance from a standard promulgated under this section. Affected employees shall be given notice of each such application and an opportunity to participate in a hearing. The Secretary shall issue such rule or order if he determines on the record, after opportunity for an inspection where appropriate and a hearing, that the proponent of the variance has demonstrated by a preponderance of the evidence that the conditions, practices, means, methods, operations, or processes used or proposed to be used by an employer will provide employment and places of employment to his employees which are as safe and healthful as those which would prevail if he complied with the standard. The rule or order so issued shall prescribe the conditions the employer must maintain, and the practices, means, methods, operations, and processes which he must adopt and utilize to the extent they differ from the standard in question. Such a rule or order may be modified or revoked upon application by an employer, employees, or by the Secretary on his own motion in the manner prescribed for its issuance under this subsection at any time after six months from its issuance.

(e) Whenever the Secretary promulgates any standard, makes any rule, order, or decision, grants any exemption or extension of time, or compromises, mitigates, or settles any penalty assessed under this Act, he shall include a statement of the reasons for such action, which shall be published in the Federal Register.

(f) Any person who may be adversely affected by a standard issued under this section may at any time prior to the sixtieth day after such standard is promulgated file a petition challenging the validity of such standard with the United States court of appeals for the circuit where-in such person resides or has his principal place of business, for a judicial review of such standard. A copy of the petition shall be forthwith transmitted by the clerk of the court to the Secretary. The filing of such petition shall not, unless otherwise ordered by the court, operate as a stay of the standard. The determinations of the Secretary shall be conclusive if supported by substantial evidence in the record considered as a whole.

(g) In determining the priority for establishing standards under this section, the Secretary shall give due regard to the urgency of the need for mandatory safety and health standards for particular industries, trades, crafts, occupations, businesses, workplaces or work environments. The Secretary shall also give due regard to the recommendations of the Secretary of Health and Human Services regarding the need for mandatory standards in determining the priority for establishing such standards.

ADVISORY COMMITTEES; ADMINISTRATION

29 USC 656
Sec. 7

(a) (1) There is hereby established a National Advisory Committee on Occupational Safety and Health consisting of twelve members appointed by the Secretary, four of whom are to be designated by the Secretary of Health and Human Services, without regard to the provisions of title 5, United States Code, governing appointments in the competitive service, and composed of representatives of management, labor, occupational safety and occupational health professions, and of the public. The Secretary shall designate one of the public members as Chairman. The members shall be selected upon the basis of their experience and competence in the field of occupational safety and health.

(2) The Committee shall advise, consult with, and make recommendations to the Secretary and the Secretary of Health and Human Services on matters relating to the administration of the Act. The Committee shall hold no fewer than two meetings during each calendar year. All meetings of the Committee shall be open to the public and a transcript shall be kept and made available for public inspection.

(3) The members of the Committee shall be compensated in accordance with the provisions of section 3109 of title 5, United States Code.

(4) The Secretary shall furnish to the Committee an executive secretary and such secretarial, clerical, and other services as are deemed necessary to the conduct of its business.

(b) An advisory committee may be appointed by the Secretary to assist him in his standard-setting functions under section 6 of this Act. Each such committee shall consist of not more than fifteen members and shall include as a member one or more designees of the Secretary of Health and Human Services, and shall include among its members an equal number of persons qualified by experience and affiliation to present the viewpoint of the employers involved, and of persons similarly qualified to present the viewpoint of the workers involved, as well as one or more representatives of health and safety agencies of the States. An advisory committee may also include such other persons as the Secretary may appoint who are qualified by knowledge and experience to make a useful contribution to the work of such commit-tee, including one or more representatives of professional organizations of technicians or professionals specializing in occupational safety or health, and one or more representatives of nationally recognized standards-producing organizations, but the number of persons so appointed to any such advisory committee shall not exceed the number appointed to such committee as representatives of Federal and State agencies. Persons appointed to advisory committees from private life shall be compensated in the same manner as consultants or experts under section 3109 of title 5, United States Code. The Secretary shall pay to any State which is the employer of a member of such a committee who is a representative of the health or safety agency of that State, reimbursement sufficient to cover the actual cost to the State resulting from such representative's membership on such committee. Any meeting of such committee shall be open to the public and an accurate record shall be kept and made available to the public. No member of such committee (other than representatives of employers and employees) shall have an economic interest in any proposed rule.

(c) In carrying out his responsibilities under this Act, the Secretary is authorized to–

(1) use, with the consent of any Federal agency, the services, facilities, and personnel of such agency, with or without reimbursement, and with the consent of any State or political subdivision thereof, accept and use the services, facilities, and personnel of any agency of such State or subdivision with reimbursement; and

(2) employ experts and consultants or organizations thereof as authorized by section 3109 of title 5, United States Code, except that contracts for such employment may be renewed annually; compensate individuals so employed at rates not in excess of the rate specified at the time of service for grade GS-18 under section 5332 of title 5, United States Code, including travel time, and allow them while away from their homes or regular places of business, travel expenses (including per diem in lieu of subsistence) as authorized by section 5703 of title 5, United States Code, for person in the Government service employed intermittently, while so employed.

INSPECTIONS, INVESTIGATIONS, AND RECORDKEEPING

29 USC 657
Sec. 8

(a) In order to carry out the purposes of this Act, the Secretary, upon presenting appropriate credentials to the owner, operator, or agent in charge, is authorized–
   (1) to enter without delay and at reasonable times any factory, plant, establishment, construction site, or other area, workplace or environment where work is performed by an employee of an employer; and
   (2) to inspect and investigate during regular working hours and at other reasonable times, and within reasonable limits and in a reasonable manner, any such place of employment and all pertinent conditions, structures, machines, apparatus, devices, equipment, and materials therein, and to question privately any such employer, owner, operator, agent or employee.

(b) In making his inspections and investigations under this Act the Secretary may require the attendance and testimony of witnesses and the production of evidence under oath. Witnesses shall be paid the same fees and mileage that are paid witnesses in the courts of the United States. In case of a contumacy, failure, or refusal of any person to obey such an order, any district court of the United States or the United States courts of any territory or possession, within the jurisdiction of which such person is found, or resides or transacts business, upon the application by the Secretary, shall have jurisdiction to issue to such person an order requiring such person to appear to produce evidence if, as, and when so ordered, and to give testimony relating to the matter under investigation or in question, and any failure to obey such order of the court may be punished by said court as a contempt thereof.

(c) (1) Each employer shall make, keep and preserve, and make available to the Secretary or the Secretary of Health and Human Services, such records regarding his activities relating to this Act as the Secretary, in cooperation with the Secretary of Health and Human Services, may prescribe by regulation as necessary or appropriate for the enforcement of this Act or for developing information regarding the causes and prevention of occupational accidents and illnesses. In order to carry out the provisions of this paragraph such regulations may include provisions requiring employers to conduct periodic inspections. The Secretary shall also issue regulations requiring that employers, through posting of notices or other appropriate means, keep their employees informed of their protections and obligations under this Act, including the provisions of applicable standards.
   (2) The Secretary, in cooperation with the Secretary of Health and Human Services, shall prescribe regulations requiring employers to maintain accurate records of, and make periodic reports on, work-related deaths, injuries and illnesses other than minor injuries requiring only first aid treatment and which do not involve medical treatment, loss of consciousness, restriction of work or motion, or transfer to another job.
   (3) The Secretary, in cooperation with the Secretary of Health and Human Services, shall issue regulations requiring employers to maintain accurate records of employee exposures to potentially toxic materials or harmful physical agents which are required to be monitored or measured under section 6. Such regulations shall provide employees or their representatives with an opportunity to observe such monitoring or measuring, and to have access to the records thereof. Such regulations shall also make appropriate provision for each employee or former employee to have access to such records as will indicate his own exposure to toxic materials or harmful physical agents. Each employer shall promptly notify any employee who has been or is being exposed to toxic materials or harmful physical agents in concentrations or at levels which exceed those prescribed by an applicable occupational safety and health

standard promulgated under section 6, and shall inform any employee who is being thus exposed of the corrective action being taken.

(d) Any information obtained by the Secretary, the Secretary of Health and Human Services, or a State agency under this Act shall be obtained with a minimum burden upon employers, especially those operating small businesses. Unnecessary duplication of efforts in obtaining information shall be reduced to the maximum extent feasible.

(e) Subject to regulations issued by the Secretary, a representative of the employer and a representative authorized by his employees shall be given an opportunity to accompany the Secretary or his authorized representative during the physical inspection of any workplace under subsection (a) for the purpose of aiding such inspection. Where there is no authorized employee representative, the Secretary or his authorized representative shall consult with a reasonable number of employees concerning matters of health and safety in the workplace.

(f) (1) Any employees or representative of employees who believe that a violation of a safety or health standard exists that threatens physical harm, or that an imminent danger exists, may request an inspection by giving notice to the Secretary or his authorized representative of such violation or danger. Any such notice shall be reduced to writing, shall set forth with reasonable particularity the grounds for the notice, and shall be signed by the employees or representative of employees, and a copy shall be provided the employer or his agent no later than at the time of inspection, except that, upon the request of the person giving such notice, his name and the names of individual employees referred to therein shall not appear in such copy or on any record published, released, or made available pursuant to subsection (g) of this section. If upon receipt of such notification the Secretary determines there are reasonable grounds to believe that such violation or danger exists, he shall make a special inspection in accordance with the provisions of this section as soon as practicable, to determine if such violation or danger exists. If the Secretary determines there are no reasonable grounds to believe that a violation or danger exists he shall notify the employees or representative of the employees in writing of such determination.

(2) Prior to or during any inspection of a workplace, any employees or representative of employees employed in such workplace may notify the Secretary or any representative of the Secretary responsible for conducting the inspection, in writing, of any violation of this Act which they have reason to believe exists in such workplace. The Secretary shall, by regulation, establish procedures for informal review of any refusal by a representative of the Secretary to issue a citation with respect to any such alleged violation and shall furnish the employees or representative of employees requesting such review a written statement of the reasons for the Secretary's final disposition of the case.

(g) (1) The Secretary and Secretary of Health and Human Services are authorized to compile, analyze, and publish, either in summary or detailed form, all reports or information obtained under this section.

(2) The Secretary and the Secretary of Health and Human Services shall each prescribe such rules and regulations as he may deem necessary to carry out their responsibilities under this Act, including rules and regulations dealing with the inspection of an employer's establishments.

(h) The Secretary shall not use the results of enforcement activities, such as the number of citations issued or penalties assessed, to evaluate employees directly involved in enforcement activities under this Act or to impose quotas or goals with regard to the results of such activities.

CITATIONS

29 USC 658
Sec. 9

(a) If, upon inspection or investigation, the Secretary or his authorized representative believes that an employer has violated a requirement of section 5 of this Act, of any standard, rule or order promulgated pursuant to section 6 of this Act, or of any regulations prescribed pursuant to this Act, he shall with reasonable promptness issue a citation to the employer. Each citation shall be in writing and shall describe with particularity the nature of the violation, including a reference to the provision of the Act, standard, rule, regulation, or order alleged to have been violated. In addition, the citation shall fix a reasonable time for the abatement of the violation. The Secretary may prescribe procedures for the issuance of a notice in lieu of a citation with respect to de minimis violations which have no direct or immediate relationship to safety or health.

(b) Each citation issued under this section, or a copy or copies thereof, shall be prominently posted; as prescribed in regulations issued by the Secretary, at or near each place a violation referred to in the citation occurred.

(c) No citation may be issued under this section after the expiration of six months following the occurrence of any violation.

PROCEDURE FOR ENFORCEMENT

29 USC 659
Sec. 10

(a) If, after an inspection or investigation, the Secretary issues a citation under section 9(a), he shall, within a reasonable time after the termination of such inspection or investigation, notify the employer by certified mail of the penalty, if any, proposed to be assessed under section 17 and that the employer has fifteen working days within which to notify the Secretary that he wishes to contest the citation or proposed assessment of penalty. If, within fifteen working days from the receipt of the notice issued by the Secretary the employer fails to notify the Secretary that he intends to contest the citation or proposed assessment of penalty, and no notice is field by any employee or representative of employees under subsection (c) within such time, the citation and the assessment, as proposed, shall be deemed a final order of the Commission and not subject to review by the court or agency.

(b) If the Secretary has reason to believe that an employer has failed to correct a violation for which a citation has been issued within the period permitted for its correction (which period shall not begin to run until the entry of a final order by the Commission in the case of any review proceedings under this section initiated by the employer in good faith and not solely for delay or avoidance of penalties), the Secretary shall notify the employer by certified mail of such failure and of the penalty proposed to be assessed under section 17 by reason of such failure, and that the employer has fifteen working days within which to notify the Secretary that he wishes to contest the Secretary's notification or the proposed assessment of penalty. If, within fifteen working days from the receipt of notification issued by the Secretary, the employer fails to notify the Secretary that he intends to contest the notification or proposed assessment of penalty, the notification and assessment, as proposed, shall be deemed a final order of the Commission and not subject to review by any court or agency.

(c) If an employer notifies the Secretary that he intends to contest a citation issued under section 9(a) or notification issued under subsection (a) or (b) of this section, or if, within fifteen working days of the issuance of a citation under section 9(a), any employee or representative of employees files a notice with the Secretary alleging that the period of time fixed in the citation for the abatement of the violation is unreasonable, the Secretary shall immediately advise the Commission of such notification, and the Commission shall afford an opportunity for a hearing (in accordance with section 554 of title 5, United States Code, but without regard to subsection (a)(3) of such section). The Commission shall thereafter issue an order, based on findings of fact, affirming, modifying, or vacating the Secretary's citation or proposed penalty, or directing other appropriate relief, and such order shall become final thirty days after its issuance. Upon a showing by an employer of a good faith effort to comply with the abatement requirements of a citation, and that abatement has not been completed because of factors beyond his reasonable control, the Secretary, after an opportunity for a hearing as provided in this subsection, shall issue an order affirming or modifying the abatement requirements in such citation. The rules of procedure prescribed by the Commission shall provide affected employees or representatives of affected employees an opportunity to participate as parties to hearings under this subsection.

JUDICIAL REVIEW

29 USC 660
Sec. 11

(a) Any person adversely affected or aggrieved by an order of the Commission issued under subsection (c) of section 10 may obtain a review of such order in any United States court of appeals for the circuit in which the violation is alleged to have occurred or where the employer has its principal office, or in the Court of Appeals for the District of Columbia Circuit, by filing in such court within sixty days following the issuance of such order a written petition praying that the order be modified or set aside. A copy of such petition shall be forthwith transmitted by the clerk of the court to the Commission and to the other parties, and thereupon the Commission shall file in the court the record in the proceeding as provided in section 2112 of title 28, United States Code. Upon such filing, the court shall have jurisdiction of the proceeding and of the question determined therein, and shall have power to grant such temporary relief or restraining order as it deems just and proper, and to make and enter upon the pleadings, testimony, and proceedings set forth in such record a decree affirming, modifying, or setting aside in whole or in part, the order of the Commission and enforcing the same to the extent that such order is affirmed or modified. The commencement of proceedings under this subsection shall not, unless ordered by the court, operate as a stay of the order of the Commission. No objection that has not been urged before the Commission shall be considered by the court, unless the failure or neglect to urge such objection shall be excused because of extraordinary circumstances. The findings of the Commission with respect to questions of fact, if supported by substantial evidence on the record considered as a whole, shall be conclusive. If any party shall apply to the court for leave to adduce additional evidence and shall show to the satisfaction of the court that such additional evidence is material and that there were reasonable grounds for the failure to adduce such evidence in the hearing before the Commission, the court may order such additional evidence to be taken before the Commission and to be made a part of the record. The Com-mission may modify its findings as to the facts, or make new findings, by reason of additional evidence so taken and filed, and it shall file such modified or new findings, which findings with respect to questions of fact, if supported by substantial evidence on the record considered as a whole, shall be conclusive, and its recommendations, if any, for the modification or setting aside of its original order. Upon the filing of the record with it, the jurisdiction of the court shall be exclusive and its judgment and decree shall be final, except that the same shall be subject to review by the Supreme Court of the United States, as provided in section 1254 of title 28, United States Code.

(b) The Secretary may also obtain review or enforcement of any final order of the Commission by filing a petition for such relief in the United States court of appeals for the circuit in which the alleged violation occurred or in which the employer has its principal office, and the provisions of subsection (a), shall govern such proceedings to the extent applicable. If no petition for review, as provided in subsection (a), is filed within sixty days after service of the Commission's order, the Commission's findings of fact and order shall be conclusive in connection with any petition for enforcement which is filed by the Secretary after the expiration of such sixty-day period. In any such case, as well as in the case of a noncontested citation or notification by the Secretary which has become a final order of the Commission under subsection (a) or (b) of section 10, the clerk of the court, unless otherwise ordered by the court, shall forthwith enter a decree enforcing the order and shall transmit a copy of such decree to the Secretary and the employer named in the petition. In any contempt proceeding brought to enforce a decree of a court of appeals entered pursuant to this sub-section or subsection (a), the court of appeals may assess the penalties provided in section 17, in addition to invoking any other available remedies.

(c) (1) No person shall discharge or in any manner discriminate against any employee because such employee has filed any complaint or instituted or caused to be instituted any proceeding under or related to this Act or has testified or is about to testify in any such proceeding or because of the exercise by such employee on behalf of himself or others of any right afforded by this Act.

(2) Any employee who believes that he has been discharged or otherwise discriminated against by any person in violation of this subsection may, within thirty days after such violation occurs, file a complaint with the Secretary alleging such discrimination. Upon receipt of such complaint, the Secretary shall cause such investigation to be made as he deems appropriate. If upon such investigation, the Secretary determines that the provisions of this subsection have been violated, he shall bring an action in any appropriate United States district court against such person. In any such action the United States district courts shall have jurisdiction, for cause shown to restrain violations of paragraph (1) of this subsection and order all appropriate relief including rehiring or reinstatement of the employee to his former position with back pay.

(3) Within 90 days of the receipt of a complaint filed under this subsection the Secretary shall notify the complainant of his determination under paragraph 2 of this subsection.

THE OCCUPATIONAL SAFETY AND HEALTH REVIEW COMMISSION

29 USC 661
Sec. 12

(a) The Occupational Safety and Health Review Commission is hereby established. The Commission shall be composed of three members who shall be appointed by the President, by and with the advice and consent of the Senate, from among persons who by reason of training, education, or experience are qualified to carry out the functions of the Commission under this Act. The President shall designate one of the members of the Commission to serve as Chairman.

(b) The terms of members of the Commission shall be six years except that (1) the members of the Commission first taking office shall serve, as designated by the President at the time of appointment, one for a term of two years, one for a term of four years, and one for a term of six years, and (2) a vacancy caused by the death, resignation, or removal of a member prior to the expiration of the term for which he was appointed shall be filled only for the remainder of such unexpired term. A member of the Commission may be removed by the President for inefficiency, neglect of duty, or malfeasance in office.

(c) (1) Section 5314 of title 5, United States Code, is amended by adding at the end thereof the following new paragraph:
"(57) Chairman, Occupational Safety and Health Review Commission."

(2) Section 5315 of title 5, United States Code, is amended by adding at the end thereof the following new paragraph:
"(94) Members, Occupational Safety and Health Review Commission."

(d) The principal office of the Commission shall be the District of Columbia. Whenever the Commission deems that the convenience of the public or the parties may be promoted, or delay or expense may be minimized it may hold hearings or conduct other proceedings at any other place.

(e) The Chairman shall be responsible on behalf of the Commission for the administrative operations of the Commission and shall appoint such administrative law judges[1] and other employees as he deems necessary to assist in the performance of the Commission's functions and to fix their compensation in accordance with the provisions of chapter 51 and subchapter III of chapter 53 of title 5, United States Code, relating to classification and General Schedule pay rates: Provided, That assignment, removal and compensation of administrative law judges shall be in accordance with sections 3105, 3344, 5362, and 7521 of title 5, United States Code.

(f) For the purpose of carrying out its functions under this Act, two members of the Commission shall constitute a quorum and official action can be taken only on the affirmative vote of at least two members.

(g) Every official act of the Commission shall be entered of record, and its hearings and records shall be open to the public. The Commission is authorized to make such rules as are necessary for the orderly transaction of its proceedings. Unless the Commission has adopted a different rule, its proceedings shall be in accordance with the Federal Rules of Civil Procedure.

(h) The Commission may order testimony to be taken by deposition in any proceedings pending before it at any state of such proceeding. Any person may be compelled to appear and depose, and to produce books, papers, or documents, in the same manner as witnesses may be compelled to appear and testify and produce like documentary evidence before the Commission. Witnesses whose depositions are taken under this subsection, and the person taking such depositions, shall be entitled to the same fees as are paid for like services in the courts of the United States.

(i) For the purpose of any proceeding before the Commission, the provisions of section 11 of the National Labor Relations Act (29 U.S.C. 161) are hereby made applicable to the jurisdiction and powers of the Commission.

(j) An administrative law judge appointed by the Commission shall hear, and make a determination upon, any proceeding instituted before the Commission and any motion in connection therewith, assigned to such administrative law judge by the Chairman of the Commission, and shall make a report of any such determination which constitutes his final disposition of the proceedings. The report of the administrative law judge shall become the final order of the Commission within thirty days after such report by the administrative law judge, unless within such period any Commission member has directed that such report shall be reviewed by the Commission.

(k) Except as otherwise provided in this Act, the administrative law judges shall be subject to the laws governing employees in the classified civil service, except that appointments shall be made without regard to section 5108 of title 5, United States Code. Each administrative law judge shall receive compensation at a rate not less than that prescribed for GS-16 under section 5332 of title 5, United States Code.

---

[1] As amended by PL 95-251, March 27, 1978.

PROCEDURES TO COUNTERACT IMMINENT DANGERS

29 USC 662
Sec. 13

(a) The United States district courts shall have jurisdiction, upon petition of the Secretary, to restrain any conditions or practices in any place of employment which are such that a danger exists which could reasonably be expected to cause death or serious physical harm immediately or before the imminence of such danger can be eliminated through the enforcement procedures otherwise provided by this Act. Any order issued under this section may require such steps to be taken as may be necessary to avoid, correct, or remove such imminent danger and prohibit the employment or presence of any individual in locations or under conditions where such imminent danger exists, except individuals whose presence is necessary to avoid, correct, or remove such imminent danger or to maintain the capacity of a continuous process operation to resume normal operations without a complete cessation of operations, or where a cessation of operations is necessary, to permit such to be accomplished in a safe and orderly manner.

(b) Upon the filing of any such petition the district court shall have jurisdiction to grant such injunctive relief or temporary restraining order pending the outcome of an enforcement proceeding pursuant to this Act. The proceeding shall be as provided by Rule 65 of the Federal Rules, Civil Procedure, except that no temporary restraining order issued without notice shall be effective for a period longer than five days.

(c) Whenever and as soon as an inspector concludes that conditions or practices described in subsection (a) exist in any place of employment, he shall inform the affected employees and employers of the danger and that he is recommending to the Secretary that relief be sought.

(d) If the Secretary arbitrarily or capriciously fails to seek relief under this section, any employee who may be injured by reason of such failure, or the representative of such employees, might bring an action against the Secretary in the United States district court for the district in which the imminent danger is alleged to exist or the employer has its principal office, or for the District of Columbia, for a writ of mandamus to compel the Secretary to seek such an order and for such further relief as may be appropriate.

REPRESENTATION IN CIVIL LITIGATION

29 USC 663
Sec. 14

Except as provided in section 518(a) of title 28, United States Code, relating to litigation before the Supreme Court, the Solicitor of Labor may appear for and represent the Secretary in any civil litigation brought under this Act but all such litigation shall be subject to the direction and control of the Attorney General.

CONFIDENTIALITY OF TRADE SECRETS

29 USC 664
Sec. 15

All information reported to or otherwise obtained by the Secretary or his representative in connection with any inspection or proceeding under this Act which contains or which might reveal a trade secret referred to in section 1905 of title 18 of the United States Code shall be considered confidential for the purpose of that section, except that such information may be disclosed to other officers or employees concerned with carrying out this Act or when relevant in any proceeding under this Act. In any such proceeding the Secretary, the Commission, or the court shall issue such orders as may be appropriate to protect the confidentiality of trade secrets.

VARIATIONS, TOLERANCES, AND EXEMPTIONS

29 USC 665
Sec. 16

The Secretary, on the record, after notice and opportunity for a hearing may provide such reasonable limitations and may make such rules and regulations allowing reasonable variations, tolerances, and exemptions to and from any or all provisions of this Act as he may find necessary and proper to avoid serious impairment of the national defense. Such action shall not be in effect for more than six months without notification to affected employees an opportunity being afforded for a hearing.

PENALTIES

29 USC 666
Sec. 17

(a) Any employer who willfully or repeatedly violates the requirements of section 5 of this Act, any standard, rule, or order promulgated pursuant to section 6 of this Act, or regulations prescribed pursuant to this Act, may be assessed a civil penalty or not more than $70,000 for each violation, but not less than $5,000 for each willful violation.

(b) Any employer who has received a citation for a serious violation of the requirements of section 5 of this Act, of any standard, rule, or order promulgated pursuant to section 6 of this Act, or of any regulations prescribed pursuant to this Act, shall be assessed a civil penalty of up to $7,000 for each such violation.

(c) Any employer who has received a citation for a violation of the requirements of section 5 of this Act, of any standard, rule, or order promulgated pursuant to section 6 of this Act, or of regulations prescribed pursuant to this Act, and such violation is specifically determined not to be of a serious nature, may be assessed a civil penalty of up to $7,000 for each such violation.

(d) Any employer who fails to correct a violation for which a citation has been issued under section 9(a) within the period permitted for its correction (which period shall not begin to run until the date of the final order of the Commission in the case of any review proceeding under section 10 initiated by the employer in good faith and not solely for delay or avoidance of penalties), may be assessed a civil penalty of not more than $7,000 for each day during which such failure or violation continues.

(e) Any employer who willfully violates any standard, rule, or order promulgated pursuant to section 6 of this Act, or of any regulations prescribed pursuant to this Act, and that violation caused death to any employee, shall, upon conviction, be punished by a fine of not more than $10,000 or by imprisonment for not more than six months, or by both; except that if the conviction is for a violation committed after a first conviction of such person, punishment shall be by a fine of not more than $20,000 or by imprisonment for not more than one year, or by both.

(f) Any person who gives advance notice of any inspection to be conducted under this Act, without authority from the Secretary or his designees, shall upon conviction, be punished by a fine of not more than $1,000 or by imprisonment for not more than six months, or by both.

(g) Whoever knowingly makes any false statement, representation, or certification in any application, record, report, plan, or other document filed or required to be maintained pursuant to this Act shall, upon conviction, be punished by a fine of not more than $10,000, or by imprisonment for not more than six months, or by both.

(h) (1) Section 1114 of title 18, United States Code, is hereby amended by striking out "designated by the Secretary of Health and Human Services to conduct investigations, or inspections under the Federal Food, Drug, and Cosmetic Act" and inserting in lieu thereof "or of the Department of Labor assigned to perform investigative, inspection, or law enforcement functions."
   (2) Notwithstanding the provisions of sections 1111 and 1114 of title 18, United States Code, whoever, in violation of the provisions of section 1114 of such title, kills a person while engaged in or on account of the performance of investigative, inspection, or law enforcement functions added to such section 1114 by paragraph (1) of this subsection, and who would otherwise be subject to the penalty provisions of such section 1111, shall be punished by imprisonment for any term of years or for life.

(i) Any employer who violates any of the posting requirements, as prescribed under the provisions of this Act, shall be assessed a civil penalty of up to $7,000 for each violation.

(j) The Commission shall have authority to assess all civil penalties provided in this section, giving due consideration to the appropriateness of the penalty with respect to the size of the business of the employer being charged, the gravity of the violation, the good faith of the employer, and the history of previous violations.

(k) For purposes of this section, a serious violation shall be deemed to exist in a place of employment if there is a substantial probability that death or serious physical harm could result from a condition which exists, or from one or more practices, means, methods, operations, or processes which have been adopted or are in use, in such place of employment unless the employer did not, and could not with the exercise of reasonable diligence, know of the presence of the violation.

(l) Civil penalties owed under this Act shall be paid to the Secretary for deposit into the Treasury of the United States and shall accrue to the United States and may be recovered in a civil action in the name of the United States brought in the United States district court for the district where the violation is alleged to have occurred or where the employer has its principal office.

STATE JURISDICTION AND STATE PLANS

29 USC 667
Sec. 18

(a) Nothing in this Act shall prevent any State agency or court from asserting jurisdiction under State law over any occupational safety or health issue with respect to which no standard is in effect under section 6.

(b) Any State which, at any time, desires to assume responsibility for development and enforcement therein of occupational safety and health standards relating to any occupational safety or health issue with respect to which a Federal standard has been promulgated under section 6 shall submit a State plan for the development of such standards and their enforcement.

(c) The Secretary shall approve the plan submitted by a State under subsection (b), or any modification thereof, if such plan in his judgment—
   (1) designates a State agency or agencies as the agency or agencies responsible for administering the plan throughout the State,
   (2) provides for the development and enforcement of safety and health standards relating to one or more safety or health issues, which standards (and the enforcement of which standards) are or will be at least as effective, in providing safety and healthful

employment and places of employment as the standards promulgated under section 6 which relate to the same issues, and which standards, when applicable to products which are distributed or used in interstate commerce, are required by compelling local conditions and do not unduly burden interstate commerce,

(3) provides for a right of entry and inspection of all workplaces subject to the Act which is at least as effective as that provided in section 8, and includes a prohibition on advance notice of inspections,

(4) contains satisfactory assurances that such agency or agencies have or will have the legal authority and qualified personnel necessary for the enforcement of such standards,

(5) gives satisfactory assurances that such State will devote adequate funds to the administration and enforcement of such standards,

(6) contains satisfactory assurances that such State will, to the extent permitted by its law, establish and maintain an effective and comprehensive occupational safety and health program applicable to all employees of public agencies of the State and its political subdivisions, which program is as effective as the standards contained in an approved plan,

(7) requires employers in the State to make reports to the Secretary in the same manner and to the same extent as if the plan were not in effect, and

(8) provides that the State agency will make such reports to the Secretary in such form and containing such information, as the Secretary shall from time to time require.

(d) If the Secretary rejects a plan submitted under subsection (b), he shall afford the State submitting the plan due notice and opportunity for a hearing before so doing.

(e) After the Secretary approves a State plan submitted under subsection (b), he may, but shall not be required to, exercise his authority under sections 8,9,10,13, and 17 with respect to comparable standards promulgated under section 6, for the period specified in the next sentence. The Secretary may exercise the authority referred to above until he determines, on the basis of actual operations under the State plan, that the criteria set forth in subsection (c) are being applied, but he shall not make such determination for at least three years after the plan's approval under subsection (c). Upon making the determination referred to in the preceding sentence, the provisions of sections 5 (a)(2), 8 (except for the purpose of carrying out subsection (f) of this section), 8, 9, 10, 13, and 17, and standards promulgated under section 6 of this Act, shall not apply with respect to any occupational safety or health issues covered under the plan, but the Secretary may retain jurisdiction under the above provisions in any proceeding commenced under section 9 or 10 before the date of determination.

(f) The Secretary shall, on the basis of reports submitted by the State agency and his own inspections make a continuing evaluation of the manner in which each State having a plan approved under this section is carrying out such plan. Whenever the Secretary finds, after affording due notice and opportunity for a hearing, that in the administration of the State plan there is a failure to comply substantially with any pro-vision of the State plan (or any assurance contained therein), he shall notify the State agency of his withdrawal of approved of such plan and upon receipt of such notice such plan shall cease to be in effect, but the State may retain jurisdiction in any case commenced before the withdrawal of the plan in order to enforce standards under the plan when-ever the issues involved do not relate to the reasons for the withdrawal of the plan.

(g) The State may obtain a review of a decision of the Secretary withdrawing approval of or rejecting its plan by the United States court of appeals for the circuit in which the State is located by filing in such court within thirty days following receipt or notice of such decision a petition to modify or set aside in whole or in part the action of the Secretary. A copy of such petition shall forthwith be served upon the Secretary, and thereupon the Secretary shall certify and file in the court the record upon which the decision complained of was issued as provided in section 2112 of title 28, United States Code. Unless the court finds that the Secretary's decision in rejecting a proposed State plan or withdrawing his approval of such a plan is not supported by substantial evidence the court shall affirm the Secretary's decision. The judgment of the court shall be subject to review by the Supreme Court of the United States upon certiorari or certification as provided in section 1254 of title 28, United States Code.

(h) The Secretary may enter into an agreement with a State under which the State will be permitted to continue to enforce one or more occupational health and safety standards in effect in such State until final action is taken by the Secretary with respect to a plan submitted by a State under subsection (b) of this section, or two years from the date of enactment of this Act, whichever is earlier.

FEDERAL AGENCY SAFETY PROGRAMS AND RESPONSIBILITIES

29 USC 668
Sec. 19

(a) It shall be the responsibility of the head of each Federal agency (not including the United States Postal Service) to establish and maintain an effective and comprehensive occupational safety and health program which is consistent with the standards promulgated under section 6. The head of each agency shall (after consultation with representatives of the employees thereof)–
   (1) provide safe and healthful places and conditions of employment, consistent with the standards set under section 6;
   (2) acquire, maintain, and require the use of safety equipment, personal protective equipment, and devices reasonably necessary to protect employees;
   (3) keep adequate records of all occupational accidents and illnesses for proper evaluation and necessary corrective action;
   (4) consult with the Secretary with regard to the adequacy as to form and content of records kept pursuant to subsection (a)(3) of this section; and
   (5) make an annual report to the Secretary with respect to occupational accidents and injuries and the agency's program under this section. Such report shall include any report submitted under section 7902 (e)(2) of title 5, United States Code.

(b) The Secretary shall report to the President a summary or digest of reports submitted to him under subsection (a)(5) of this section, together with his evaluations of and recommendations derived from such reports.

(c) Section 7902(c)(1) of title 5, United States Code, is amended by inserting after "agencies" the following: "and of labor organizations representing employees."

(d) The Secretary shall have access to records and reports kept and filed by Federal agencies pursuant to subsections (a) (3) and (5) of this section unless those records and reports are specifically required by Executive order to be kept secret in the interest of the national defense or foreign policy, in which case the Secretary shall have access to such information as will not jeopardized national defense or foreign policy.

RESEARCH AND RELATED ACTIVITIES

29 USC 669
Sec. 20

(a) (1) The Secretary of Health and Human Services, after consultation with the Secretary and with other appropriate Federal departments or agencies, shall conduct (directly or by grants or contracts) research, experiments, and demonstrations relating to occupational safety and health, including studies of psychological factors involved, and relating to innovative methods, techniques, and approaches for dealing with occupational safety and health problems.

(2) The Secretary of Health and Human Services shall from time to time consult with the Secretary in order to develop specific plans for such research, demonstrations, and experiments as are necessary to produce criteria, including criteria identifying toxic substances, enabling the Secretary to meet his responsibility for the formulation of safety and health standards under this Act; and the Secretary of Health and Human Services, on the basis of such research, demonstrations, and experiments and any other information available to him, shall develop and publish at least annually such criteria as will effectuate the purposes of this Act.

(3) The Secretary of Health and Human Services, on the basis of such research, demonstrations, and experiments, and any other information available to him, shall develop criteria dealing with toxic materials and harmful physical agents and substances which will describe exposure levels that are safe for various periods of employment, including but not limited to the exposure levels at which no employee will suffer impaired health or functional capacities or diminished life expectancy as a result of his work experience.

(4) The Secretary of Health and Human Services shall also conduct special research, experiments, and demonstrations relating to occupational safety and health, which may require ameliorative action beyond that which is otherwise provided for in the operating provisions of this Act. The Secretary of Health and Human Services shall also conduct research into the motivational and behavioral factors relating to the field of occupational safety and health.

(5) The Secretary of Health and Human Services, in order to comply with his responsibilities under paragraph (2), and in order to develop needed information regarding potentially toxic substances or harmful physical agents, may prescribe regulations requiring employers to measure, record, and make reports on the exposure of employees to substances or physical agents which the Secretary of Health and Human Services reasonably believes may endanger the health or safety of employees. The Secretary of Health and Human Services also is authorized to establish such programs of medical examinations and tests as may be necessary for determining the incidence of occupational illnesses and the susceptibility of employees to such illnesses. Nothing in this or any other provision of this Act shall be deemed to authorize or require medical examination, immunization, or treatment for those who object thereto on religious grounds, except where such is necessary for the protection of the health or safety of others. Upon the request of any employer who is required to measure and record exposure of employees to substances or physical agents as provided under this subsection, the Secretary of Health and Human Services shall furnish full financial or other assistance to such employer for the purpose of defraying any additional expense incurred by him in carrying out the measuring and recording as provided in this subsection.

(6) The Secretary of Health and Human Services shall publish within six months of enactment of this Act and thereafter as needed but at least annually a list of all known toxic substances by generic family or other useful grouping, and the concentrations at which such toxicity is known to occur. He shall determine following a written request by any employer or authorized representative of employees, specifying with reasonable particularity the grounds on which the request is made, whether any substance normally found in the place of employment has potentially toxic effects in such concentrations as used or found; and shall submit such determination both to employers and affected employees as soon as possible. If the Secretary of Health and Human Services determines that any substance is potentially toxic at the concentrations in which it is used or found in a place of employment, and such substance is not covered by an occupational safety and health standard promulgated under section 6, the Secretary of Health and Human Services shall immediately submit such determination to the Secretary, together with all pertinent criteria.

(7) Within two years of enactment of this Act, and annually thereafter the Secretary of Health and Human Services shall conduct and publish industry-wide studies of the

effect of chronic or low-level exposure to industrial materials, processes, and stresses on the potential for illness, disease, or loss of functional capacity in aging adults.

(b) The Secretary of Health and Human Services is authorized to make inspections and question employers and employees as provided in section 8 of this Act in order to carry out his functions and responsibilities under this section.

(c) The Secretary is authorized to enter into contracts, agreements, or other arrangements with appropriate public agencies or private organizations for the purpose of conducting studies relating to his responsibilities under this Act. In carrying out his responsibilities under this subsection, the Secretary shall cooperate with the Secretary of Health and Human Services in order to avoid any duplication of efforts under this section.

(d) Information, obtained by the Secretary and the Secretary of Health and Human Services under this section shall be disseminated by the Secretary to employers and employees and organizations thereof.

(e) The functions of the Secretary of Health and Human Services under this Act shall, to the extent feasible, be delegated to the Director of the National Institute for Occupational Safety and Health established by section 22 of this Act.

TRAINING AND EMPLOYEE EDUCATION

29 USC 670
Sec. 21

(a) The Secretary of Health and Human Services, after con-sultation with the Secretary and with other appropriate Federal depart-ments and agencies, shall conduct, directly or by grants or contracts (1) education programs to provide an adequate supply of qualified personnel to carry out the purposes of this Act, and (2) informational programs on the importance of and proper use of adequate safety and health equipment.

(b) The Secretary is also authorized to conduct, directly or by grants or contracts, short-term training of personnel engaged in work related to his responsibilities under this Act.

(c) The Secretary, in consultation with the Secretary of Health and Human Services, shall (1) provide for the establishment and supervision of programs for the education and training of employers and employees in the recognition, avoidance, and prevention of unsafe or unhealthful working conditions in employments covered by this Act, and (2) consult with and advise employers and employees, and organizations representing employers and employees, as to effective means of preventing occupational injuries and illnesses.

(d) (1) The Secretary shall establish and support cooperative agreements with the States under which employers subject to this Act may consult with State personnel with respect to–
    (A) the application of occupational safety and health requirements under this Act or under State plans approved under section 18; and
    (B) voluntary efforts that employers may undertake to establish and maintain safe and healthful employment and places of employment.
Such agreements may provide, as a condition of receiving funds under such agreements, for contributions by States towards meeting the costs of such agreements.
(2) Pursuant to such agreements the State shall provide on-site consultation at the employer's worksite to employers who request such assistance. The State may also provide other education and training programs for employers and employees in the State. The State shall ensure that on-site consultations conducted pursuant to such agreements include provision for the participation by employees.
(3) Activities under this subsection shall be conducted independently of any enforcement activity. If an employer fails to take immediate action to eliminate employee exposure to an imminent danger identified in a consultation or fails to correct a serious hazard

so identified within a reasonable time, a report shall be made to the appropriate enforcement authority for such action as is appropriate.

(4) The Secretary shall, by regulation after notice and opportunity for comment, establish rules under which an employer–
   (A) which requests and undergoes an on-site consultative visit provided under this subsection;
   (B) which corrects the hazards that have been identified during the visit within the time frames established by the State and agrees to request a subsequent consultative visit if major changes in working conditions or work processes occur which introduce new hazards in the workplace; and
   (C) which is implementing procedures for regularly identifying and preventing hazards regulated under this Act and maintains appropriate involvement of, and training for, management and non-management employees in achieving safe and healthful working conditions, may be exempt from an inspection (except an inspection requested under section 8(f) or an inspection to determine the cause of a workplace accident which resulted in the death of one or more employees or hospitalization for three or more employees) for a period of 1 year from the closing of the consultative visit.

(5) A State shall provide worksite consultations under paragraph(2) at the request of an employer. Priority in scheduling such consultations shall be assigned to requests from small businesses which are in higher hazard industries or have the most hazardous conditions at issue in the request.

OCCUPATIONAL SAFETY AND HEALTH ADMINISTRATION

29 USC 670 Note

For necessary expenses for the Occupational Safety and Health Administration, $336,480,000, including not to exceed $77,941,000 which shall be the maximum amount available for grants to States under section 23(g) of the Occupational Safety and Health Act, which grants shall be no less than 50 percent of the costs of State occupational safety and health programs required to be incurred under plans approved by the Secretary under section 18 of the Occupational Safety and Health Act of 1970; and, in addition, notwithstanding 31 U.S.C. 3302, <<NOTE: 29 USC 670 note.>> the Occupational Safety and Health Administration may retain up to $750,000 per fiscal year of training institute course tuition fees, otherwise authorized by law to be collected, and may utilize such sums for occupational safety and health training and education grants: Provided, That, notwithstanding 31 U.S.C. 3302, the Secretary of Labor is authorized, during the fiscal year ending September 30, 1998, to collect and retain fees for services provided to Nationally Recognized Testing Laboratories, and may utilize such sums, in accordance with the provisions of 29 U.S.C. 9a, to administer national and international laboratory recognition programs that ensure the safety of equipment and products used by workers in the workplace: Provided further, That none of the funds appropriated under this paragraph shall be obligated or expended to prescribe, issue, administer, or enforce any standard, rule, regulation, or order under the Occupational Safety and Health Act of 1970 which is applicable to any person who is engaged in a farming operation which does not maintain a temporary labor camp and employs ten or fewer employees: Provided further, That no funds appropriated under this paragraph shall be obligated or expended to administer or enforce any standard, rule, regulation, or order under the Occupational Safety and Health Act of 1970 with respect to any employer of ten or fewer employees who is included within a category having an occupational injury lost workday case rate, at the most precise Standard Industrial Classification Code for which such data are published, less than the national average rate as such rates are most recently published by the Secretary, acting through the Bureau of Labor Statistics, in accordance with section 24 of that Act (29 U.S.C. 673), except–

(1) to provide, as authorized by such Act, consultation, technical assistance, educational and training services, and to conduct surveys and studies;

(2) to conduct an inspection or investigation in response to an employee complaint, to issue a citation for violations found during such inspection, and to assess a penalty for violations which are not corrected within a reasonable abatement period and for any willful violations found;

(3) to take any action authorized by such Act with respect to imminent dangers;

(4) to take any action authorized by such Act with respect to health hazards;

(5) to take any action authorized by such Act with respect to a report of an employment accident which is fatal to one or more employees or which results in hospitalization of two or more employees, and to take any action pursuant to such investigation authorized by such Act; and

(6) to take any action authorized by such Act with respect to complaints of discrimination against employees for exercising rights under such Act: Provided further, That the foregoing proviso shall not apply to any person who is engaged in a farming operation which does not maintain a temporary labor camp and employs ten or fewer employees.

NATIONAL INSTITUTE FOR OCCUPATIONAL SAFETY AND HEALTH

29 USC 671
Sec. 22

(a) It is the purpose of this section to establish a National Institute for Occupational Safety and Health in the Department of Health and Human Services in order to carry out the policy set forth in section 2 of this Act and to perform the functions of the Secretary of Health and Human Services under sections 20 and 21 of this Act.

(b) There is hereby established in the Department of Health and Human Services a National Institute for Occupational Safety and Health. The Institute shall be headed by a Director who shall be appointed by the Secretary of Health and Human Services, and who shall serve for a term of six years unless previously removed by the Secretary of Health and Human Services.

(c) The Institute is authorized to–
  (1) develop and establish recommended occupational safety and health standard; and
  (2) perform all functions of the Secretary of Health and Human Services under sections 20 and 21 of this Act.

(d) Upon his own initiative, or upon the request of the Secretary or the Secretary of Health and Human Services, the Director is authorized (1) to conduct such research and experimental programs as he determines are necessary for the development of criteria for new and improved occupational safety and health standards, and (2) after consideration of the results of such research and experimental programs make recommendations concerning new or improved occupational safety and health standards. Any occupational safety and health standard recommended pursuant to this section shall immediately be forwarded to the Secretary of Labor, and to the Secretary of Health and Human Services.

(e) In addition to any authority vested in the Institute by the other provisions of this section, the Director, in carrying out the functions of the Institute, is authorized to–
  (1) prescribe such regulations as he deems necessary governing the manner in which its functions shall be carried out;
  (2) receive money and other property donated, bequeathed, or devised, without condition or restriction other than that it be used for the purposes of the Institute and to use, sell, or to otherwise dispose of such property for the purpose of carrying out its functions;
  (3) receive (and use, sell, or otherwise dispose of, in accordance with paragraph (2)), money and other property donated, bequeathed, or devised to the Institute with a

condition or restriction, including a condition that the Institute use other funds of the Institute for the purposes of the gift;
(4) in accordance with the civil service laws, appoint and fix the compensation of such personnel as may be necessary to carry out the provisions of this section;
(5) obtain the services of experts and consultants in accordance with the provisions of section 3109 of title 5, United States Code;
(6) accept and utilize the services of voluntary and noncompensated personnel and reimburse them for travel expenses, including per diem, as authorized by section 5703 of title 5, United States Code;
(7) enter into contracts, grants or other arrangements, or modifications thereof to carry out the provisions of this section, and such contracts or modifications thereof may be entered into without performance of other bonds, and without regard to section 3709 of the Revised Statutes, as amended (41 U.S.C. 5), or any other provision of law relating to competitive bidding;
(8) make advance, progress, and other payments which the Director deems necessary under this title without regard to the provisions of section 3648 of the Revised Statutes, as amended (31 U.S.C. 529); and
(9) make other necessary expenditures.

(f) The Director shall submit to the Secretary of Health and Human Services, to the President, and to the Congress an annual report of the operations of the Institute under this Act, which shall include a detailed statement of all private and public funds received and expended by it, and such recommendations as he deems appropriate.

(g) Lead-Based Paint Activities.–
  (1) Training Grant Program
    (A) The Institute, in conjunction with the Administrator of the Environmental Protection Agency, may make grants for the training and education of workers and supervisors who are or may be directly engaged in lead-based paint activities.
    (B) Grants referred to in subparagraph (A) shall be awarded to nonprofit organizations (including colleges and universities, joint labor-management trust funds, States, and nonprofit government employee organizations)–
      (i) which are engaged in the training and education of workers and supervisors who are or who may be directly engaged in lead-based paint activities (as defined in title IV of the Toxic Substances Control Act),
      (ii) which have demonstrated experience in implementing and operating health and safety training and education programs, and
      (iii) with a demonstrated ability to reach, and involve in lead-based paint training programs, target populations of individuals who are or will be engaged in lead-based paint activities.
    Grants under this subsection shall be awarded only to those organizations that fund at least 30 percent of their lead-based paint activities training programs from non-Federal sources, excluding in-kind contributions. Grants may also be made to local governments to carry out such training and education for their employees.
    (C) There are authorized to be appropriated, at a minimum, $10,000,000 to the Institute for each of the fiscal years 1994 through 1997 to make grants under this paragraph.
  (2) Evaluation of Programs.–The Institute shall conduct periodic and comprehensive assessments of the efficacy of the worker and supervisor training programs developed and offered by those receiving grants under this section. The Director shall prepare reports on the results of these assessments addressed to the Administrator of the Environmental Protection Agency to include recommendations as may be appropriate for the revision of these programs. The sum of $500,000 is authorized to be appropriated to the Institute for each of the fiscal years 1994 through 1997 to carry out this paragraph.

GRANTS TO THE STATES

29 USC 672
Sec. 23

(a) The Secretary is authorized, during the fiscal year ending June 30, 1971, and the two succeeding fiscal years, to make grants to the States which have designated a State agency under section 18 to assist them–
  (1) in identifying their needs and responsibilities in the area of occupational safety and health,
  (2) in developing State plans, under section 18, or
  (3) in developing plans for–
    (A) establishing systems for the collection of information concerning the nature and frequency of occupational injuries and diseases;
    (B) increasing the expertise and enforcement capabilities of their personnel engaged in occupational safety and health programs; or
    (C) otherwise improving the administration and enforcement of State occupational safety and health laws, including standards, thereunder, consistent with the objectives of this Act.

(b) The Secretary is authorized, during the fiscal year ending June 30, 1971, and the two succeeding fiscal years, to make grants to the States for experimental and demonstration projects consistent with the objectives set forth in subsection (a) of this section.

(c) The Governor of the State shall designate the appropriate State agency for receipt of any grant made by the Secretary under this section.

(d) Any State agency designated by the Governor of the State desiring a grant under this section shall submit an application therefor to the Secretary.

(e) The Secretary shall review the application, and shall, after consultation with the Secretary of Health and Human Services, approve or reject such application.

(f) The Federal share for each State grant under subsection (a) or (b) of this section may not exceed 90 per centum of the total cost of the application. In the event the Federal share for all States under either such subsection is not the same, the differences among the States shall be established on the basis of objective criteria.

(g) The Secretary is authorized to make grants to the States to assist them in administering and enforcing programs for occupational safety and health contained in State plans approved by the Secretary pursuant to section 18 of this Act. The Federal share for each State grant under this subsection may not exceed 50 per centum of the total cost to the State of such a program. The last sentence of subsection (f) shall be applicable in determining the Federal share under this subsection.

(h) Prior to June 30, 1973, the Secretary shall, after consultation with the Secretary of Health and Human Services, transmit a report to the President and to the Congress, describing the experience under the grant programs authorized by this section and making any recommendations he may deem appropriate.

STATISTICS

29 USC 673
Sec. 24

(a) In order to further the purposes of this Act, the Secretary, in consultation with the Secretary of Health and Human Services, shall develop and maintain an effective program of collection, compilation, and analysis of occupational safety and health statistics. Such program may cover all employments whether or not subject to any other provisions of this

Act but shall not cover employments excluded by section 4 of the Act. The Secretary shall compile accurate statistics on work injuries and illnesses which shall include all disabling, serious, or significant injuries and illnesses, whether or not involving loss of time from work, other than minor injuries requiring only first aid treatment and which do not involve medical treatment, loss of consciousness, restriction of work or motion, or transfer to another job.

(b) To carry out his duties under subsection (a) of this section, the Secretary may-
   (1) promote, encourage, or directly engage in programs of studies, information and communication concerning occupational safety and health statistics;
   (2) make grants to States or political subdivisions thereof in order to assist them in developing and administering programs dealing with occupational safety and health statistics; and
   (3) arrange, through grants or contracts, for the conduct of such research and investigations as give promise of furthering the objectives of this section.

(c) The Federal share for each grant under subsection (b) of this section may be up to 50 per centum of the State's total cost.

(d) The Secretary may, with the consent of any State or political subdivision thereof, accept and use the services, facilities, and employees of the agencies of such State or political subdivision, with or without reimbursement, in order to assist him in carrying out his functions under this section.

(e) On the basis of the records made and kept pursuant to section 8(c) of this Act, employers shall file such reports with the Secretary as he shall prescribe by regulation, as necessary to carry out his functions under this Act.

(f) Agreements between the Department of Labor and States pertaining to the collection of occupational safety and health statistics already in effect on the effective date of this Act shall remain in effect until superseded by grants or contracts under this Act.

AUDITS

29 USC 674
Sec. 25

(a) Each recipient of a grant under this Act shall keep such records as the Secretary or the Secretary of Health and Human Services shall prescribe, including records which fully disclose the amount and disposition by such recipient of the proceeds of such grant, the total cost of the project or undertaking in connection with which such grant is made or sued, and the amount of that portion of the cost of the project or undertaking supplied by other sources, and such other records as will facilitate an effective audit.

(b) The Secretary or the Secretary of Health and Human Services, and the Comptroller General of the United States, or any of their duly authorized representatives, shall have access for the purpose of audit and examination to any books, documents, papers, and records of the recipients of any grant under this Act that are pertinent to any such grant.

ANNUAL REPORT

29 USC 675
Sec. 26

Within one hundred and twenty days following the convening of each regular session of each Congress, the Secretary and the Secretary of Health and Human Services shall each prepare and submit to the President for transmittal to the Congress a report upon the subject matter of this Act, the progress toward achievement of the purpose of this Act, the needs and requirements in the field of occupational safety and health, and any other relevant information.

Such reports shall include information regarding occupational safety and health standards, and criteria for such standards, developed during the preceding year; evaluation of standards and criteria previously developed under this Act, defining areas of emphasis for new criteria and standards; and evaluation of the degree of observance of applicable occupational safety and health standards, and a summary of inspection and enforcement activity undertaken; analysis and evaluation of research activities for which results have been obtained under governmental and nongovernmental sponsorship; an analysis of major occupational diseases; evaluation of available control and measurement technology for hazards for which standards or criteria have been developed during the preceding year; description of cooperative efforts undertaken between Government agencies and other interested parties in the implementation of this Act during the preceding year; a progress report on the development of an adequate supply of trained manpower in the field of occupational safety and health, including estimates of future needs and the efforts being made by Government and others to meet those needs; listing of all toxic substances in industrial usage for which labeling requirements, criteria, or standards have not yet been established; and such recommendations for additional legislation as are deemed necessary to protect the safety and health of the worker and improve the administration of this Act.

NATIONAL COMMISSION ON STATE WORKMEN'S COMPENSATION LAWS

29 USC 676
Sec. 27

(a) (1) The Congress hereby finds and declares that–
- (A) the vast majority of American workers, and their families, are dependent on workmen's compensation for their basic economic security in the event such workers suffer disabling injury or death in the course of their employment; and that the full protection of American workers from job-related injury or death requires an adequate, prompt, and equitable system of workmen's compensation as well as an effective program of occupational health and safety regulation; and
- (B) in recent years serious questions have been raised concerning the fairness and adequacy of present workmen's compensation laws in the light of the growth of the economy, the changing nature of the labor force, increases in medical knowledge, changes in the hazards associated with various types of employment, new technology creating new risks to health and safety, and increases in the general level of wages and the cost of living.

(2) The purpose of this section is to authorize an effective study and objective evaluation of State workmen's compensation laws in order to determine if such laws provide an adequate, prompt, and equitable system of compensation for injury or death arising out of or in the course of employment.

(b) There is hereby established a National Commission on State Workmen's Compensation Laws.

(c) (1) The Workmen's Compensation Commission shall be composed of fifteen members to be appointed by the President from among members of State workmen's compensation boards, representatives of insurance carriers, business, labor, members of the medical profession having experience in industrial medicine or in workmen's compensation cases, educators having special expertise in the field of workmen's compensation, and representatives of the general public. The Secretary, the Secretary of Commerce, and the Secretary of Health and Human Services shall be ex officio members of the Workmen's Compensation Commission:

(2) Any vacancy in the Workmen's Compensation Commission shall not affect its powers.

(3) The President shall designate one of the members to serve as Chairman and one to serve as Vice Chairman of the Workmen's Compensation Commission.

(4) Eight members of the Workmen's Compensation Commission shall constitute a quorum.

(d) (1) The Workmen's Compensation Commission shall undertake a comprehensive study and evaluation of State workmen's compensation laws in order to determine if such laws provide an adequate, prompt, and equitable system of compensation. Such study and evaluation shall include, without being limited to, the following subject: (A) the amount and duration of permanent and temporary disability benefits and the criteria for determining the maximum limitations thereon, (B) the amount and duration of medical benefits and provisions insuring adequate medical care and free choice of physician, (C) the extent of coverage of workers, including exemptions based on numbers or type of employment, (D) standards for determining which injuries or diseases should be deemed compensable, (E) rehabilitation, (F) coverage under second and subsequent injury funds, (G) time limits on filing claims, (H) waiting periods, (I) compulsory or elective coverage, (J) administration, (K) legal expenses, (L) the feasibility and desirability of a uniform system of reporting information concerning job-related injuries and diseases and the operation of workmen's compensation laws, (M) the resolution of conflict of laws, extraterritoriality and similar problems arising from claims with multistate aspects, (N) the extent to which private insurance carriers are excluded from supplying workmen's compensation coverage and the desirability of such exclusionary practices, to the extent they are found to exist, (O) the relationship between workmen's compensation on the one hand, and old-age, disability, and survivors insurance and other types of insurance, public or private, on the other hand, (P) methods of implementing the recommendations of the Commission.

(2) The Workmen's Compensation Commission shall transmit to the President and to the Congress not later than July 31, 1972, a final report containing a detailed statement of the findings and conclusions of the Commission, together with such recommendations as it deems advisable.

(e) (1) The Workmen's Compensation Commission or, on the authorization of the Workmen's Compensation Commission, any subcommittee or members thereof, may, for the purpose of carrying out the provisions of this title, hold such hearings, take such testimony, and sit and act at such times and places as the Workmen's Compensation Commission deems advisable. Any member authorized by the Workmen's Compensation Commission may administer oaths or affirmations to witnesses appearing before the Workmen's Compensation Commission or any subcommittee or members thereof.

(2) Each department, agency, and instrumentality of the executive branch of the Government, including independent agencies, is authorized and directed to furnish to the Workmen's Compensation Commission, upon request made by the Chairman or Vice Chairman, such information as the Workmen's Compensation Commission deems necessary to carry out its functions under this section.

(f) Subject to such rules and regulations as may be adopted by the Workmen's Compensation Commission, the Chairman shall have the power to–
  (1) appoint and fix the compensation of an executive director, and such additional staff personnel as he deems necessary, without regard to the provisions of title 5, United States Code, governing appointments in the competitive service, and without regard to the provisions of chapter 51 and subchapter III of chapter 53 of such title relating to classification and General Schedule pay rates, but at rates not in excess of the maximum rate for GS-18 of the General Schedule under section 5332 of such title, and
  (2) procure temporary and intermittent services to the same extent as is authorized by section 3109 of title 5, United States Code.

(g) The Workmen's Compensation Commission is authorized to enter into contract with Federal or State agencies, private firms, institutions, and individuals for the conduct of research or surveys, the preparation of reports, and other activities necessary to the discharge of its duties.

(h) Members of the Workmen's Compensation Commission shall receive compensation for each day they are engaged in the performance of their duties as members of the Workmen's Compensation Commission at the daily rate prescribed for GS-18 under section 5332 of title 5, United States Code, and shall be entitled to reimbursement for travel, subsistence, and other necessary expenses incurred by them in the performance of their duties as members of the Workmen's Compensation Commission.

(i) There are hereby authorized to be appropriated such sums as may be necessary to carry out the provisions of this section.

(j) On the ninetieth day after the date of submission of its final report to the President, the Workmen's Compensation Commission shall cease to exist.

SEPARABILITY

29 USC 677
Sec. 32

If any provision of this Act, or the application of such provision to any person or circumstance, shall be held invalid, the remainder of this Act, or the application of such provision to persons or circumstances other than those as to which it is held invalid, shall not be affected thereby.

APPROPRIATIONS

29 USC 678
Sec. 33

There are authorized to be appointed to carry out this Act for each fiscal year such sums as the Congress shall deem necessary.

·   ·   ·

PROVISIONS THAT DO NOT AMEND

FROM PL 99-499, OCTOBER 17, 1986

WORKER PROTECTION STANDARDS

Sec. 126

(a) Promulgation.–Within one year after the date of the enactment of this section, the Secretary of Labor shall, pursuant to section 6 of the Occupational Safety and Health Act of 1970, promulgate standards for the health and safety protection of employees engaged in hazardous waste operations.

(b) Proposed Standards.–The Secretary of Labor shall issue proposed regulations on such standards which shall include, but need not be limited to, the following worker protection provisions:
  (1) Site Analysis.–Requirements for a formal hazard analysis of the site and development of a site specific plan for worker protection.
  (2) Training.–Requirements for contractors to provide initial and routine training of workers before such workers are permitted to engage in hazardous waste operations which would expose them to toxic substances.
  (3) Medical Surveillance.–A program of regular medical examination, monitoring, and surveillance of workers engaged in hazardous waste operations which would expose them to toxic substances.

(4) Protective Equipment.–Requirements for appropriate personal protective equipment, clothing, and respirators for work in hazardous waste operations.
(5) Engineering Controls.–Requirements for engineering controls concerning the use of equipment and exposure of workers engaged in hazardous waste operations.
(6) Maximum Exposure Limits.–Requirements for maximum exposure limitations for workers engaged in hazardous waste operations, including necessary monitoring and assessment procedures.
(7) Informational Program.–A program to inform workers engaged in hazardous waste operations of the nature and degree of toxic exposure likely as a result of such hazardous waste operations.
(8) Handling.–Requirements for the handling, transporting, labeling, and disposing of hazardous wastes.
(9) New Technology Program.–A program for the introduction of new equipment or technologies that will maintain worker protections.
(10) Decontamination Procedures.–Procedures for decontamination.
(11) Emergency Response.–Requirements for emergency response and protection of workers engaged in hazardous waste operations.

(c) Final Regulations.–Final regulations under subsection (a) shall take effect one year after the date they are promulgated. In promulgating final regulations on standards under subsection (a), the Secretary of Labor shall include each of the provisions listed in paragraphs (1) through (11) of subsection (b) unless the Secretary determines that the evidence in the public record considered as a whole does not support inclusion of any such provision.

(d) Specific Training Standards.–
(1) Offsite Instruction; Field Experience.–Standards promulgated under subsection (a) shall include training standards requiring that general site workers (such as equipment operators, general laborers, and other supervised personnel) engaged in hazardous substance removal or other activities which expose or potentially expose such workers to hazardous substances receive a minimum of 40 hours of initial instruction off the site, and a minimum of three days of actual field experience under the direct supervision of a trained, experienced supervisor, at the time of assignment. The requirements of the pre-ceding sentence shall not apply to any general site worker who has received the equivalent of such training. Workers who may be exposed to unique or special hazards shall be provided additional training.
(2) Training of Supervisors.–Standards promulgated under subsection (a) shall include training standards requiring that onsite managers and supervisors directly responsible for the hazardous waste operations (such as foremen) receive the same training as general site workers set forth in paragraph (1) of this subsection and at least eight additional hours of specialized training on managing hazardous waste operations. The requirements of the preceding sentence shall not apply to any person who has received the equivalent of such training.
(3) Certification; Enforcement.–Such training standards shall contain provisions for certifying that general site workers, onsite managers, and supervisors have received the specified training and shall prohibit any individual who has not received the specified training from engaging in hazardous waste operations covered by the standard.
(4) Training of Emergency Response Personnel.–Such training standards shall set forth requirements for the training of workers who are responsible for responding to hazardous emergency situations who may be exposed to toxic substances in carrying out their responsibilities.

(e) Interim Regulations.–The Secretary of Labor shall issue interim final regulations under this section within 60 days after the enactment of this section which shall provide no less protection under this section for workers employed by contractors and emergency response workers than the protections contained in the Environmental Protection Agency

Manual (1981) "Health and Safety Requirements for Employees Engaged in Field Activities" and existing standards under the Occupational Safety and Health Act of 1970 found in subpart C of part 1926 of title 29 of the Code of Federal Regulations. Such interim final regulations shall take effect upon issuance and shall apply until final regulations become effective under subsection (c).

(f) Coverage of Certain State and Local Employees.–Not later than 90 days after the promulgation of final regulations under subsection (a), the Administrator shall promulgate standards identical to those promulgated by the Secretary of Labor under subsection (a). Standards promulgated under this subsection shall apply to employees of State and local governments in each State which does not have in effect an approved State plan under section 18 of the Occupational Safety and Health Act of 1970 providing for standards for the health and safety protection of employees engaged in hazardous waste operations.

(g) Grant Program.–
   (1) Grant Purposes.–Grants for the training and education of workers who are or may be engaged in activities related to hazardous waste removal or containment or emergency response may be made under this subsection.
   (2) Administration.–Grants under this subsection shall be administered by the National Institute of Environmental Health Sciences.
   (3) Grant Recipients.–Grants shall be awarded to nonprofit organizations which demonstrate experience in implementing and operating worker health and safety training and education programs and demonstrate the ability to reach and involve in training programs target populations of workers who are or will be engaged in hazardous waste removal or containment or emergency response operations.

FROM PL 100-202, DECEMBER 22, 1987

OCCUPATIONAL SAFETY AND HEALTH ADMINISTRATION
SALARIES AND EXPENSES

For necessary expenses for the Occupational Safety and Health Administration, $235,474,000 including not to exceed $40,524,000, which shall be the maximum amount available for grants to States under section 23(g) of the Occupational Safety and Health Act, which grants shall be no less than fifty percent of the costs of State occupational safety and health programs required to be incurred under plans approved by the Secretary under section 18 of the Occupational Safety and Health Act of 1970; Provided, That none of the funds appropriated under this paragraph shall be obligated or expended for the assessment of civil penalties issued for first instance violations of any standard, rule, or regulation promulgated under the Occupational Safety and Health Act of 1970 (other than serious, willful, or repeated violations under section 17 of the Act) resulting from the inspection of any establishment or workplace subject to the Act, unless such establishment or workplace is cited, on the basis of such inspection, for ten or more violations: Provided further, That none of the funds appropriated under this paragraph shall be obligated or expended to prescribe, issue, administer, or enforce any standard, rule, regulation, or order under the Occupational Safety and Health Act of 1970 which is applicable to any person who is engaged in a farming operation which does not maintain a temporary labor camp and employs ten or fewer employees; Provided further, That none of the funds appropriated under this paragraph shall be obligated or expended to prescribe, issue, administer, or enforce any standard, rule, regulation, order or administrative action under the Occupational Safety and Health Act of 1970 affecting any work activity by reason of recreational hunting, shooting, or fishing; Provided further, That no funds appropriated under this paragraph shall be obligated or expended to administer or enforce any standard, rule, regulation, or order under the Occupational Safety and Health Act of 1970 with respect to any employer of ten or fewer employees who is included within a category having an occupational injury lost work day case rate, at the most precise Standard Industrial Classification Code for which such data are published, less than the national average rate as such rates are most recently published by the Secretary, acting through the Bureau of Labor Statistics, in accordance with section 24 of that Act (29 U.S.C. 673), except–

(1) to provide, as authorized by such Act, consultation, technical assistance, educational and training services, and to conduct surveys and studies;

(2) to conduct an inspection or investigation in response to an employee complaint, to issue a citation for violations found during such inspection, and to assess a penalty for violations which are not corrected within a reasonable abatement period and for any willful violations found;

(3) to take any action authorized by such Act with respect to imminent dangers;

(4) to take any action authorized by such Act with respect to health hazards;

(5) to take any action authorized by such Act with respect to a report of an employment accident which is fatal to one or more employees or which results in hospitalization of five or more employees, and to take any action pursuant to such investigation authorized by such Act; and

(6) to take any action authorized by such Act with respect to complaints of discrimination against employees for exercising rights under such Act; Provided further, That the foregoing proviso shall not apply to any person who is engaged in a farming operation which does not maintain a temporary labor camp and employs ten or fewer employees; Provided further, That none of the funds appropriated under this paragraph shall be obligated or expended for the proposal or assessment of any civil penalties for the violation or alleged violation by an employer of ten or fewer employees of any standard, rule, regulation, or order promulgated under the Occupational Safety and Health Act of 1970 (other than serious, willful or repeated violations and violations which pose imminent danger under section 13 of the Act) if, prior to the inspection which gives rise to the alleged violation, the employer cited has (1) voluntarily requested consultation under a pro-gram operated pursuant to section 7(c)(1) or section 18 of the Occupational Safety and Health Act of 1970 or from a private consultative source approved by the Administration and (2) had the consultant examine the condition cited and (3) made or is in the process of making a reasonable good faith effort to eliminate the hazard created by the condition cited as such, which was identified by the aforementioned consultant, unless changing circumstances or workplace conditions render inapplicable the advice obtained from such consultants; Provided further, That none of the funds appropriated under this paragraph may be obligated or expended for any State plan monitoring visit by the Secretary of Labor under section 18 of the Occupational Safety and Health Act of 1970, of any factory, plant, establishment, construction site, or other area, workplace or environment where such a workplace or environment has been inspected by an employee of a State acting pursuant to section 18 of such Act within the six months preceding such inspection: Provided further, That this limitation does not prohibit the Secretary of Labor from conducting such monitoring visit at the time and place of an inspection by an employee of a State acting pursuant to section 18 of such Act, or in order to investigate a complaint about State program administration including a failure to respond to a worker complaint regarding a violation of such Act, or in order to investigate a discrimination complaint under section 11(c) of such Act, or as part of a special study monitoring program, or to investigate a fatality or catastrophe; Provide further, That none of the funds appropriated under this paragraph may be obligated or expended for the inspection, investigation, or enforcement of any activity occurring on the Outer Continental Shelf which exceeds the authority granted to the Occupational Safety and Health Administration by any provision of the Outer Continental Shelf Lands Act, or the Outer Continental Shelf Lands Act Amendments of 1978.

FROM PL 102-550, OCTOBER 28, 1992

This subtitle [Subtitle B–Lead Exposure Reduction] may be cited as the "Lead-Based Paint exposure Reduction Act."

Subtitle C—Worker Protection

Sec. 1031

Worker Protection. Not later than 180 days after the enactment of this Act, the Secretary of Labor shall issue an interim final regulation regulating occupational exposure to lead in the construction industry. Such interim final regulation shall provide employment and places of employment to employees which are as safe and healthful as those which would prevail under the Department of Housing and Urban Development guidelines published at Federal Register 55, page 38973 (September 28, 1990) (Revised Chapter 8). Such interim final regulations shall take effect upon issuance (except that such regulations may include a reasonable delay in the effective date), shall have the legal effect of an Occupational Safety and Health Standard, and shall apply until a final standard becomes effective under section 6 of the Occupational Safety and Health Act of 1970.

Sec. 1032

Coordination Between Environmental Protection Agency and Department of Labor. The Secretary of Labor, in promulgating regulations under section 1031, shall consult and coordinate with the Administrator of the Environmental Protection Agency for the purpose of achieving the maximum enforcement of title IV of the Toxic Substances Control Act and the Occupational Safety and Health Act of 1970 while imposing the least burdens of duplicative requirements on those subject to such title and Act and for other purposes.

From PL 104-1, JANUARY 23, 1995

CONGRESSIONAL ACCOUNTABILITY ACT OF 1995

PART C—OCCUPATIONAL SAFETY AND HEALTH ACT OF 1970

RIGHTS AND PROTECTIONS UNDER THE OCCUPATIONAL SAFETY AND HEALTH ACT OF 1970; PROCEDURES FOR REMEDY OF VIOLATIONS

USC1341
Sec. 215

(a) Occupational Safety and Health Protections.–
   (1) In General.–Each employing office and each covered employee shall comply with the provisions of section 5 of the Occupational Safety and Health Act of 1970 (29 U.S.C. 654).
   (2) Definitions.–For purposes of the application under this section of the Occupational Safety and Health Act of 1970–
      (A) the term "employer" as used in such Act means an employing office;
      (B) the term "employee" as used in such Act means a covered employee;
      (C) the term "employing office" includes the General Accounting Office, the Library of Congress, and any entity listed in subsection (a) of section 210 that is responsible for correcting a violation of this section, irrespective of whether the entity has an employment relationship with any covered employee in any employing office in which such a violation occurs; and
      (D) the term "employee" includes employees of the General Accounting Office and the Library of Congress.

(b) Remedy.–The remedy for a violation of subsection (a) shall be an order to correct the violation, including such order as would be appropriate if issued under section 13(a) of the Occupational Safety and Health Act of 1970 (29 U.S.C. 662(a)).

(c) Procedures.–
  (1) Requests for Inspections.–Upon written request of any employing office or covered employee, the General Counsel shall exercise the authorities granted to the Secretary of Labor by subsections (a), (d), (e), and (f) of section 8 of the Occupational Safety and Health Act of 1970 (29 U.S.C. 657 (a), (d), (e), and (f)) to inspect and investigate places of employment under the jurisdiction of employing offices.
  (2) Citations, Notices, and Notifications.–For purposes of this section, the General Counsel shall exercise the authorities granted to the Secretary of Labor in sections 9 and 10 of the Occupational Safety and Health Act of 1970 (29 U.S.C. 658 and 659), to issue–
    (A) a citation or notice to any employing office responsible for correcting a violation of subsection (a); or
    (B) a notification to any employing office that the General Counsel believes has failed to correct a violation for which a citation has been issued within the period permitted for its correction.
  (3) Hearings and Review.–If after issuing a citation or notification, the General Counsel determines that a violation has not been corrected, the General Counsel may file a complaint with the Office against the employing office named in the citation or notification. The complaint shall be submitted to a hearing officer for decision pursuant to subsections (b) through (h) of section 405, subject to review by the Board pursuant to section 406.
  (4) Variance Procedures.–An employing office may request from the Board an order granting a variance from a standard made applicable by this section. For the purposes of this section, the Board shall exercise the authorities granted to the Secretary of Labor in sections 6(b)(6) and 6(d) of the Occupational Safety and Health Act of 1970 (29 U.S.C. 655(b)(6) and 655(d)) to act on any employing office's request for a variance. The Board shall refer the matter to a hearing officer pursuant to subsections (b) through (h) of section 405, subject to review by the Board pursuant to section 406.
  (5) Judicial Review.–The General Counsel or employing office aggrieved by a final decision of the Board under paragraph (3) or (4), may file a petition for review with the United States Court of Appeals for the Federal Circuit pursuant to section 407.
  (6) Compliance Date.–If new appropriated funds are necessary to correct a violation of subsection (a) for which a citation is issued, or to comply with an order requiring correction of such a violation, correction or compliance shall take place as soon as possible, but not later than the end of the fiscal year following the fiscal year in which the citation is issued or the order requiring correction becomes final and not subject to further review.
(d) Regulations to Implement Section.–
  (1) In General.–The Board shall, pursuant to section 304, issue regulations to implement this section.
  (2) Agency Regulations.–The regulations issued under paragraph (1) shall be the same as substantive regulations promulgated by the Secretary of Labor to implement the statutory provisions referred to in subsection (a) except to the extent that the Board may determine, for good cause shown and stated together with the regulation, that a modification of such regulations would be more effective for the implementation of the rights and protections under this section.
  (3) Employing Office Responsible for Correction.–The regulations issued under paragraph (1) shall include a method of identifying, for purposes of this section and for different categories of violations of subsection (a), the employing office responsible for correction of a particular violation.
(e) Periodic Inspections; Report to Congress.–
  (1) On a regular basis, and at least once each Congress, the General Counsel, exercising the same authorities of the Secretary of Labor as under subsection (c)(1), shall conduct periodic inspections of all facilities of the House of Representatives, the

Senate, the Capitol Guide Service, the Capitol Police, the Congressional Budget Office, the Office of the Architect of the Capitol, the Office of the Attending Physician, the Office of Compliance, the Office of Technology Assessment, the Library of Congress, and the General Accounting Office to report on compliance with subsection (a).
(2) Report.–On the basis of each periodic inspection, the General Counsel shall prepare and submit a report–
  (A) to the Speaker of the House of Representatives, the President pro tempore of the Senate, and the Office of the Architect of the Capitol or other employing office responsible for correcting the violation of this section uncovered by such inspection, and
  (B) containing the results of the periodic inspection, identifying the employing office responsible for correcting the violation of this section uncovered by such inspection, describing any steps necessary to correct any violation of this section, and assessing any risks to employee health and safety associated with any violation.
(3) Action after Report.–If a report identifies any violation of this section, the General Counsel shall issue a citation or notice in accordance with subsection (c)(2)(A).
(4) Detailed Personnel.–The Secretary of Labor may, on request of the Executive Director, detail to the Office such personnel as may be necessary to advise and assist the Office in carrying out its duties under this section.

(f) Initial Period for Study and Corrective Action.–The period from the date of the enactment of this Act until December 31, 1996, shall be available to the Office of the Architect of the Capitol and other employing offices to identify any violations of subsection (a), to determine the costs of compliance, and to take any necessary corrective action to abate any violations. The Office shall assist the Office of the Architect of the Capitol and other employing offices by arranging for inspections and other technical assistance at their request. Prior to July 1, 1996, the General Counsel shall conduct a thorough inspection under subsection (e)(1) and shall submit the report under subsection (e)(2) for the One Hundred Fourth Congress.

(g) Effective Date.–
  (1) In General.–Except as provided in paragraph (2), subsections (a), (b), (c), and (e)(3) shall be effective on January 1, 1997.
  (2) General Accounting Office and Library of Congress.–This section shall be effective with respect to the General Accounting Office and the Library of Congress 1 year after transmission to the Congress of the study under section 230.

From PL 104-208, SEPTEMBER 30, 1996

SALARIES AND EXPENSES

For necessary expenses for the Occupational Safety and Health Administration, $325,734,000, including not to exceed $77,354,000 which shall be the maximum amount available for grants to States under section 23(g) of the Occupational Safety and Health Act, which grants shall be no less than fifty percent of the costs of State occupational safety and health programs required to be incurred under plans approved by the Secretary under section 18 of the Occupational Safety and Health Act of 1970; and, in addition, notwithstanding 31 U.S.C. 3302, the Occupational Safety and Health Administration may retain up to $750,000 per fiscal year of training institute course tuition fees, otherwise authorized by law to be collected, and may utilize such sums for occupational safety and health training and education grants: Provided, That, notwithstanding 31 U.S.C. 3302, the Secretary of Labor is authorized, during the fiscal year ending September 30, 1997, to collect and retain fees for services provided to Nationally Recognized Testing Laboratories, and may utilize such sums, in accordance with the provisions of 29 U.S.C. 9a, to administer national and international laboratory recognition programs that ensure the safety of equipment and products used by workers in the workplace: Provided further, That none of the funds appropriated under this paragraph shall be obligated or expended to prescribe, issue, administer, or enforce any standard, rule, regulation, or order under the Occupational Safety and Health Act of 1970 which is applicable to any person who is engaged in a farming operation which does not maintain a temporary labor camp and employs ten or fewer employees: Provided further, That no funds appropriated under this paragraph shall be obligated or expended to administer or enforce any standard, rule, regulation, or order under the Occupational Safety and Health Act of 1970 with respect to any employer of ten or fewer employees who is included within a category having an occupational injury lost workday case rate, at the most precise Standard Industrial Classification Code for which such data are published, less than the national average rate as such rates are most recently published by the Secretary, acting through the Bureau of Labor Statistics, in accordance with section 24 of that Act (29 U.S.C. 673), except–

(1) to provide, as authorized by such Act, consultation, technical assistance, educational and training services, and to conduct surveys and studies;

(2) to conduct an inspection or investigation in response to an employee complaint, to issue a citation for violations found during such inspection, and to assess a penalty for violations which are not corrected within a reasonable abatement period and for any willful violations found;

(3) to take any action authorized by such Act with respect to imminent dangers;

(4) to take any action authorized by such Act with respect to health hazards;

(5) to take any action authorized by such Act with respect to a report of an employment accident which is fatal to one or more employees or which results in hospitalization of two or more employees, and to take any action pursuant to such investigation authorized by such Act; and

(6) to take any action authorized by such Act with respect to complaints of discrimination against employees for exercising rights under such Act: Provided further, That the foregoing proviso shall not apply to any person who is engaged in a farming operation which does not maintain a temporary labor camp and employs ten or fewer employees.

FROM PUBLIC LAW 105-78, NOVEMBER 13, 1997

For necessary expenses for the Occupational Safety and Health Administration, $336,480,000, including not to exceed $77,941,000 which shall be the maximum amount available for grants to States under section 23(g) of the Occupational Safety and Health Act, which grants shall be no less than 50 percent of the costs of State occupational safety and health programs required to be incurred under plans approved by the Secretary under section 18 of the Occupational Safety and Health Act of 1970; and, in addition, notwithstanding 31 USC 3302, the Occupational Safety and Health Administration may retain up to $750,000 per fiscal year of training institute course tuition fees, otherwise authorized by law to be collected, and may utilize such sums for occupational safety and health training and education grants: Provided, That, notwithstanding 31 USC 3302, the Secretary of Labor is authorized, during the fiscal year ending September 30, 1998, to collect and retain fees for services provided to Nationally Recognized Testing Laboratories, and may utilize such sums, in accordance with the provisions of 29 USC 9a, to administer national and international laboratory recognition programs that ensure the safety of equipment and products used by workers in the workplace: Provided further, That none of the funds appropriated under this paragraph shall be obligated or expended to prescribe, issue, administer, or enforce any standard, rule, regulation, or order under the Occupational Safety and Health Act of 1970 which is applicable to any person who is engaged in a farming operation which does not maintain a temporary labor camp and employs ten or fewer employees: Provided further, That no funds appropriated under this paragraph shall be obligated or expended to administer or enforce any standard, rule, regulation, or order under the Occupational Safety and Health Act of 1970 with respect to any employer of ten or fewer employees who is included within a category having an occupational injury lost workday case rate, at the most precise Standard Industrial Classification Code for which such data are published, less than the national average rate as such rates are most recently published by the Secretary, acting through the Bureau of Labor Statistics, in accordance with section 24 of that Act (29 USC 673), except–

(1) to provide, as authorized by such Act, consultation, technical assistance, educational and training services, and to conduct surveys and studies;
(2) to conduct an inspection or investigation in response to an employee complaint, to issue a citation for violations found during such inspection, and to assess a penalty for violations which are not corrected within a reasonable abatement period and for any willful violations found;
(3) to take any action authorized by such Act with respect to imminent dangers;
(4) to take any action authorized by such Act with respect to health hazards;
(5) to take any action authorized by such Act with respect to a report of an employment accident which is fatal to one or more employees or which results in hospitalization of two or more employees, and to take any action pursuant to such investigation authorized by such Act; and
(6) to take any action authorized by such Act with respect to complaints of discrimination against employees for exercising rights under such Act: Provided further, That the foregoing proviso shall not apply to any person who is engaged in a farming operation which does not maintain a temporary labor camp and employs ten or fewer employees.

Index

A
A/C Electric Co. v. OSHRC, 247
Abatement period, extension of, 242
Access to information, for employees, 149-150
Access to records, 30-31, 136-139
Accident reports, 28-29
Accord, Peterson Bros. Steel Erection Co. v. Secretary of Labor, 88
Acetylaminofluorene, recordkeeping requirement, 116
Actual, constructive knowledge, 84-85
Administrative law judge
  Department of Labor, 56
  proceedings before, 244-255
    answer, 244-245
    complaint, 244
    defenses, 245-249
      borrowed servant defense, 248
      greater hazard defense, 247
      impossibility of compliance, 246
      multi-employer worksite doctrine, 248
      procedural defenses, 245-246
      promptness, 246
      substantive defenses, 246-249
      unpreventable employee misconduct, 247
    discovery, 249-254
      depositions, 252
      exhibits, 253
      interrogatories, 251-252
      production of documents, 250-251
        protection of confidential information, 253-254
      requests for admissions, 251
      subpoenas, 252-253
      trade secrets, 253-254
    hearing, 254-255
    simplified proceedings, 255
    "E-Z trial" procedures, 255
Administrative officer, OSHA, assistant Secretary of Labor for Occupational Safety and Health, 2
Administrative Procedures Act, 54, 275
Administrative remedies, exhaustion of, 271-273
Advanced notice of proposed rulemaking, 55

Advisory committees, in rulemaking process, 54-55
Affirmative defenses, to duty to comply, 86-92
  de minimis violations, 92
  general duty clause, 89
  greater hazard defense, 89-90
  infeasibility, 86-89
    alternative measures element, 89
  invalidity of standard, 91-92
  unpreventable employee misconduct, 90-91
AFL-CIO v. OSHA, 9
Agencies, federal, grading of, Maxwell School of Government, 1
Agency inspections, use of audit, 204
Air contaminants standard, permissible exposure limits, 69
Alaska, occupational safety, health plan, 306
Alaska Department of Labor and Workforce Development, 307
Alpha-naphthylamine, 116
American Airlines, Inc. v. Secretary of Labor, 86
American Conference of Governmental Industrial Hygienists, threshold limit values, 44
American National Standards Institute, 101
American National Standards Institute (ANSI), 101
American Smelting and Refining Co. v. OSHRC, 98
American Textile Manufacturers Institute, Inc. v. Donovan, 19
American Textile Mfrs. Inst., Inc. v. Donovan, 67
Aminodiphenyl, 116
Annual Report on Carcinogens published by National Toxicology Program (NTP), 44
Annual summary of accidents, illnesses, 29
ANSI. *See* American National Standards Institute
Answer, proceedings before administrative law judge, 244-245
Anthony Crane Rental, Inc. v. Reich, 83
Appeal process, Occupational Safety and Health Review Commission, 39
Arbitration, 159
Arizona, occupational safety, health plan, 306
ASARCO v. NLRB, 210
Asbestos, 6, 70

recordkeeping requirement, 116
Asbestos Information Association/North America v. OSHA, 70
Assistant Secretary of Labor for Occupational Safety and Health, administration of OSHA, 2
Associated Indus. of New York State, Inc. v. U.S. Dep't of Labor, 144
Atlantic & Gulf Stevedores, Inc. v. OSHRC, 88, 273
Audit, recommendations, 202
Audit privilege, determining whether to apply privilege, 209-210
Auditors, censoring of, 202
Austin Road Co. v. OSHRC, 277
Authority to inspect places of employment, 216

B
B & B Insulation, Inc. v. OSHRC, 86
Balancing, feasibility, 16-19
Barlow case
  warrantless inspections, 27-28
Barlow case, warrantless inspections, 27-28
Baroid Div. v. OSHRC, 98
Barrentine v. Arkansaw Freight Sys., 158
Beatty Equip. Leasing, Inc. v. Secretary of Labor, 83
Bell v. Dynamite Foods, 163
Benzene, 66
  recordkeeping requirement, 116
Benzidine, 116
Beta-naphthylamine, recordkeeping requirement, 116
Bis-chloromethyl ether, recordkeeping requirement, 116
Bloodborne pathogens, recordkeeping requirement, 116
Borrowed servant defense, as defense in proceedings before administrative law judge, 248
Bratton Corp. v. OSHRC, 84
Brennan v. OSHRC, 80, 83, 99, 103, 249, 269, 277
Brennan v. Smoke-Craft, Inc., 103
Brennan v. Winters Battery Mfg. Co., 271
Brevik v. Kite Painting, Inc., 160
Brisk Waterproofing, 106
Bristol Steel & Iron Works, Inc. v. OSHRC, 86
Brock v. Dun-Par Engineered Form Co., 89
Brock v. Emerson Electric Co., 228
Brock v. L. E. Myers Company, 27
Buckeye Indus., Inc. v. Secretary of Labor, 273

Builders Steel Co. v. Marshall, 277
Building and Constr. Trades Dep't, AFL-CIO v. Brock, 66
Bunge Corp. v. Secretary of Labor, 80
Butadiene, recordkeeping requirement, 116
Butler Lime & Cement Co. v. OSHRC, 278

C
Cabesuela v. Browning-Ferris Industries of Calif. Inc., 160
Cadmium, recordkeeping requirement, 116
Caldwell v. U.S., 289
California, occupational safety, health plan, 306
California Department of Industrial Relations, 307
Cancer, International Agency for Research on, 44
Cape and Vineyard Division v. OSAHRC, 102, 103
Carlyle Compressor Co. v. OSHRC, 106
Carpal tunnel syndrome, 47
Caterpillar Inc. v. Herman, 89, 90, 92
CDC. See Centers for Disease Control
Censoring of auditors, 202
Centers for Disease Control, 40
CERCLA. See Comprehensive Environmental Response, Compensation, and Liability Act
Cerracchio v. Alden Leeds, Inc., 163
Cerro Metal Products, 106
CH2M Hill Central, Inc. v. OSHRC, 272
CH2M Hill Central, Inv. v. OSHRC, 270
Challenges, 240-243
  abatement period, extending, 242
  commissioners, review by, 255-257
  employee participation, 242-243
  notice of contest, 241
  settlement, 240-241
    after filing notice of contest, 243
Challenging standards, 73
Chamber of Commerce v. Herman, 60
Chambers of Commerce v. OSHA, 148
Champlin Petroleum Co. v. OSHRC, 106
Chao v. Symms Fruit Ranch, Inc., 92
Chemical-specific standards, recordkeeping, 114-118
Citation, 23-24. See also Violation
  challenging, 236, 240-243
    abatement period, extending, 242
    employee participation, 242-243
    notice of contest, 241
    settlement, after filing notice of contest, 243
    settlement agreement, 240-241

patterns, 24-25
Classification, OSHA violations, 236-240
Clean Air Act, 4, 38, 108
Closing conference, inspection, 232-233
Coal tar pitch volatiles, recordkeeping requirement, 116
Coast Guard regulations, maritime workers subject to, exemption from OSHA, 6
Coke oven emissions, 116
Collective bargaining agreements, 158-159
Commission order
 final, judicial review, enforcement actions, 270-271
 stay of, 257
Commission review, 256-257
Commissioners, review of decision by, 255-257
 commission review, 256-257
 discretionary review, 256
 final decisions, 256
 interlocutory review, 255-256
 reports of judges, 256
 stay of commission order, 257
Common law audit privilege, 208-210
Common sense' recognition, recognized hazards, general duty clause, 102
Communication of company rules, 26-27
Communication of hazard
 assessing hazard, 168-173
  hazard determination, 168-171
 employee exposure, 171
 HAZCOM standard, 165
 labelling, 174-176
 mixtures, 171-173
 multiple employer worksites, 190-191
 National Toxicology Program, annual report, 171
 regulations, 41-46
 trade secrets, 189-190
 warning, forms of, 174-176
 written hazard communication program, 173
Company rules, communication, enforcement of, 26-27
Compare American Textile Mfgrs. Inst. v. Donovan, 106
Competence of inspectors, 22-23
Complaint, proceedings before administrative law judge, 244
Completion, standards, deletion processes, 12-13
Compliance, 20-28
Compliance safety and health officers, 215, 285

Comprehensive Environmental Response, Compensation, and Liability Act, 187
Comprehensive Occupational Safety and Health Reform Act, 296
Conclusions of law, standard of review for, 275-276
Confidentiality, of audit information, 201
Confined space entry, 120
Congress, general duty clause guidance, 94-96
Conn-OSHA, 307
Connecticut, occupational safety, health plan, 306
Connecticut Department of Labor, 307
Consensual searches, warrant requirement, contrasted, 219-222
Consensus standards, Section 6(a), 11-12
Consolidated Andy, Inc. v. Donovan, 270
Constitutional challenges, judicial review, enforcement actions, 273-274
Construction
 contractors, 82-83
 sites, inspection, 233-234
Constructive knowledge, 84-85
Consultation, 36-37
Consumer Product Safety Act, 186
Contest rate, 26
Continental Can Co. v. Marshall, 249
Corrected conditions, no citation for, 204
Cotter & Co. v. OSHRC, 86
Cotton dust, 67, 116
Coverage, of OSHA, 5
Cranes, recordkeeping requirement, 118, 120
Creation of OSHA, 3
Criminal, violation, defined, 23
Criminal enforcement, 259-266
 employee, definition of, 260-261
 environmental statutes, prosecution under, 264
 false statements, advance notice, 262
 federal prosecution, 260-262
 Kennedy, Ted, Senator, 265
 memorandums of understanding, 264
 multi-employer doctrine, 261
 Nader, Ralph, 265
 recent legislation, 265
 Resource Conservation and Recovery Act, 264
 state enforcement, 263-264
 Wellstone, Paul, Senator, 265
 willful violations, causing death to employee, 261-262
Cuyahoga Valley Ry. v. United Transp. Union, 149

D

Dale Madden Constr., Inc. v. Hodgson, 269
Daniel Intern Corporation v. OSHRC, 262
Davidson v. Light, 209
Dayton Tire Plant, 279
DBCP. *See* Dibromochloropropane
De minimis violation, 92, 237
  defined, 23
Deering Milliken, Inc. v. OSHRC, 73, 273
Defenses
  affirmative, to duty to comply, 86-92
    de minimis violations, 92
    general duty clause, 89
    greater hazard defense, 89-90
    infeasibility, 86-89
      alternative measures element, 89
    invalidity of standard, 91-92
    unpreventable employee misconduct, 90-91
  in proceedings before administrative law judge, 245-249
    borrowed servant defense, 248
    greater hazard defense, 247
    impossibility of compliance, 246
    multi-employer worksite doctrine, 248
    procedural defenses, 245-246
    promptness, 246
    substantive defenses, 246-249
    unpreventable employee misconduct, 247
Deletion processes, standards, 12-13
Demonstration of serious harm, general duty clause and, 104-105
Department of Health and Human Services, 4
Department of Labor, 21
  administrative law judge, 56
Department of Transportation, 37
Depositions, proceedings before administrative law judge, 252
Derricks, recordkeeping requirement, 118, 120
DeTrae Enterprises, Inc. v. Secretary of Labor, 84
Diamond Roofing Co. v. OSHRC, 269
Dibromo-3-chloropropane, 116
Dibromochloropropane, 70
Dichlorobenzidine, 116
Dimethylaminoazobenzene, recordkeeping requirement, 116
Disclosure, of audit information, 207-213
  common law audit privilege, 208-210
  determining whether to apply privilege, 209-210
  self-audit privilege, 207-211
  statutory audit privilege, 210-211

Discovery, proceedings before administrative law judge, 249-254
  depositions, 252
  exhibits, 253
  interrogatories, 251-252
  production of documents, 250-251
  requests for admissions, 251
  subpoenas, 252-253
  trade secrets, 253-254
Discretionary review, 256
  petition for, 271
Discrimination, employee protection from, 147-148
Documentation of steps respond to hazards, 201-202
Documents, inspection of, 230-231
Dole v. Williams Enterprises, Inc., 90
Donovan v. A.A. Beiro Constr. Co., 222
Donovan v. Burlington Northern, Inc., 222
Donovan v. Castle & Cooke Foods and OSHRC, 18
Donovan v. Daniel Constr. Co., 92
Donovan v. Hackney, Inc., 221
Donovan v. Hahner, Foreman & Harness, Inc., 153
Donovan v. Lone Steer, Inc., 229
Donovan v. Missouri Farmers Association, 102
Donovan v. R.D. Andersen Constr. Co., 162
Donovan v. Southwest Electric Co., 221
Donovan v. Union Packing, 229
Donovan v. Williams Enterprises, Inc., 88
Donovan v. Wollaston Alloys, 221, 229
Dowling v. American Hawaii Cruises, Inc., 209
Dray v. New Market Poultry Products, Inc., 164
Dry Color Manufacturers' Association v. Department of Labor, 71
D&S Grading Co. v. Secretary of Labor, 278
Dun-Par Engineered Form Co. v. Marshall, 84, 87, 89
Duty to comply, 77-92
  applicability
    OSHA standards, 77-79
    problems of, 78-79
  employer responsibility, 81-84
    actual, constructive knowledge, 84-85
    affirmative defenses, 86-92
      de minimis violations, 92
      general duty clause, 89
      greater hazard defense, 89-90
      infeasibility, 86-89
        alternative measures element, 89
      invalidity of standard, 91-92

unpreventable employee misconduct, 90-91
general construction contractors, 82-83
multi-employer liability rules, legal status of, 83
multi-employer worksite defense rules, 84
multi-employer worksite liability rules, 81-82
non-construction applications, multi-employer liability rules, 83
exposure rule, 79-80
general principles, 79-84
preemption, 77-78
reasonably predictable exposure, 80
zone of danger, 80

E
"E-Z trial" procedures, 255
Egregious penalty policy, 238-239
Egregious violation, defined, 23
Electric Line Builders of Kansas, Inc. v. Marshall, 247
Electric Smith, Inc. v. Secretary of Labor, 84
Electrical systems, recordkeeping requirement, 120
Ellis v. Chase Communications, Inc., 153
Emergency temporary standards, 15, 70-71
Empire-Detroit Steel v. OSHRC, 97
Employee, definition of, 260-261
Employee alarms, 120
Employee information, training, 181-183
Employee interviews, inspection, 231-232
Employee misconduct, general duty clause, 103-104
Employee right-to-know, 165-196
Employee rights, 146-150
 access to information, 149-150
 complaints, 146-147
 discrimination, protection from, 147-148
 enforcement, participation in, 148-149
 imminent danger inspection, 288-291
  dangerous work, refusing, 289-291
  right to inform, assist with inspection, 288-289
  writ of mandamus, 289
 inspections, participation in, 148-149
 refusal to work, 147
Employer, defined, 5
Employer recognition of hazard, general duty clause and, 99-100
Employer responsibility, 81-84
 actual, constructive knowledge, 84-85

affirmative defenses, 86-92
 de minimis violations, 92
 general duty clause, 89
 greater hazard defense, 89-90
 infeasibility, 86-89
  alternative measures element, 89
 invalidity of standard, 91-92
 unpreventable employee misconduct, 90-91
general construction contractors, 82-83
multi-employer liability rules, legal status of, 83
multi-employer worksite defense rules, 84
multi-employer worksite liability rules, 81-82
non-construction applications, multi-employer liability rules, 83
Employers' rights, 142-146
 citations, challenging, 143-144
 civil penalties, challenging, 143-144
 inspections, 142-143
 judicial review, 144
 notice of contest, 143
 rulemaking, participation in, 144-145
 trade secrets, protection of, 145-146
 warrants, 142-143
Enactment, of OSHA, 3
Enforcement of company rules, 26-27
English v. General Electric Co., 162
Ensign-Bickford Corporation v. OSHRC, 262
Enterprises, Inc. v. Marshall, 222
Environmental Protection Agency, 38
 OSHA, contrasted, 1, 2
Environmental statutes, criminal prosecution under, 264
EPA's Emergency Planning and Community Right-to-Know Act, 32
Ergonomics, 47-49
Ergonomics Program Standard, 57, 58
Ethylene oxide, recordkeeping requirement, 116
Ethyleneimine, recordkeeping requirement, 116
ETS. *See* Emergency temporary standard
Evidence from audit, use of information to supplement, 206
Executive Order 12866, 64-66
Executive orders, 61-66
Exemptions, from OSHA, 5-6
Exhaustion of administrative remedies, 271-273
Exhibits, proceedings before administrative law judge, 253
Exposure rule, 79-80

F

Failure to correct, violation, defined, 23
False statements, advance notice, 262
Farm workers, 6
Faultless Div., Bliss & Laughlin Indus. v. Secretary, 88
Feasibility, balancing debate, 16-19
Feasible hazard abatement, general duty clause, 105-106
Federal agencies, 34
  grading of, Maxwell School of Government, 1
Federal Alcohol Administration Act, 186
Federal Aviation Administration, 38
Federal employees, exemption from OSHA, 6
Federal Food, Drug, and Cosmetic Act, 186
Federal Insecticide, Fungicide, and Rodenticide Act, 186
Federal Mine Safety and Health Act of 1977, 294, 295
Federal preemption, 45-46
Federal prosecution, 260-262
Federal Railroad Administration, 37
  railroad workers under, exemption from OSHA, 6
Federal Railroad Safety Act, 152
Federal Seed Act, 187
Federal Water Pollution Control Act, 38
Field Inspection Reference Manual, 102
Field Operations Manual, 99, 102, 107
Field structure, 21
  inspections, 21
Final commission orders, judicial review, enforcement actions, 270-271
Final decisions, review of by commissioners, 256
Final rule, in rulemaking process, 57-58
Findings of fact, standard of review for, 276-278
Fines, 23-24. *See also* Penalties; Violation
Fire brigades, recordkeeping requirement, 120
Fire extinguishers, 118, 120
Firestone case, 288
First aid, recordkeeping requirement, 120
Flammable/combustible liquids, recordkeeping requirement, 120
Florida Peach Growers Assn. v. Dept. of Labor, 38
Florida Peach Growers Ass'n v. U.S. DOL, 71
Form 101, 29
Form 200, 28
Formaldehyde, 66
  recordkeeping requirement, 116
Fourth amendment, inspections and, 216-219

G

Gade v. National Solid Waste Management Association, 46, 260
Gade v. National Solid Wastes Management Association, 192
Gateway Coal Co. v. U.M.W., 157
General construction contractors, 82-83
General duty clause, 15-16, 89, 93-110
  5(a)(1), 15-16
  American National Standards Institute, 101
  Clean Air Act, 108
  Congress, guidance provided by, 94-96
  employee misconduct, 103-104
  feasible hazard abatement, 105-106
  Field Inspection Reference Manual, 102
  Field Operations Manual, 99, 102, 107
  framework of, 96-98
  inspection and, 224
  recognized hazards, 98-102
    common sense' recognition, 102
    employer recognition, 99-100
    industry recognition, 100-101
  replacement of, with OSHA standards, 106-108
  serious harm, demonstration of, 104-105
  unforeseeable hazards, 102-103
General Electric Co. v. OSHRC, 269
General Electric Co. v. Secretary, 90
Georgia Pacific Corp. v. OSHRC, 86
Globe Contractors, Inc. v. Herman, 272
Good faith, penalty reduction for, 205
Grading of federal agencies, Maxwell School of Government, 1
Granite City Terminals Corp., 86
Granite Construction Co. v. Superior Court, 220
Graphic Communications International Union v. Salem Gravure Division of World Color Press, Inc., 278
Greater hazard defense, 89-90
  as defense in proceedings before administrative law judge, 247
Gutierrez v. Sundance Indian Jewelry, 164

H

Hawaii, occupational safety, health plan, 306
Hawaii Department of Labor and Industrial Relations, 307
Hazard communication, 30, 41-46, 165-196
  components of regulation, 42-43
  hazard evaluation, 43-44
  material safety data sheet, 43

recordkeeping requirement, 120
scope, 42-43
Hazard communication standard, 43, 132-135, 236
Hazard evaluation, 43-44
Hazardous Waste Operations and Emergency Response regulations, 305
HAZWOPER. *See* Hazardous Waste Operations and Emergency Response regulations
Health inspector, 215
Health standards
 overview of, 9
 safety standards, contrasted, 68
Hickman v. Taylor, 212
Hines v. Elf Atochem N. Am. Inc., 163
Home workplaces, 6-7
Horizontal standards, vertical standards, contrasted, 69-70

I
IARC. *See* International Agency for Research on Cancer
IBP, Inc. v. Herman, 83, 248
ICR. *See* Information collection request
Illinois v. Chicago Magnet Wire Corp., 220
Illness/injury recordkeeping requirements, 120-131
 access, 125-126
 exemptions, 122-123
 information to be recorded, 126-129
 maintaining, retaining records, 124-125
 musculoskeletal disorders, 129-131
 posting, 124
 recordkeeping forms, 123
 those subject to rule, 121-122
Illnesses, accidents, annual summary of, 29
Imminent danger, defined, 282-283
Imminent danger inspection, 279-297
 Area Director, Compliance Safety and Health Officer, contact to employer, 285
 employee rights, 288-291
  dangerous work, refusing, 289-291
  right to inform, assist with inspection, 288-289
  writ of mandamus, 289
 employer response, 285-286
 Federal Mine Safety and Health Act of 1977, 294, 295
 imminent danger, defined, 282-283
 injunction proceedings, 286-288
 inspection procedure, 283-286

legislation, 296-297
Mine Safety and Health Agency, 294
penalties, 291-292
Secretary of Labor, response of, 284-285
temporary restraining order, 286
Imminent Danger Restraint, 295
Impossibility of compliance, as defense in proceedings before administrative law judge, 246
In re Establishment Inspection of Jeep Corp., 219
In re Grand Jury Matter, 212
In re Key Energy Services, Inc., 191
Indiana, occupational safety, health plan, 306
Indiana Department of Labor, 307
Industrial Commission of Arizona, 307
Industrial hygiene officer, 215
Industrial Truck Ass'n, Inc. v. Henry, 193
Industrial trucks, recordkeeping requirement, 120
Industrial Union Department, AFL-CIO v. American Petroleum Institute, 66
Industrial Union Department, AFL-CIO v. Bingham, 70
Industrial Union Department, AFL v. Hodgson, 17, 53
Industry recognition, recognized hazards, general duty clause, 100-101
Infeasibility, 86-89
 alternative measures element, 89
Information collection request, 62
Injunction proceedings, imminent danger, 286-288
Inorganic arsenic, recordkeeping requirement, 116
Inspection, 20-28, 215-234
 authority, to inspect places of employment, 216
 compliance safety and health officers, 215
 consensual searches, warrant requirement, contrasted, 219-222
 construction sites, 233-234
 field structure, 21
 fourth amendment, 216-219
 health inspector, 215
 imminent danger, 279-297
  Area Director, Compliance Safety and Health Officer, contact to employer, 285
  employee rights, 288-291
   dangerous work, refusing, 289-291
   right to inform, assist with inspection, 288-289
   writ of mandamus, 289
  employer response, 285-286

358 ❖ OSHA Law Handbook

Federal Mine Safety and Health Act of 1977, 294, 295
  imminent danger, defined, 282-283
  injunction proceedings, 286-288
  inspection procedure, 283-286
  legislation, 296-297
  Mine Safety and Health Agency, 294
  penalties, 291-292
  Secretary of Labor, response of, 284-285
  temporary restraining order, 286
industrial hygiene officer, 215
multi-party worksites, 233-234
number per year, 24
overview, 3
probable cause, 218-219
  corroboration of information, 219
  specificity of information, 218
  staleness, 219
  violation, likelihood of, 219
procedures, 223-233
  closing conference, 232-233
  documents, 230-231
  employee interviews, 231-232
  general duty clause, 224
  general requirements, 223-224
  opening conference, 224-226
  pictures, 230-231
  records inspection, subpoenas, 227-229
  subpoenas, 227-229
  walkaround, 229-230
role of, 21
safety inspector, 215
safety officer, 215
specific evidence test, 218
training, competence of inspectors, 22-23
types of inspections, 223
violations, by date, 22
warrant
  challenging of, 221-222
  requirement, exceptions, 222
Interagency Regulatory Liaison Group, 38
Interim order, 75
Interlocutory review of decision, 255-256
International Agency for Research on Cancer, 44
International Union, UAW v. General Dynamics Land System Division, 16
International Union, UAW v. OSHA, 66, 67, 68
International Union, UAW v. Pendergrass, 66
Interrogatories, proceedings before administrative law judge, 251-252

Interviews, employee, inspection, 231-232
Invalidity of standard, 91-92
Ionizing radiation, 116, 120
Iowa, occupational safety, health plan, 306
Iowa Division of Labor, 307
IPB, Inc. v. Herman, 261
IRLG. *See* Interagency Regulatory Liaison Group

J
Jeffress, Charles N., 2
J.L. Foti Const. Co. v. Donovan, 222
Jones & Laughlin Steel Corp. v. OSAHRC, 103
Judicial decisions, precedential effect, judicial review, enforcement actions, 278
Judicial review, enforcement actions, 267-278
  Administrative Procedure Act, 275
  constitutional challenges, 273-274

  findings of fact, standard of review for, 276-278
  jurisdiction, 268-270
    standing to bring appeal, 268-270
  notice of contest, 271
  Occupational Safety and Health Act, 267
  petition for discretionary review, 271
  precedential effect, judicial decisions, 278
  scope, 274-278
    procedural matters, 274-275
    standard of review for conclusions of law, 275-276
  timing, 270-274
  exhaustion of administrative remedies, 271-273
  final commission orders, 270-271
Jurisdiction
  judicial review, enforcement actions, 268-270
  standing to bring appeal, 268-270
  overlapping, 37-38
  state, 71-72

K
Kelly Springfield Tire Co. v. Donovan, 105
Kennedy, Ted, Senator, 265
Kentucky, occupational safety, health plan, 306
Kentucky Labor Cabinet, 308
Keystone Roofing Co. v. OSHRC, 272
Knowledge, actual, constructive, 84-85
Kropp Forge Co. v. Secretary of Labor, 86

L
Labelling, communication of hazard, 174-176
Laboratories, recordkeeping requirement, 120

Lake Butler Apparel Co. v. Secretary of Labor, 142, 222
Lance Roofing v. Hodgson, 271
Lead
  recordkeeping requirement, 116
  standard, provisions, 117-118
Legislative framework, OSHA, 3-7
Lockout/tagout, recordkeeping requirement, 120
Loomis Cabinet Co. v. OSHRC, 73
Louisiana Chemical Ass'n b. Bingham, 60
Lundgrin v. Claytor, 286

M
Machine guarding, recordkeeping requirement, 120
Magma Copper Co. v. Marshall, 97
Magna Copper Co. v. Marshall, 99
Maritime workers, subject to Coast Guard regulations, exemption from OSHA, 6
Marquette Cement, 102
Marshall v. Barlow's Inc., 11, 27, 28, 142, 216, 217, 218
Marshall v. Firestone Tire & Rubber Co., 148
Marshall v. Grand River Dam Authority, 221
Marshall v. Klug & Smith, Inc., 162
Marshall v. Knutson Constr. Co., 83
Marshall v. Natl. Indus. Constructors, 153
Marshall v. OSHRC, 269
Marshall v. P & Z Co., 162
Marshall v. Power City Elec., 162
Marshall v. Springville Poultry, 162
Marshall v. Sun Petroleum Prods., 149, 269
Marshall v. Union Oil Co., 73
Marshall v. Wallace Bros., Mfg. Co., 162
Martin Painting & Coating Co. v. Marshall, 276
Martin v. Bally's Park Place Hotel & Casino, 212
Martin v. OSHRC, 131, 276
Maryland, occupational safety, health plan, 306
Maryland Division of Labor and Industry, 308
Material safety data sheet, 43, 176-180
Maxwell School of Government, Syracuse University, grading of federal agencies, 1
McComb v. Jacksonville Paper Co., 221
McGowan v. Marshall, 274
McLaughlin v. A.B. Chance Co., 228
McLaughlin v. Kings Island, 138, 145, 228
Meatpacking plants, ergonomics, 47
Mechanical power presses, recordkeeping requirement, 118, 120
Medical records, 29-30

Medical services, recordkeeping requirement, 120
Melerine v. Avondale Shipyards, Inc., 83
Memorandums of understanding, 264
Methyl chloromethyl ether, recordkeeping requirement, 116
Metzlen v. Arcadian Corp., 239
Michigan, occupational safety, health plan, 306
Michigan Department of Consumer and Industry Services, 308
"Mickey Mouse" standards, 10, 12
Miles v. Martin Marietta Corp., 163
Mine Safety and Health Agency, 294
Mineral Industries v. OSHRC, 80
Minnesota, occupational safety, health plan, 306
Minnesota Department of Labor and Industry, 308
Modern Drop Forge Co. v. Secretary of Labor, 90, 247
Mohawk Excavating, Inc. v. OSHRC, 273
Monitoring, 29-30
Morgan & Culpepper, Inc. v. OSHRC, 80
MSDS. *See* Material safety data sheet
Multi-employer doctrine, 261
Multi-employer worksite defense, 84
  in proceedings before administrative law judge, 248
Multi-employer worksite liability rules, 81-82
Multi-party worksites, inspection, 233-234
Multiple employer worksites, communication of hazard, 190-191
Musculoskeletal disorders, 48
  recordkeeping requirements, 129-131

N
N-nitrosodimethylamine, recordkeeping requirement, 116
Nader, Ralph, 265
National Association of Manufacturers, 7
National Constructors Ass'n v. Marshall, 92
National Grain and Feed Association, Inc. v. OSHA, 68
National Industrial Constructors, Inc. v. OSHRC, 73
National Industrial Contractors, Inc. v. OSHRC, 273
National Institute of Occupational Safety and Health, 4, 13, 40-41
National Labor Relations Act, 137-138, 155-157
  cooperation, between OSHA, NLRB, 157
  Section 7, 155-157

Section 502, compared, 157
National Realty, 100, 101, 103, 104
National Toxicology Program, annual report, 171
Negotiated Rulemaking Act, 59
Nevada, occupational safety, health plan, 306
Nevada Division of Industrial Relations, 308
New Jersey, occupational safety, health plan, 306
New Jersey Chamber of Commerce v. Hughey, 46
New Jersey Department of Labor, 308
New Mexico, occupational safety, health plan, 306
New Mexico Environment Department, 309
New York, occupational safety, health plan, 306
New York Department of Labor, 309
New York State Electric & Gas Corp. v. Sec. of Labor and OSHRC, 247
Newport News Shipbuilding v. Marshall, 160
Nitrobiphenyl, 116
NLRA. *See* National Labor Relations Act
NLRB v. C & C Plywood Corp., 160
NLRB v. City Disposal Systems, 155
NLRB v. Jasper Seating Co., 156
NLRB v. Maine Caterers, Inc., 222
No-effect levels, standards, 11
Noblecraft Indus., Inc. v. Secretary of Labor, 273
Noise, recordkeeping requirement, 120
Non-serious, violation, defined, 23
North Carolina, occupational safety, health plan, 306
North Carolina Department of Labor, 309
Northeast Erectors Ass'n v. Secretary of Labor, 272
Notice of contest, 143, 241, 271
 settlement after filing of, 243
Notice of intent to appear, in rulemaking process, 56
NRA. *See* Negotiated Rulemaking Act
NTP. *See* National Toxicology Program

O
Occupational Safety and Health Review Commission, 4, 39-40
 appeal process, 39
 limitations of, 39-40
Ocean Electric Corp. v. Sec. of Labor, 247
Office of Management and Budget, 41, 62
Oil, Chem. & Atomic Workers v. OSHRC, 269
Opening conference, inspection, 224-226
Oregon, occupational safety, health plan, 306
Oregon Occupational Safety and Health Division, 309
Other-than-serious violations, 237
Overlapping jurisdiction, 37-38
Owens-Corning Fiberglass Corp. v. Donovan, 86

P
P. Gioioso & Sons, Inc. v. OSHRC, 247, 272
Paperwork Reduction Act of 1995, 61-62
PBR, Inc. v. Secretary of Labor and OSHRC, 89
Peabody Southwest, Inc. v. Texas, 220
PELs. *See* Permissible exposure limits
Penalties, 23-24. *See also* Violation
 contesting
  increasing size of, 239
  OSHA budget and, 239-240
 with date, 25
 patterns, 24-25
 proposed, overview, 3
 reduction, audit and, 207
Penn-Dixie Steel Corp. v. OSHRC, 271
Pennsylvania Power & Light Co. v. OSAHRC, 103
Pennsylvania Power & Light Co. v. OSHRC, 247
People v. Chicago Magnet Wire Corp., 263
People v. Pymm Thermometer Company, 263
Performance standards, specification standards, contrasted, 69
Permanent standards
 content of, 68-69
 Section 6(b), 13-15
Permanent variances, 20, 74
Permissible exposure limits, air contaminants standard, 69
Perry v. O'Donnell, 222
Personal protective equipment, recordkeeping requirement, 120
Pesticides, 6
Peterson Bros. Steel Erec. Co., 88
Peterson v. Chesapeake & Ohio Ry. Co., 209
Petition for discretionary review, 271
Phoenix Roofing, Inc. v. Dole, 92
Pictures, inspection, 230-231
Pitt-Des Moines, 261
Poultry plants, ergonomics, 47
Power Plant Div., Brown & Root, Inc. v. OSHRC, 86
Power tools, recordkeeping requirement, 120
Pratt & Whitney Aircraft Div. v. Secretary, 99
Precedential effect, judicial decisions, judicial review, enforcement actions, 278
Preemption, 77-78
 federal, 45-46
 whistleblowing, 162

Price v. County of San Diego, 210
Prill v. NLRB, 156
Privilege, self-audit, 207-211
Probable cause, 216-219
  inspections and, 218-219
    corroboration of information, 219
    specificity of information, 218
    staleness, 219
    violation, likelihood of, 219
Procedural defenses, in proceedings before administrative law judge, 245-246
Process safety management, recordkeeping requirement, 120
Production of documents, proceedings before administrative law judge, 250-251
Programmatic standards, 31
Prohibition against unreasonable search, seizure, 216-219
Promptness, as defense in proceedings before administrative law judge, 246
Proposed rulemaking, advanced notice of, 55
Prudential Ins. Co. v. Lai, 159
Puerto Rico, occupational safety, health plan, 306
Puerto Rico Department of Labor and Human Resources, 309
Purpose of OSHA, 4

Q
Quality Stamping Products v. OSHRC, 88

R
R. P. Carbone Constr. Co. v. OSHRC, 83
Railroad workers, under Federal Railroad Administration, exemption from OSHA, 6
Ray Evars Welding Co. v. OSAHRC, 103
Ray Evers Welding Co. v. OSHRC, 86
Reasonably predictable exposure, 80
Recordkeeping, 28-31, 111-139
  access to records, 30-31, 136-139
  accident reports, 28-29
  chemical-specific standards, 114-118
  enforcement, 135-136
  hazard communication, 30
    standard, 132-135
  illness/injury recordkeeping requirements, 120-131
    access, 125-126
    exemptions, 122-123
    information to be recorded, 126-129
    maintaining, retaining, 124-125

musculoskeletal disorders, 129-131
    posting, 124
    recordkeeping forms, 123
    those subject to rule, 121-122
  list of safety standards, types of records, 120
  medical records, 29-30
  monitoring, 29-30
  programmatic standards, 31
  reasons for requiring, 112-113
    for safety standards, 118-120
  statutory basis, 111-112
  types of records, 114
Records, access to, 30-31
Records inspection, subpoenas, 227-229
Refusal to work, 31-33, 147, 152-160
  arbitration, 159
  collective bargaining agreements, 158-159
  common law, 160
  enforcing rights, 153-154
  Federal Railroad Safety Act, 152
  federal statutes, 152-160
  National Labor Relations Act, 155-157
    cooperation, between OSHA, NLRB, 157
    Section 7, 155-157
    Section 502, compared, 157
  referral, 159-160
  remedies, 154
  Secretary's burden, in litigation, 154
  shifting burden analysis, 154
  state statutes, 160
  Surface Transportation Assistance Act, 152
Regulatory Flexibility Act, 63
Regulatory impact analyses, 64-66
Reich v. Cambridgport Air Systems, Inc., 154
Reich v. Dayton Tire, 286, 287, 288
Reich v. Hercules, Inc., 210
Reich v. Hoy Shoe Co., 154
Reich v. Simpson, Gumpertz & Heger, Inc., 276
Reich v. Sysco Corp., 159
Reich v. Trinity Industries, Inc., 90
Repeat violations, 238
  defined, 23
Request for information, advanced notice of proposed rulemaking, 55
Requests for admissions, proceedings before administrative law judge, 251
Resource Conservation and Recovery Act, 187, 264
Respiratory protection, recordkeeping requirement, 120

362 ❖ OSHA Law Handbook

Restraining order, with imminent danger, 286
Review of decision, interlocutory, 255-256
Riggers & Erectors, Inc. v. OSHRC, 86
Rockwell International Corp. v. NLRB, 157
Rosenberg v. Merrill Lynch, 159
RSR Corp. v. Donovan, 73, 270
R.T. Vanderbilt Co. v. OSHRC, 269
Rulemaking process, 53-76
  advanced notice of proposed rulemaking, 55
  advisory committees, 54-55
  decision to initiate, 54
  final rule, 57-58
  hearing, 56-57
  negotiated, 59
  notice of intent to appear, 56
  notice of proposed rulemaking, 55-56
  request for information, 55
  traditional, 54-58

S
Sabine Consolidated, Inc. v. State, 220
Safety, health standards, 59-71
  basic framework, 59-60
  characterization of, 66-70
  controlling statutes, 61-66
  emergency temporary standards, 70-71
  Executive Order 12866, 64-66
  executive orders, 61-66
   horizontal standards, vertical standards, contrasted, 69-70
  information collection request, 62
  Office of Management and Budget, 62
  Paperwork Reduction Act of 1995, 61-62
  permanent standards, content of, 68-69
  principles, OSHA standards, 66-68
  Regulatory Flexibility Act, 63
  regulatory impact analyses, 64-66
  safety standards, health standards, contrasted, 68
  Small Business Regulatory Enforcement Fairness Act, 63-64
  specification standards, performance standards, contrasted, 69
  statutory authority to issue standards, 60-61
  temporary standards, emergency, 70-71
Safety inspector, 215
Safety officer, 215
Safety standards, health standards, contrasted, 68
SBREFA. *See* Small Business Regulatory Enforcement Fairness Act
Scope of audit, 199-200

Search
  consensual, warrant requirement, contrasted, 219-222
  seizure, prohibition against, 216-219
Sec. of Labor v. Dun-Par Engineered Form Co., 247
Secretary of Labor
  for Occupational Safety and Health, Assistant, administration of OSHA, 2
  response of, with imminent danger, 284-285
Secretary of Labor v. Catapano, 289
Secretary of Labor v. Gene T. Jones Tire & Battery Distributors, 293
Secretary of Labor v. H.M.S. Direct Mail, 154
Secretary of Labor v. L.E. Myers Co., 90
Secretary of Labor v. Manganas, 293
Secretary of Labor v. Master Metals, Inc., 293
Secretary of Labor v. OSHRC, 92
Secretary of Labor v. Pete Barrios, et al., 231
Secretary v. American Sterilizer Co., 115
Secretary v. Bechtel Power Corp., 105
Secretary v. Beird-Poulan, 101
Secretary v. Cerro Metal Products, 103, 106
Secretary v. Copperweld Steel Co, 99
Secretary v. Dayton Tire, 245
Secretary v. FMC Corp., 103, 106
Secretary v. General Dynamics Land Systems Div. Inc., 100
Secretary v. General Electric Co., 99
Secretary v. Granite City Terminals Corp., 103
Secretary v. IBP, Inc., 248
Secretary v. Monfort of Colorado, Inc., 139, 145
Secretary v. Northwest Airlines, Inc., 104
Secretary v. Nova Fedrick/AJV, 246
Secretary v. Price Chopper Supermarkets, 125
Secretary v. Rockwell International Corporation, 103
Secretary v. Scafar Contracting, Inc., 246
Secretary v. Southern Nuclear Operating Co., Inc., 246
Secretary v. Wyman-Gordon Co., 138
See v. City of Seattle, 228
Self-audit, 197-214
  in agency inspections, 204
  audit information, 200
  audit team, 199
  auditing tips, 201-202
  censoring of auditors, 202
  confidentiality, 201
  corrected conditions, no citation for, 204
  disclosure, audit information, 207-213
    common law audit privilege, 208-210

determining whether to apply privilege, 209-210
self-audit privilege, 207-211
statutory audit privilege, 210-211
documentation, steps responding to hazards, 201-202
evidence, use of information to supplement, 206
findings of, 202
good faith, penalty reduction for, 205
overview, 199-200
penalty reduction, 207
policy, 203-207
  provisions, 204-205
  purpose, 203
  scope, 203-204
privilege, 207-211
scope of audit, 199-200
timely respond to hazard identified, 201
tips, 201-202
voluntary safety, health auditing, 198-202
willfulness, protection from use of to show, 205
Serious violation, 237
defined, 23
Servicing truck tires, recordkeeping requirement, 120
Settlement
after filing notice of contest, 243
agreement of, 240-241
Simplified proceedings, 255
"E-Z trial" procedures, 255
Six Month Statute of Limitations, 245
Slings, recordkeeping requirement, 120
Small Business Regulatory Enforcement Fairness Act, 51, 63-64
Smith Steel Casting Co. v. Donovan, 278
Society of Plastics Industry v. OSHA, 17
Solid Waste Disposal Act, 187
South Carolina, occupational safety, health plan, 306
South Carolina Department of Labor, Licensing, and Regulation, 309
Southern Colo. Prestress Co. v. OSHRC, 88
Spancrete Northeast Inc. v. OSHRC, 89
Specific evidence test, inspection and, 218
Specification standards, performance standards, contrasted, 69
Spinelli v. United States, 222
Sprinklers/hoses standpipes, recordkeeping requirement, 120
Staleness, inspection, probable cause and, 219

Standard of review for conclusions of law, 275-276
Standards, OSHA, 7-11
applicability of, 77-79
areas covered by, 7-8
challenging of, 73
completion, 12-13
consensus, Section 6(a), 11-12
deletion processes, 12-13
emergency temporary, 15
no-effect levels, 11
overview of, 8-9
permanent, Section 6(b), 13-15
programmatic, 31
setting of, 11-19
procedures for, 11
Standing to bring appeal, 268-270
State employees, 34
exemption from OSHA, 6
State jurisdiction, 71-72
State occupational safety, health plan, 34-36, 299-310
approval of, 299-302
  program content, 300
  steps for approval, 300-302
approved state plans, 303, 306
complaints, 302-303
preemption, 305-306
state standards, 303-305
State Sheet Metal Co., 88
Statutory audit privilege, 210-211
Statutory authority to issue standards, 60-61
Stay of commission order, 257
Stephenson Enters. V. Marshall, 222
Subpart Z list, 44
Subpoenas
inspection, 227-229
proceedings before administrative law judge, 252-253
Substantive defenses, as defense in proceedings before administrative law judge, 246-249
Surface Transportation Assistance Act, 152
Syracuse University, Maxwell School of Government, grading of federal agencies, 1

T

Taylor Diving & Salvage Co., Inc. v. U.S. Department of Labor, 71
Telecommuting, 6-7
Temporary, emergency, standards, 15

Temporary restraining order, with imminent danger, 286
Temporary standards, emergency, 70-71
Temporary variances, 20, 74
Ten workers, or fewer, workplaces employing, exemption from OSHA, 5
Tennessee, occupational safety, health plan, 306
Tennessee Department of Labor, 310
Threshold limit values, chemical substances, physical agents, American Conference of Governmental Industrial Hygienists, 44
Timing, judicial review, enforcement actions, 270-274
  exhaustion of administrative remedies, 271-273
  final commission orders, 270-271
Titanium Metals Corp. of Am. v. Usery, 97
Titanium Metals Corp. v. Usery, 104
Toxic Substances Control Act, 186
Trade secrets, 44-45
  communication of hazard and, 189-190
  proceedings before administrative law judge, 253-254
  protection of, 145-146
Traditional rulemaking process, 54-58
Training, competence of inspectors, 22-23
Turner Co. v. Secretary of Labor, 18
Types of records, 114
  safety standards, 120

U

Unforeseeable hazards, general duty clause, 102-103
United Auto Workers v. General Dynamics, 107
United States v. Matlock, 234
United States v. Morton Salt Co., 229
United States v. Pitt-Des Moines, Inc., 83
United States v. Thriftmarts, Inc., 142, 222
United States v. Westinghouse Electric Corp., 138
United Steelworkers of America v. Auchter, et al., 45
United Steelworkers of America v. Marshall, 58, 105
United Steelworkers of America v. Pendergrass, 42
Universal Construction Corp., 83
Unpreventable employee misconduct, 90-91
  as defense in proceedings before administrative law judge, 247
Unreasonable search, seizure, prohibition against, 216-219
Upjohn v. United States, 211
Upper extremity cumulative trauma disorders, 131

US v. Ladish Malting, 261, 262
US v. Pitt-Des Moines, Inc., 261
Usery v. Marquette Cement Mfg. Co., 101, 102
Utah, occupational safety, health plan, 306
Utah Labor Commission, 310

V

Variances, 19-20, 73-75
  permanent, 20, 74
  temporary, 20, 74
Ventilation systems, recordkeeping requirement, 120
Vermont, occupational safety, health plan, 306
Vermont Department of Labor and Industry, 310
Vertical standards, horizontal standards, contrasted, 69-70
Vinyl chloride, 70
  recordkeeping requirement, 116
Violations
  classification of, 236-240
  criminal enforcement, 259-266
  de minimis, defined, 23
  inspections, by date, 22
  serious, as proportion of total violations, 24
Virgin Islands, occupational safety, health plan, 306
Virgin Islands Department of Labor, 310
Virginia, occupational safety, health plan, 306
Virginia Department of Labor and Industry, 310
Virus-Serum-Toxin Act of 1913, 186
Voegele Company, Inc. v. OSHRC, 86
Voluntary safety, health auditing, 198-202

W

Walkaround, inspection, 229-230
Walking working surfaces, recordkeeping requirement, 120
Warning of hazards, forms of, 174-176
Warrant, 216-219
  challenging of, 221-222
  requirement for
    consensual searches, contrasted, 219-222
    exceptions, 222
Warrantless inspections
  Barlow case, 27-28
Warrantless inspections, *Barlow* case, 27-28
Washington, occupational safety, health plan, 306
Washington Department of Labor and Industries, 310
Welding, recordkeeping requirement, 120

Wellstone, Paul, Senator, 265
West Point-Pepperell, Inc. v. Donovan, 222
Weyerhauser Co. v. Marshall, 218
Whirlpool Corp. v. Marshall, 31, 147, 153, 288, 290, 291
Whistleblowing, 31-33, 161-164
  common law, 163
  enforcement procedure, 162
  federal statutes, 161-162
  preemption, 162
  protection of, 32-33
  state statutes, 162-163
Wickard v. Filburn, 277
Willful violation, 237-238
  causing death to employee, 261-262
  defined, 23

Willfulness, evidence, protection from audit use, 205
Wood ladders, recordkeeping requirement, 120
Worker right-to-know rule, 41
Written hazard communication program, 173
Wyoming, occupational safety, health plan, 306
Wyoming Department of Employment, 310

Y
Young Sales Corp., 101

Z
Zone of danger, 80